Computational Drug Discovery

Computational Drug Discovery

Methods and Applications

*Edited by Vasanthanathan Poongavanam and
Vijayan Ramaswamy*

Volume 1

WILEY VCH

Editors

Dr. Vasanthanathan Poongavanam
Uppsala University
Department of Chemistry-BMC
751 05 Uppsala
Sweden

Dr. Vijayan Ramaswamy
University of Texas MD Anderson Cancer Center
Institute for Applied Cancer Science
TX
United States

Cover: © Vasanthanathan Poongavanam

All books published by **WILEY-VCH** are carefully produced. Nevertheless, authors, editors, and publisher do not warrant the information contained in these books, including this book, to be free of errors. Readers are advised to keep in mind that statements, data, illustrations, procedural details or other items may inadvertently be inaccurate.

Library of Congress Card No.: applied for

British Library Cataloguing-in-Publication Data
A catalogue record for this book is available from the British Library.

Bibliographic information published by the Deutsche Nationalbibliothek
The Deutsche Nationalbibliothek lists this publication in the Deutsche Nationalbibliografie; detailed bibliographic data are available on the Internet at <http://dnb.d-nb.de>.

© 2024 WILEY-VCH GmbH, Boschstraße 12, 69469 Weinheim, Germany

All rights reserved (including those of translation into other languages). No part of this book may be reproduced in any form – by photoprinting, microfilm, or any other means – nor transmitted or translated into a machine language without written permission from the publishers. Registered names, trademarks, etc. used in this book, even when not specifically marked as such, are not to be considered unprotected by law.

Print ISBN: 978-3-527-35374-3
ePDF ISBN: 978-3-527-84072-4
ePub ISBN: 978-3-527-84073-1
oBook ISBN: 978-3-527-84074-8

Typesetting   Straive, Chennai, India

Contents

Volume 1

Preface *xv*
Acknowledgments *xix*
About the Editors *xxi*

Part I   Molecular Dynamics and Related Methods in Drug Discovery *1*

1   Binding Free Energy Calculations in Drug Discovery *3*
Anita de Ruiter and Chris Oostenbrink
1.1   Introduction *3*
1.1.1   Free Energy and Thermodynamic Cycles *4*
1.2   Endpoint Methods *5*
1.2.1   MM/PBSA and MM/GBSA *5*
1.2.2   Linear Response Approximations *7*
1.3   Alchemical Methods *9*
1.3.1   Free Energy Perturbation *9*
1.3.2   Thermodynamic Integration *10*
1.3.3   Bennett's Acceptance Ratio *10*
1.3.4   Nonequilibrium Methods *11*
1.3.5   Multiple Compounds *11*
1.3.6   One-Step Perturbation Approaches *12*
1.3.7   Challenges in Alchemical Free Energy Calculations *13*
1.4   Pathway Methods *15*
1.5   Final Thoughts *17*
         References *17*

2   Gaussian Accelerated Molecular Dynamics in Drug Discovery *21*
Hung N. Do, Jinan Wang, Keya Joshi, Kushal Koirala, and Yinglong Miao
2.1   Introduction *21*
2.2   Methods *22*

2.2.1	Gaussian Accelerated Molecular Dynamics	22
2.2.2	Ligand Gaussian Accelerated Molecular Dynamics	24
2.2.3	Energetic Reweighting of GaMD for Free Energy Calculations	25
2.2.4	GLOW: A Workflow Integrating Gaussian Accelerated Molecular Dynamics and Deep Learning for Free Energy Profiling	26
2.2.5	Binding Kinetics Obtained from Reweighting of GaMD Simulations	26
2.2.6	Gaussian Accelerated Molecular Dynamics Implementations and Software	27
2.3	Applications	28
2.3.1	G-Protein-Coupled Receptors	28
2.3.1.1	Characterizing the Binding and Unbinding of Caffeine in Human Adenosine A_{2A} Receptor	28
2.3.1.2	Unraveling the Allosteric Modulation of Human A_1 Adenosine Receptor	29
2.3.1.3	Ensemble Based Virtual Screening of Allosteric Modulators of Human A_1 Adenosine Receptor	32
2.3.2	Nucleic Acids	33
2.3.2.1	Exploring the Binding of Risdiplam Splicing Drug Analog to Single-Stranded RNA	33
2.3.2.2	Uncovering the Binding of RNA to a Musashi RNA-Binding Protein	33
2.3.3	Human Angiotensin-Converting Enzyme 2 Receptor	35
2.3.4	Discovery of Novel Small-Molecule Calcium Sensitizers for Cardiac Troponin C	37
2.3.5	Binding Kinetics Prediction from GaMD Simulations	37
2.4	Conclusions	39
	References	40
3	**MD Simulations for Drug-Target (Un)binding Kinetics**	**45**
	Steffen Wolf	
3.1	Introduction	45
3.1.1	Preface	45
3.1.2	Motivation for Predicting (Un)binding Kinetics	45
3.1.3	The Time Scale Problem of MD Simulations	46
3.2	Theory of Molecular Kinetics Calculation	47
3.2.1	Nonequilibrium Statistical Mechanics in a Nutshell	47
3.2.2	Kramers Rate Theory	48
3.2.3	Biased MD Methods	49
3.2.3.1	Temperature- and Barrier-Scaling	49
3.2.3.2	Bias Potential-Based Methods	49
3.2.3.3	Bias Force-Based Methods	50
3.2.3.4	Knowledge-Biased Methods	50
3.2.3.5	Coarse-graining and Master Equation Approaches	51
3.3	Challenges and Caveats in Rate Prediction	51
3.3.1	Finding Reaction Coordinates and Pathways	51
3.3.2	Error Ranges of Estimates	52

3.3.3	A Need for Reliable Benchmarking Systems	*53*
3.3.4	Problems with Force Fields	*53*
3.4	Methods for Rate Prediction	*53*
3.4.1	Unbinding Rate Prediction	*53*
3.4.1.1	Empirical Predictions	*53*
3.4.1.2	Prediction of Absolute Unbinding Rates	*54*
3.4.2	Binding Rate Prediction	*56*
3.5	State-of-the-Art in Understanding Kinetics	*57*
3.6	Conclusion	*57*
	References	*58*

4   Solvation Thermodynamics and its Applications in Drug Discovery *65*
Kuzhanthaivelan Saravanan and Ramesh K. Sistla

4.1	Introduction	*65*
4.1.1	Protein Folding	*65*
4.1.2	Protein–Ligand Interactions	*66*
4.2	Tools to Assess the Solvation Thermodynamics	*70*
4.2.1	Watermap	*71*
4.2.2	GIST	*72*
4.2.3	3D-RISM	*74*
4.3	Case Studies	*75*
4.3.1	Watermap	*75*
4.3.1.1	Background and Approach	*75*
4.3.1.2	Results and Discussion	*75*
4.3.2	Grid Inhomogeneous Solvation Theory (GIST)	*76*
4.3.2.1	Objective and Approach	*76*
4.3.2.2	Results and Discussion	*77*
4.3.3	Three-Dimensional Reference Interaction-Site Model (3D-RISM)	*78*
4.3.3.1	Objective and Background	*78*
4.3.3.2	Results and Discussion	*79*
4.4	Conclusion	*80*
	References	*80*

5   Site-Identification by Ligand Competitive Saturation as a Paradigm of Co-solvent MD Methods *83*
Asuka A. Orr and Alexander D. MacKerell Jr

5.1	Introduction	*83*
5.2	SILCS: Site Identification by Ligand Competitive Saturation	*90*
5.3	SILCS Case Studies: Bovine Serum Albumin and Pembrolizumab	*97*
5.3.1	SILCS Simulations	*98*
5.3.2	FragMap Construction	*99*
5.3.3	SILCS-MC	*100*
5.3.4	SILCS-Hotspots	*102*
5.3.5	SILCS-PPI	*103*

5.3.6	SILCS-Biologics *105*
5.4	Conclusion *106*
	Conflict of Interest *106*
	Acknowledgments *107*
	References *107*

Part II Quantum Mechanics Application for Drug Discovery *119*

6 QM/MM for Structure-Based Drug Design: Techniques and Applications *121*
Marc W. van der Kamp and Jaida Begum
- 6.1 Introduction *121*
- 6.2 QM/MM Approaches *122*
- 6.2.1 Combined Quantum Mechanical/Molecular Mechanical Energy Calculations *122*
- 6.2.2 QM/MM Methods for the Evaluation of Non-Covalent Inhibitor Binding *124*
- 6.2.3 QM/MM Reaction Modeling *125*
- 6.3 Applications of QM/MM for Covalent Drug Design and Evaluation *128*
- 6.3.1 Covalent Tyrosine Kinase Inhibitors for Cancer Treatment *128*
- 6.3.2 Evaluation of Antibiotic Resistance Conferred by β-Lactamases *133*
- 6.3.3 Covalent SARS-CoV-2 Inhibitors: Mechanism and Insights for Design *138*
- 6.4 Conclusions and Outlook *143*
- References *144*

7 Recent Advances in Practical Quantum Mechanics and Mixed-QM/MM-Driven X-Ray Crystallography and Cryogenic Electron Microscopy (Cryo-EM) and Their Impact on Structure-Based Drug Discovery *157*
Oleg Borbulevych and Lance M. Westerhoff
- 7.1 Introduction *157*
- 7.2 Feasibility of Routine and Fast QM-Driven X-Ray Refinement *159*
- 7.3 Metrics to Measure Improvement *160*
- 7.3.1 Ligand Strain Energy *160*
- 7.3.2 ZDD of Difference Density *161*
- 7.3.3 Overall Crystallographic Structure Quality Metrics: MolProbity Score and Clashscore *162*
- 7.4 QM Region Refinement *162*
- 7.5 ONIOM Refinement *165*
- 7.6 XModeScore: Distinguish Protomers, Tautomers, Flip States, and Docked Ligand Poses *168*

7.7	Impact of the QM-Driven Refinement on Protein–Ligand Affinity Prediction *169*	
7.7.1	Impact of Structure Inspection and Modification *172*	
7.7.2	Impact of Selecting Protomer States: Implications of XModeScore on SBDD *174*	
7.8	Conclusion *175*	
	Acknowledgments *177*	
	References *177*	
8	**Quantum-Chemical Analyses of Interactions for Biochemical Applications** *183*	
	Dmitri G. Fedorov	
8.1	Introduction *183*	
8.2	Introduction to FMO *184*	
8.3	Pair Energy Decomposition Analysis (PIEDA) *186*	
8.3.1	Formulation of PIEDA *186*	
8.3.2	Applications of PIEs and PIEDA *189*	
8.3.3	Example of PIEDA *189*	
8.4	Partition Analysis (PA) *190*	
8.4.1	Formulation of PA *193*	
8.4.2	Applications and an Example of PA *194*	
8.5	Partition Analysis of Vibrational Energy (PAVE) *195*	
8.5.1	Formulation of PAVE *196*	
8.5.2	Applications of PAVE *196*	
8.6	Subsystem Analysis (SA) *197*	
8.6.1	Formulation of SA *197*	
8.6.2	Examples of SA and PAVE *200*	
8.7	Fluctuation Analysis (FA) *201*	
8.8	Free Energy Decomposition Analysis (FEDA) *202*	
8.9	Other Analyses of Chemical Reactions *202*	
8.10	Conclusions *203*	
	References *203*	
Part III	**Artificial Intelligence in Pre-clinical Drug Discovery** *211*	
9	**The Role of Computer-Aided Drug Design in Drug Discovery** *213*	
	Storm van der Voort, Andreas Bender, and Bart A. Westerman	
9.1	Introduction to Drug–Target Interactions, Hit Identification *213*	
9.2	Lead Identification and Optimization: QSAR and Docking-Based Approaches *215*	
9.3	DTI Machine Learning Methods *215*	
9.4	Supervised, Non-supervised and Semi-supervised Learning Methods *216*	

9.5	Graph-Based Methods to Label Data for DTI Prediction	*217*
9.6	The Importance of Explainable ML Methods: Linking Molecular Properties to Effects	*218*
9.7	Predicting Therapeutic Responses	*219*
9.8	ADMET-tox Prediction	*220*
9.9	Challenging Aspects of Using Computational Methods in Drug Discovery	*220*
9.9.1	What are Those Limitations?	*221*
	References	*223*
10	**AI-Based Protein Structure Predictions and Their Implications in Drug Discovery**	*227*
	Tahsin F. Kellici, Dimitar Hristozov, and Inaki Morao	
10.1	Introduction	*227*
10.2	Impact of AI-Based Protein Models in Structural Biology	*229*
10.2.1	Combination of AI-Based Predictions with Cryo-EM and X-Ray Crystallography	*229*
10.2.2	Combination of AI-Based Predictions with NMR Structures	*232*
10.2.3	Combination of AI-Based Predictions with Other Experimental Restraints	*234*
10.2.4	Impact of Deep Learning Models in Other Areas of Structural Biology	*235*
10.3	Combination of AI-Based Methods with Computational Approaches	*236*
10.3.1	Combination of Structure Prediction with Other Computational Approaches	*242*
10.4	Current Challenges and Opportunities	*243*
10.5	Conclusions	*246*
	References	*246*
11	**Deep Learning for the Structure-Based Binding Free Energy Prediction of Small Molecule Ligands**	*255*
	Venkatesh Mysore, Nilkanth Patel, and Adegoke Ojewole	
11.1	Introduction	*255*
11.2	Deep Learning Models for Reasoning About Protein–Ligand Complexes	*257*
11.2.1	Datasets	*258*
11.2.2	Convolutional Neural Networks	*258*
11.2.2.1	Background	*258*
11.2.2.2	Voxelized Grid Representation	*258*
11.2.2.3	Descriptors	*259*
11.2.2.4	Applications	*259*
11.2.3	Graph Neural Networks	*260*
11.2.3.1	Background	*260*
11.2.3.2	Graph Representation	*260*

11.2.3.3	Descriptors *260*	
11.2.3.4	Applications *260*	
11.2.3.5	Extension to Attention Based Models *261*	
11.2.3.6	Geometric Deep Learning and Other Approaches *261*	
11.3	Deep Learning Approaches Around Molecular Dynamics Simulations *261*	
11.3.1	Enhanced Sampling *262*	
11.3.2	Physics-inspired Neural Networks *262*	
11.3.3	Modeling Dynamics *263*	
11.3.3.1	Applications *263*	
11.4	Modifying AlphaFold2 for Binding Affinity Prediction *264*	
11.4.1	Modifying AlphaFold2 Input Protein Database for Accurate Free Energy Predictions *265*	
11.4.2	Modifying Multiple Sequence Alignment for AlphaFold2-Based Docking *265*	
11.5	Conclusion *266*	
11.5.1	New Models for Binding Affinity Prediction *266*	
11.5.2	Retrospective from the Compute Industry *266*	
11.5.2.1	Future DL-Based Binding Affinity Computation will Require Massive Scalability *267*	
11.5.2.2	Single GPU Optimizations for DL *267*	
11.5.2.3	Distributed DL Training and Inference *267*	
	References *268*	
12	**Using Artificial Intelligence for *de novo* Drug Design and Retrosynthesis** *275*	
	Rohit Arora, Nicolas Brosse, Clarisse Descamps, Nicolas Devaux, Nicolas Do Huu, Philippe Gendreau, Yann Gaston-Mathé, Maud Parrot, Quentin Perron, and Hamza Tajmouati	
12.1	Introduction *275*	
12.1.1	Traditional Drug Design and Discovery Process Is Slow and Expensive *275*	
12.1.2	Success and Limitations of Standard Computational Methods *276*	
12.1.3	AI-Based Methods can Accelerate Medicinal Chemistry *277*	
12.2	Quantitative Structure-Activity Relationship Models *278*	
12.2.1	Introduction to QSAR Models *278*	
12.2.2	QSAR Machine Learning Methods *278*	
12.2.3	QSAR Deep Neural Networks Methods *280*	
12.3	Modes of Generative AI in Chemistry *281*	
12.3.1	General Introduction *281*	
12.3.2	Generative AI in Lead Optimization *281*	
12.3.3	Fragment Growing *283*	
12.3.4	Novelty Generation *283*	
12.3.4.1	The Model *283*	
12.3.4.2	Optimization of the Novelty Generator *285*	

12.4	Importance of Synthetic Accessibility	*285*
12.4.1	Overview	*285*
12.4.2	Synthetic Scores	*286*
12.4.3	Integration of Synthetic Scores in Generative AI	*288*
12.4.3.1	An Example of a Lead Optimization Use Case	*288*
12.5	The Road Ahead	*290*
	References	*290*

13 Reliability and Applicability Assessment for Machine Learning Models *299*
Fabio Urbina and Sean Ekins

13.1	Introduction	*299*
13.2	Challenges for Modeling	*300*
13.3	Example 1: BBB Applicability Domain Comparison	*302*
13.4	Example 2: Models for Uncertainty Estimation for Multitask Toxicity Predictions	*303*
13.5	Example 3: Class-Conditional Conformal Predictors	*307*
13.6	Conclusions	*308*
	Funding	*309*
	Competing Interests	*309*
	References	*309*

Volume 2

Preface *xv*
Acknowledgments *xix*
About the Editors *xxi*

Part IV Chemical Space and Knowledge-Based Drug Discovery *315*

14 Enumerable Libraries and Accessible Chemical Space in Drug Discovery *317*
Tim Knehans, Nicholas A. Boyles, and Pieter H. Bos

15 Navigating Chemical Space *337*
Ákos Tarcsay, András Volford, Jonathan Buttrick, Jan-Constantin Christopherson, Máté Erdős, and Zoltán B. Szabó

16 Visualization, Exploration, and Screening of Chemical Space in Drug Discovery *365*
José J. Naveja, Fernanda I. Saldívar-González, Diana L. Prado-Romero, Angel J. Ruiz-Moreno, Marco Velasco-Velázquez, Ramón Alain Miranda-Quintana, and José L. Medina-Franco

17 SAR Knowledge Bases for Driving Drug Discovery *395*
Nishanth Kandepedu, Anil Kumar Manchala, and Norman Azoulay

18 Cambridge Structural Database (CSD) – Drug Discovery Through Data Mining & Knowledge-Based Tools *419*
Francesca Stanzione, Rupesh Chikhale, and Laura Friggeri

Part V Structure-Based Virtual Screening Using Docking *441*

19 Structure-Based Ultra-Large Virtual Screenings *443*
Christoph Gorgulla

20 Community Benchmarking Exercises for Docking and Scoring *471*
Bharti Devi, Anurag TK Baidya, and Rajnish Kumar

Part VI In Silico ADMET Modeling *495*

21 Advances in the Application of In Silico ADMET Models – An Industry Perspective *497*
Wenyi Wang, Fjodor Melnikov, Joe Napoli, and Prashant Desai

Part VII Computational Approaches for New Therapeutic Modalities *537*

22 Modeling the Structures of Ternary Complexes Mediated by Molecular Glues *539*
Michael L. Drummond

23 Free Energy Calculations in Covalent Drug Design *561*
Levente M. Mihalovits, György G. Ferenczy, and György M. Keserű

Part VIII Computing Technologies Driving Drug Discovery *579*

24 Orion® A Cloud-Native Molecular Design Platform *581*
Jesper Sørensen, Caitlin C. Bannan, Gaetano Calabrò, Varsha Jain, Grigory Ovanesyan, Addison Smith, She Zhang, Christopher I. Bayly, Tom A. Darden, Matthew T. Geballe, David N. LeBard, Mark McGann, Joseph B. Moon, Hari S. Muddana, Andrew Shewmaker, Jharrod LaFon, Robert W. Tolbert, A. Geoffrey Skillman, and Anthony Nicholls

25 Cloud-Native Rendering Platform and GPUs Aid Drug Discovery *617*
Mark Ross, Michael Drummond, Lance Westerhoff, Xavier Barbeu, Essam Metwally, Sasha Banks-Louie, Kevin Jorissen, Anup Ojah, and Ruzhu Chen

26 The Quantum Computing Paradigm *627*
Thomas Ehmer, Gopal Karemore, and Hans Melo

Index *679*

Preface

Computer-aided drug design (CADD) techniques are used in almost every stage of the drug discovery continuum, given the need to shorten discovery timelines, reduce costs, and improve the odds of clinical success. CADD integrates modeling, simulation, informatics, and artificial intelligence (AI) to design molecules with desired properties. Briefly, the application of CADD methodologies in drug discovery dates back to the 1960s, tracing its origin to the development of quantitative structure–activity relationship (QSAR) approaches. Between the 1970s and 1980s, computer graphics programs to visualize macromolecules began to take off together with advancements in computational power. This coincided with the emergence of more sophisticated techniques, including mapping energetically favorable binding sites on proteins, molecular docking, pharmacophore modeling, and modeling the dynamics of biomolecules. Since then, CADD has evolved as a powerful technique opening new possibilities, leading to increased adoption within the pharmaceutical industry and contributing to the discovery of several approved drugs.

Recent developments in CADD have been propelled by advancements in computing, breakthroughs in related fields such as structural biology, and the emergence of new therapeutic modalities. Notably, the advent of highly parallelizable GPUs and cloud computing have significantly increased computing power, while quantum computing holds promise to simulate complex systems at an unprecedented scale and speed. Advances in AI technologies, particularly generative AI for molecule design, are reducing cycle times during lead optimization. Meanwhile, the resolution revolution in cryo-electron microscopy (cryo-EM) and AI-powered structure biology are shedding light on the three-dimensional structure of many therapeutically relevant drug targets, thereby expanding our ability to carry out structure-based drug design against these targets. Other exciting breakthroughs that offer new opportunities include the explosion in the size of "make-on-demand" chemical libraries that enable ultra-large-scale virtual screening for hit identification, the big data phenomena in medicinal chemistry with the advent of bioactivity databases like ChEMBL and GOSTAR that provide access to millions of SAR data points useful for building predictive models and for knowledge-based compound, the emergence of new therapeutic modalities like targeted protein degradation like PROTACs and molecular glues, and viable approaches for targeting various reactive amino acid side chains beyond cysteine for developing covalent inhibitors. These developments are also now enabling drug discovery scientists to tackle high-value drug targets previously considered undruggable.

The changing paradigm in drug discovery, complemented by technological advancements, has significantly expanded the toolbox available for computational chemists to enable drug discovery in recent years. Against this backdrop, we felt a need for a book that offers up-to-date information on the most important developments in the field of CADD. This book, titled "Computational Drug Discovery," is meant to be a valuable resource for readers seeking a comprehensive account of the latest developments in CADD methods and technologies that are transforming small-molecule drug discovery. The intended target audience for this book is medicinal chemists, computational chemists, and drug discovery professionals from industry and academia.

The book is organized into eight thematic sections, each dedicated to a cutting-edge computational method, or a technology utilized in computational drug discovery. In total, it comprises 26 chapters authored by renowned experts from academia, pharma, and major drug discovery software providers, offering a broad overview of the latest advances in computational drug discovery.

Part I explores the role of molecular dynamics simulation and related approaches in drug discovery. It encompasses various topics such as the utilization of physics-based methods for binding free energy estimation, the theory and application of enhanced sampling methods like Gaussian Accelerated MD to facilitate efficient sampling of the conformational space, understanding binding and unbinding kinetics of compound binding through molecular dynamics simulation, the application of computational approaches like WaterMap and 3D-RISM framework to understand the location and thermodynamic properties of solvents that solvate the binding pocket which offers rich physical insights compound design, and the use of mixed solvent MD simulations for mapping binding hotspots on protein surfaces based on the SILCS technology.

Part II focuses on the role of quantum mechanical approaches in drug discovery, covering topics such as the use of hybrid QM/MM method for modeling reaction mechanisms and covalent inhibitor design, refinement of X-ray and cryo-EM structures integrating QM and QM/MM approaches for accurate assignment of tautomer, protomers, and amide flip rotamers for downstream structure-based design, and quantifying protein–ligand interaction energies using QM methods at a reduced computational cost like the fragment molecular orbital (FMO) framework

Part III focuses on the application of AI in preclinical drug discovery, highlighting its growing importance across different stages of the drug discovery process. Given the recent advancements in AI and related technologies, we have chapters that outline advancements in deep learning for protein structure prediction, in particular the significant breakthrough achieved by AlphaFold2, the use of deep learning architectures such as Convolutional Neural Networks (CNNs), Graph Neural Networks (GNNs), and physics-inspired neural networks for predicting protein–ligand binding affinity, the emergence of generative modeling techniques for *de novo* design of synthetically tractable drug-like molecules that satisfy a defined set of constraints. In order to offer readers guidance on effectively applying machine learning (ML) models and ensuring their validity and usefulness, this section includes a chapter that discusses different approaches for evaluating the reliability and domain applicability of ML models.

Part IV of this book focuses on how the concept of chemical space and the big data phenomenon are driving drug discovery. It includes chapters describing innovative approaches in reaction-based enumerations that enable the generation of virtual libraries containing tangible compounds, followed by computational solutions for visualizing and navigating this vast chemical space. Additionally, this section also highlights the use of SAR *knowledge bases like GOSATR* for extracting valuable insights and generating robust design ideas based on medicinal chemistry precedence. Wrapping up the section is a chapter highlighting how the wealth of knowledge gained by mining the data in CSD is proving valuable in various stages of drug discovery.

The ever-expanding size of compound libraries and the advent of make-on-demand compound libraries have elevated virtual screening to a whole new level.

Part V focuses on ultra-large-scale virtual screening using approaches that scale virtual screening methods to match the size of these massively large compound libraries. Although virtual screening using docking is a well-established approach for hit finding in drug discovery, the ability of docking programs to generate the correct binding mode and accurately estimate binding affinity is still a challenge. Hence, we have a chapter that reviews collaborative efforts within the scientific community for evaluating and comparing the performance of docking methods, establishing standardized metrics for assessing the efficiency of virtual screening techniques through rigorous competitive evaluations.

Early profiling of absorption, distribution, metabolism, excretion, and toxicity (ADMET) endpoints in early drug discovery is essential for designing and selecting compounds with superior ADMET properties. Consequently, major pharmaceutical companies have developed and implemented predictive models within their organizations for predicting multiple endpoints to enhance compound design. **Part VI** of the book chapter offers an overview of in silico ADMET methods and their practical applications in facilitating compound design within an industrial context. **Part VII** explores the role of computational techniques in accelerating the design of cutting-edge therapeutic modalities. This section provides a comprehensive focus on two key areas: the design of molecular glues and the design of covalent inhibitors.

In addition to the aforementioned methods and approaches that revolutionize the drug discovery process, computing technologies are further accelerating drug discovery with enhanced speed and accuracy.

Part VIII is dedicated to exploring how cloud computing and quantum computing significantly expand the range of drug discovery opportunities. Particularly, there is great hope and excitement surrounding the potential applications of quantum computing in drug discovery. *"The Quantum Computing Paradigm"* provides a comprehensive review on quantum computing from the perspective of drug discovery. In addition to discussing several drug discovery applications, including peptide design, the chapter also addresses challenges associated with this emerging drug discovery technology.

In conclusion, we believe that this book provides a thorough overview of the recent advancements in computational drug discovery, making it an engaging and captivating read. We would like to express our deepest appreciation to all the authors for their invaluable contributions to this book. Their expertise, insights, and unwavering commitment have greatly enriched its content and overall significance.

14 September 2023

Vasanthanathan Poongavanam, Uppsala, Sweden
Vijayan Ramaswamy, Texas, USA

Acknowledgments

First and foremost, we would like to extend our sincerest gratitude and profound appreciation to all the contributing authors. Their unwavering commitment, tireless efforts, and remarkable enthusiasm have been instrumental in bringing this book to fruition. It is their willingness to share their knowledge and experience that has greatly enriched its content, resulting in a truly valuable and comprehensive book that provides an account of the latest advancements in the field of computer-aided drug design.

We also extend our sincere gratitude to the external reviewers for their timely feedback and insightful suggestions that helped improve the quality of the book and shape the final outcome. Our special thanks to the following individuals for their invaluable contributions in reviewing the book chapters: Dr. Andreas Tosstorff (F. Hoffmann-La Roche, Switzerland), Dr. Sagar Gore, Dr. Suneel Kumar BVS (Molecular Forecaster, Canada), Dr. Pandian Sokkar, Dr. Ono Satoshi (Mitsubishi Tanabe Pharma, Japan), Dr. Octav Caldararu (Zealand Pharma, Denmark), Dr. Sundarapandian Thangapandian (HotSpot Therapeutics, Inc, USA), Dr. Vigneshwaran Namasivayam (Dewpoint Therapeutics, Germany), Dr. Yinglong Miao (University of Kansas, USA), Nanjie Deng (Pace University, USA), and Dr. Ansuman Biswas (Ernst & Young, India).

In conclusion, we would like to express our gratitude to the publisher Wiley for entrusting us with an opportunity to edit this book and for the fruitful collaboration. Especially, we convey our appreciation to Katherine Wong (Senior Managing Editor) and Dr. Lifen Yang (Program Manager) at Wiley for their unwavering support, encouragement throughout the editing process, and their commitment to ensuring the quality and excellence of this book. The editors also extend their thanks to Prof. Jan Kihlberg and Dr. Jason B. Cross for their continuous support, which made this project possible.

About the Editors

Vasanthanathan Poongavanam is a senior scientist in the Department of Chemistry-BMC, Uppsala University, Sweden. Before starting at Uppsala University in 2016, he was a postdoctoral fellow at the University of Vienna, Austria, and at the University of Southern Denmark. He obtained his PhD degree in Computational Medicinal Chemistry as a Drug Research Academy (DRA) Fellow at the University of Copenhagen, Denmark, on computational modeling of cytochrome P450. He has published more than 65 scientific articles, including reviews and book chapters. His scientific interests focus on in silico ADMET modeling, including cell permeability and solubility, and he has worked extensively on understanding the molecular properties that govern the pharmacokinetic profile of molecules bRo5 property space, including macrocycles and PROTACs.

Vijayan Ramaswamy (R.S.K. Vijayan) is a senior research scientist affiliated with the Structural Chemistry division at the Institute for Applied Cancer Science, The University of Texas MD Anderson Cancer Center, TX, USA. In 2016, he joined MD Anderson Cancer after a brief tenure as a scientist in computational chemistry at PMC Advanced Technologies, New Jersey, USA. He undertook postdoctoral training at Rutgers University in New Jersey, USA, and Temple University in Pennsylvania, USA. He received his PhD as a CSIR senior research fellow from the Indian Institute of Chemical Biology, Kolkata, India. He is a named co-inventor on seven issued US patents, including an ATR kinase inhibitor that has advanced to clinical trials. He has published more than 20 scientific articles and authored one book chapter. His research focuses on applying computational chemistry methods to drive small-molecule drug discovery programs, particularly in oncology and neurodegenerative diseases.

Part I

Molecular Dynamics and Related Methods in Drug Discovery

1

Binding Free Energy Calculations in Drug Discovery

Anita de Ruiter[1] and Chris Oostenbrink[1,2]

[1] Institute for Molecular Modeling and Simulation, Department of Material Sciences and Process Engineering, University of Natural Resources and Life Sciences, Vienna, Muthgasse 18, 1190 Vienna, Austria
[2] Christian Doppler Laboratory for Molecular Informatics in the Biosciences, University of Natural Resources and Life Sciences, Vienna, Muthgasse 18, 1190 Vienna, Austria

1.1 Introduction

This chapter attempts to provide an overview of the different approaches and methods that are available to compute binding free-energy in drug design and drug discovery. We do not provide an exhaustive list of available methods and do not rigorously derive all of the methods from first principles. Instead, we aim to give a overview of available methods and to point at the intrinsic limitations and challenges of these methods, such that researchers applying these methods can make a fair estimate of the most appropriate methods for their aims.

Numerous methods for the calculation of binding free energies have been developed over the years [1]. Which method is the best choice depends on how many free energies need to be determined, the available computational resources, the accuracy one wishes to obtain, and other specific properties of the system under study. Let us start by separating the available methods into three classes. Binding free energies can be calculated with endpoint, alchemical, or pathway methods. These methods are very different, not only in terms of their underlying theory but also in their accuracy and efficiency. The endpoint methods are very efficient in terms of computational requirements, but, unfortunately, not very accurate. Alchemical methods, on the other hand, are considered one of the most accurate but also slow methods. Pathway methods are also computationally demanding but can give important information about the binding pathways. Which method is the best choice will mostly depend on the stage at which the drug discovery/design is currently at. In the very early stages, where whole databases of compounds need to be screened, one can likely not afford the computational costs of alchemical approaches. However, since the range of binding free energies that are to be predicted may also be rather large, the faster methods will be sufficient to pick up some hit compounds. In the lead optimization stage, where rather similar compounds are studied, a more accurate method is required that can detect smaller differences

Computational Drug Discovery: Methods and Applications, First Edition.
Edited by Vasanthanathan Poongavanam and Vijayan Ramaswamy.
© 2024 WILEY-VCH GmbH. Published 2024 by WILEY-VCH GmbH.

in the binding free energies. Because the optimization stage also focuses on fewer leads, the higher computational demand for the more accurate method can actually be afforded.

1.1.1 Free Energy and Thermodynamic Cycles

First of all, we should look into the definition of free energy to find out what kind of property we are trying to determine. The definition of free energy in statistical mechanics is

$$G = -k_B T \ln Q_{NPT} \tag{1.1}$$

where G is the Gibbs free energy, k_B is the Boltzmann constant, T is the temperature, and Q_{NPT} is the partition function for a system with a constant number of particles, pressure, and temperature. The partition function is defined as

$$Q_{NPT} = \iint e^{-H(\mathbf{r},\mathbf{p})/k_B T} d\mathbf{r} d\mathbf{p} \tag{1.2}$$

with $\mathbf{r}$ and $\mathbf{p}$ as the positions and momenta of all atoms in the system, respectively, and $H(\mathbf{r},\mathbf{p})$ as the Hamiltonian of the system, giving the total energy. It is clear from the integration of all positions and momenta in Eq. (1.2) that the free energy is intrinsically a property of a statistical mechanical ensemble and not something that can be estimated from a single configuration. Free energy is a property of all (relevant) configurations of the system together. Any effort to estimate the free energy from a single configuration will likely miss out on some relevant aspects of the ensemble, such as conformational changes and their entropic effects.

In the field of drug discovery, one is not interested in the absolute free energy of a certain state, but rather in the binding free energy of, e.g. a small molecule to a protein;

$$\Delta G_{bind} = -k_B T \ln \frac{Q_{bound}}{Q_{free}} \tag{1.3}$$

where Q_{bound} and Q_{free} are the partition functions for the system where the small molecule is bound to the protein and when both partners are free in solution, respectively.

Furthermore, during hit-to-lead or lead optimization stages, one is mostly interested in the relative binding free energy, i.e. which of the two ligands binds stronger to the protein than the other. This, together with the fact that free energy is a state function, makes it possible to design thermodynamic cycles to make it easier to calculate the free energies. Consider a thermodynamic cycle like that in Figure 1.1.

There are four states, one with ligand A bound to the protein, one with ligand A unbound from the protein, one with ligand B bound to the protein, and one with ligand B unbound from the protein. Since free energy is a state function, following the full thermodynamic cycle will lead to a free energy difference of 0:

$$\Delta G_{BA}(prot) - \Delta G_{bind}(B) - \Delta G_{BA}(free) + \Delta G_{bind}(A) = 0 \tag{1.4}$$

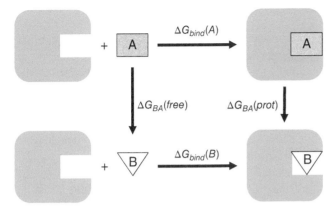

Figure 1.1 Thermodynamic cycle for the calculation of relative binding free energies of two small molecules A and B binding to a common receptor.

From this, it follows that $\Delta G_{bind}(B) - \Delta G_{bind}(A) = \Delta G_{BA}(prot) - \Delta G_{BA}(free)$. This means that we can determine the difference in binding free energy without performing a tedious simulation of the actual binding process. Although the modification of ligand A to ligand B is not something that is physically possible in the laboratory, it is possible with computer simulations and alchemical free-energy calculations. In fact, it is often easier to obtain converged results for these unphysical processes because modifying the ligands will most likely lead to much less reorganization of the protein than the binding process would. Modifying the ligand requires intermediate states, which will be discussed in more detail in the section on alchemical methods.

1.2 Endpoint Methods

As the name implies, endpoint methods only require the simulation of the endpoints of the system of interest. For binding free energy calculations, the endpoints would be the protein–ligand complex and the separate protein and ligand. That is, we explicitly simulate the states of the corners of the thermodynamic cycle of Figure 1.1. Their efficiency and reasonable accuracy make the endpoint free energy methods very popular in the early stages of drug discovery. Here, we will discuss two kinds of endstate methods: the molecular mechanics Poisson–Boltzmann surface area (MM/PBSA) methods and methods derived from linear response theory.

1.2.1 MM/PBSA and MM/GBSA

The most commonly used methods are MM/PBSA and the closely related molecular mechanics generalized Born surface area (MM/GBSA) [2, 3].

In MM/PBSA, the free energy of a state is composed of several contributions;

$$G = E_{MM} + G_{pol} + G_{np} - TS \tag{1.5}$$

$$E_{MM} = E_{bnd} + E_{el} + E_{vdW} \tag{1.6}$$

Here, E_{MM} is the molecular mechanics potential energy term, which consists of bonded interactions (E_{bnd}), electrostatic interactions (E_{el}), and van der Waals interactions (E_{vdW}). G_{pol} and G_{np} are the polar and nonpolar contributions to the solvation free energy, respectively. T represents the temperature of the system, and S is the entropy. Note that, although the free energy is a property of the ensemble and not an average over the ensemble, these methods assume that these terms together approximate the free energy of the state reasonably well and can be computed from individual configurations of the ensemble.

In order to calculate the absolute binding free energy of a system, the free energy of the free ligand (L), the unbound protein (P), as well as the complex (PL) needs to be computed;

$$\Delta G_{bind} = \langle G_{PL} \rangle_{PL} - \langle G_P \rangle_P - \langle G_L \rangle_L \tag{1.7}$$

Here, the angular brackets indicate an ensemble average from the simulation of the system indicated in the subscript. Equation (1.7) is the so-called three-average MM/PBSA (3A-MM/PBSA) since three different simulations need to be performed. The ensembles in Eq. (1.7) are generated from snapshots of molecular dynamics simulations with an explicit solvation model. Once these snapshots are generated, they are stripped from all solvent molecules and ions, and an implicit solvation model is used for further analysis.

G_{pol} is determined either by solving the Poisson–Boltzmann (PB) equation or by using the generalized Born equation (in which case the method would be called MM/GBSA). GB uses an analytical expression for the polar solvation energy and is thus much faster, but also likely to be less accurate, although this is system-dependent. G_{np} is estimated by using the solvent accessible surface area (SA). The assumption that G_{pol} and G_{np} can be approximated from an implicit solvation model means that solvent degrees of freedom are no longer treated explicitly in Eq. (1.2) and lead to the use of simple ensemble averages in Eq. (1.7). The calculation of G_{pol} furthermore depends strongly on the implicit solvation model that is used. Usually, the implicit solvation model requires a single dielectric constant to be chosen to describe the very complex electrostatic environment within the protein. This either makes the results unreliable, or the user can choose the constant such that the results are in agreement with known binding free energies for the system. In the latter case, MM/PBSA becomes more of an empirical method, where the parameters are optimized to reproduce experimental data. Finally, as a result of the implicit solvation model, MM/PBSA is not very well-suited when the binding site involves a highly charged environment or when critical water molecules are within the binding site.

The second reason that the ensemble property of Eqs. (1.1) and (1.2) may be approximated by a simple ensemble average in Eq. (1.7) is the explicit separation of the protein and ligand degrees of freedom into an energetic contribution (E_{MM}) and an entropic contribution (TS). The energetic term is computed from a force field, which is indeed well captured by an ensemble average. The entropy term is most commonly estimated with normal mode analysis (NMA). However, this method, which estimates the curvature of the energy landscape and approximates the entropy based on the expected sampling on this surface, is rather time-consuming and therefore not suitable for larger systems. More efficient methods have been explored over the years, but especially when the interest lies with relative binding free energies (like in drug discovery), the entropy term is often simply ignored. The underlying assumption would be that similar ligands will have similar entropy terms. Effectively, however, this means that the free energy is approximated by an energy.

A further, very popular, approximation is to use the single-trajectory MM/PBSA (1A-MM/PBSA), where only the complex is simulated

$$\Delta G_{bind} = \langle G_{PL} - G_P - G_L \rangle_{PL} \tag{1.8}$$

G_P and G_L are determined from the ensemble of the complex by just removing the atoms that are not part of the state of interest, i.e. for G_P, the ligand atoms are removed from the complex simulations, and for G_L, the protein atoms are removed. There are two main advantages of 1A-MM/PBSA with respect to 3A-MM/PBSA. The most obvious one is that only a single simulation needs to be performed instead of three simulations, and therefore it is computationally more efficient. The second one comes from the fact that E_{bnd} and all intramolecular contributions to E_{el} and E_{vdW} cancel in Eq. (1.8) because these energies for the apo protein and isolated ligand are calculated from exactly the same configuration as the complex. Also, the entropy estimate will seemingly become negligible since the ligand does not sample different conformations in the bound or in the free state. This significantly reduces the noise in the free energies, allowing for faster convergence of the results.

However, we need to consider what additional assumptions are being made with the 1A-MM/PBSA approach. It basically assumes that the protein and ligand visit the exact same conformations when they are in complex with each other as when they are each free in solution. This is not very likely to be the case. For example, the ligand can be forced to be more rigid (and/or bend) within a tight binding site, and a protein side chain or loop can be pushed aside upon binding of the ligand. The energetic and entropic effects of such conformational changes can be significant, but are entirely absent from the 1A-MM/PBSA approach. Unfortunately, the fact that the 1A-MM/PBSA often leads to less noise in the calculation does not make it more appropriate.

Recent advances try to address several of the approximations in the MM/PBSA and MM/GBSA methods with some promising results, but the optimal solutions remain rather system-dependent. For further reading, we suggest some recent reviews on the topic [3–5].

1.2.2 Linear Response Approximations

Other endpoint methods are based on the linear response theory. In the linear response approximation (LRA) framework [6, 7], two additional states need to be simulated, which are the neutralized states of the ligand when it is bound to the

1 Binding Free Energy Calculations in Drug Discovery

protein and when it is free in solution. Any partial charges of the ligand are set to 0 in these simulations. The charging free energy difference is then calculated with

$$\Delta G_{N \to Q}^{LRA} = G_Q - G_N = \frac{1}{2}[\langle H_Q - H_N \rangle_N + \langle H_Q - H_N \rangle_Q]$$
$$= \frac{1}{2} \left(\langle V_{ls}^{el} \rangle_N + \langle V_{ls}^{el} \rangle_Q \right) \quad (1.9)$$

where H_Q is the Hamiltonian of the charged state and H_N is the Hamiltonian of the neutralized state; subscripts after the angular brackets indicate which Hamiltonian was used to obtain the ensemble; and V_{ls}^{el} are the electrostatic interactions of the ligand with its surroundings.

In the Linear Interaction Energy (LIE) method [8], it is assumed that the electrostatic interactions of the charged ligand obtained from the ensemble of neutral states, $\langle V_{ls}^{el} \rangle_N$, will average to 0. Although this assumption is reasonable for the ligand in solution, it might not hold for the ligand bound to a protein. The protein is not likely to have a random electrostatic distribution around the neutral ligand and $\langle V_{ls}^{el} \rangle_N$ corresponds to the preorganization energy of the protein. The free energy difference of charging a ligand using LIE is then calculated with

$$\Delta G_{N \to Q}^{LIE} \approx \beta \langle V_{ls}^{el} \rangle_Q \quad (1.10)$$

where β is theoretically 1/2. The nonpolar interactions are also assumed to have a linear relationship with the free energy difference, even though this is only based on observations that the free energy of solvation for nonpolar particles and the interaction energy both seem to be linearly correlated with the size of the molecule. The binding free-energy difference based on LIE can thus be calculated with

$$\Delta G_{bind}^{LIE} = \alpha \Delta \langle V_{ls}^{vdW} \rangle_Q + \beta \Delta \langle V_{ls}^{el} \rangle_Q + \gamma \quad (1.11)$$

where α and γ are empirical parameters, which can be used to fit to experimental data from a data set. Δ indicates the difference between the ensemble averages obtained from the simulation of the free ligand and when bound to the protein. Even though β has a theoretical value of 1/2, it is also often used as an empirical parameter. Scaling the interactions with α, β, and adding γ helps compensate for the missing factors in the LIE approach, such as intramolecular energies, entropic confinement, and desolvation effects.

As an alternative to LIE, we have developed third power fitting (TPF), in which we do not assume linearity for the charging free energy [9]. Instead, the neutral and charged states are simulated, and the curvature of the charging curve is estimated by a third-order polynomial of a coupling parameter λ. Four constraints are used to find the best fit, which are based on the first $(dG/d\lambda)$ and second $(d^2G/d\lambda^2)$ derivatives of the free energy with respect to λ, from simulations in the N and Q states of LRA. It can be shown using the cumulant expansion that $d^2G/d\lambda^2$ is equal to the negative of the fluctuations of $dH/d\lambda$

$$\left. \frac{d^2 G}{d\lambda^2} \right|_S = \frac{1}{k_B T} \left(\left\langle \frac{\partial H}{\partial \lambda} \right\rangle_S^2 - \left\langle \left(\frac{\partial H}{\partial \lambda} \right)^2 \right\rangle_S \right) = \frac{1}{k_B T} \left(\langle V_{ls}^{el} \rangle_S^2 - \left\langle (V_{ls}^{el})^2 \right\rangle_S \right)$$
$$(1.12)$$

With the subscript S corresponding to either N or Q. The advantage of TPF is that there is that the nonlinearity is captured without additional simulations or empirical parameters. One should keep in mind, though, that the fluctuations in Eq. (1.12) are slower to converge.

1.3 Alchemical Methods

Once one or more lead compounds have been discovered during the early stages of drug discovery, lead optimization can be performed with more rigorous alchemical methods. Especially relative binding free energies between compounds that do not differ too much can be calculated very accurately with these methods. As mentioned above, relative binding free energies can be determined by morphing ligand A into ligand B, when bound and when free in solution. This means that simulations are performed in the direction represented by the vertical arrows of the thermodynamic cycle in Figure 1.1. Molecular properties that are changed during this process can include atom type, (partial) charges, bond lengths, angles, and dihedrals. All these changes are usually performed with several intermediate steps since convergence of the simulations would otherwise not be reached. Since free energy is a state function, any intermediate state can be chosen to make the calculations more efficient, even if it is unphysical. The intermediate states are defined by a coupling parameter λ, where at $\lambda = 0$, ligand A is represented, and at $\lambda = 1$, ligand B is represented. This means that at intermediate values of λ, the ligand is a nonphysical representation of a mixture of both ligands.

1.3.1 Free Energy Perturbation

Thermodynamic integration (TI) [10] and Bennett acceptance ratio (BAR) [11] are the two most commonly used alchemical free energy methods. All of these methods are often referred to as free energy perturbation, even though we prefer to reserve that term for the perturbation equation of Zwanzig [12]. This equation shows that relative free energy can be expressed in terms of the ratio of the partition functions of the two states.

$$\Delta G_{BA}^{FEP} = G_B - G_A = -k_B T \ln \frac{\iint e^{-H_B/k_B T}}{\iint e^{-H_A/k_B T}} \tag{1.13}$$

Multiplying the first exponential by 1 written in the form of $e^{+H_A/k_B T}e^{-H_A/k_B T}$, we find an ensemble average

$$\Delta G_{BA}^{FEP} = -k_B T \ln \left\langle e^{-\frac{H_B - H_A}{k_B T}} \right\rangle_A \tag{1.14}$$

A simulation of only state A is thus predicting the free energy difference toward state B. Accurate results are only obtained if the simulation of state A also samples the relevant conformational states for state B. If this is not the case, additional intermediate states can be used to increase the phase space overlap.

1.3.2 Thermodynamic Integration

In TI, the free energy difference between two states A and B is calculated with

$$\Delta G_{BA}^{TI} = \int_0^1 \frac{dG(\lambda)}{d\lambda} = \int_0^1 \left\langle \frac{\partial H(\lambda)}{\partial \lambda} \right\rangle_\lambda d\lambda \tag{1.15}$$

where the term with the angular brackets indicates the ensemble average of the derivative of the Hamiltonian with respect to λ obtained from the simulation at that λ-value. After all simulations at a number (N) of intermediate states are finished, the free energy difference between the two end states is obtained by numerical integration. In most cases, the trapezoidal rule is used for integration, although other integration schemes have been suggested as well. The number and spacing of intermediate states that are required to get a reliable result are strongly system-dependent. Since the integration is done numerically, areas with large curvature need to have a denser spacing of λ. We ourselves have suggested the use of extended TI, in which the values of $<\partial H/\partial \lambda>$ at many intermediate λ-values are predicted from simulations at a smaller number of λ-values, leading to smooth curves that explicitly capture the curvature [13].

1.3.3 Bennett's Acceptance Ratio

Another method to calculate the free energy difference between two states is BAR. This method uses ensemble averages obtained from both states to get a minimum statistical error of the free energy difference.

$$\Delta G_{ji}^{BAR} = k_B T \ln \left(\frac{\langle f(H_i - H_j + C) \rangle_j}{\langle f(H_j - H_i - C) \rangle_i} \right) + C \tag{1.16}$$

where $f(x) = 1/[1 + \exp(x/k_B T)]$ is the Fermi function and C is a constant. The optimal statistical estimate of the free energy difference is obtained when one solves for C such that the numerator and denominator of the logarithm are equal. It follows that

$$C = -k_B T \ln \left(\frac{N_j Q_i}{N_i Q_j} \right) \tag{1.17}$$

where N_j and N_i are the number of configurations in states j and i, respectively. Q_i and Q_j are the partition functions for states i and j, respectively. Since the partition functions are included in C, they need to be solved iteratively to get a self-consistent result. Convergence of the iterative process will only be reached if sufficient overlap between the forward and backward energy differences ($H_i - H_j$ and $H_j - H_i$ as obtained in the ensemble averages of j and i, respectively) is achieved. The final free energy difference between states A and B is determined with

$$\Delta G_{BA}^{BAR} = \sum_{i=1}^{n-1} \Delta G_{i+1,i}^{BAR} \tag{1.18}$$

BAR can be extended to using the configurations of all states to contribute to the free energy determination; this is referred to as Multistate Bennett Acceptance Ratio (MBAR) [14].

1.3.4 Nonequilibrium Methods

TI and BAR are both methods that rely on the simulation of states, which are at equilibrium. However, nonequilibrium simulations can also be used to determine free energy differences. According to the Jarzysnki equation [15], the ensemble average over the exponential distribution of the work relates to the free energy;

$$\Delta G = -k_B T \ln \left\langle e^{-W/k_B T} \right\rangle \quad (1.19)$$

This means that alchemical perturbations can be performed in a single simulation, where λ is changing from 0 to 1, continuously. Since the system does not need to be in equilibrium, these simulations can be rather short. However, they must be performed many times, and initial configurations must be sampled from a proper ensemble, in order to get a reasonable estimate of the work distribution.

Another nonequilibrium method is the Crooks Gaussian intersection (CGI) [16]. Similar to the Jarzynski equation, it requires the work obtained from a number of nonequilibrium simulations. In addition, simulations are performed in the inverse direction (i.e. changing λ from 1 to 0). The forward and backward work distributions are fitted with Gaussian distributions, and the intersection of them leads to the free energy estimate;

$$\frac{P_f(W)}{P_b(W)} = e^{W - \Delta G/k_B T} \quad (1.20)$$

Here, $P_f(W)$ and $P_b(W)$ are the work distributions for the forward and backward simulations, respectively. CGI will give results with reasonable accuracy when the distributions are sufficiently sampled and there is an overlap between the forward and backward work distributions.

1.3.5 Multiple Compounds

The alchemical methods described above are used to determine a single relative binding free energy between two compounds. However, during the lead optimization stage, there are many more compounds of interest. Just performing all possible combinations of compounds will be extremely time-consuming and will likely lead to many non-converged simulations because the differences between the compounds are too large. So, how do we decide which combinations of the compounds are good to determine the relative binding free energies? For smaller sets of compounds, this can be done manually, but for larger sets, the lead optimization mapper (LOMAP) [17] has been introduced. LOMAP uses the maximal common substructure to group structurally similar compounds together. It makes sure that rings are preserved as much as possible and that the compounds have the same net charge when relative binding free energies are being calculated (see below for a short description of why such perturbations are particularly challenging). It then determines optimal thermodynamic cycles, like in Figure 1.2, which can be used as an internal validation. When the free energy along the thermodynamic cycle is close to 0, the free energy calculations are likely to have converged. LOMAP designs these cycles automatically and keeps them rather small in order to prevent cancellation of

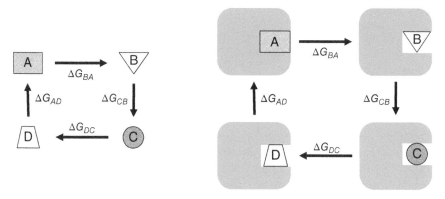

Figure 1.2 Thermodynamic cycles designed for internal validation of the calculated relative binding free energies.

errors. It also makes sure that all compounds (of the same net charge) are connected to each other.

1.3.6 One-Step Perturbation Approaches

Another approach to include multiple compounds is to generate simulation setups from which the binding free energies of more than two compounds can be estimated in one go. We have already discussed that intermediate states do not necessarily have to be physical. This means we can also use methods that rely on the simulation of only a single nonphysical (intermediate) state. An alternative thermodynamic cycle can be created, where one goes from state A to a reference state REF and then from REF to state B.

$$\Delta G_{A \to B} = \Delta G_{A \to REF} + \Delta G_{REF \to B} \tag{1.21}$$

The free energy difference can still be estimated with the Zwanzig relationship, based on the simulation of the REF state.

$$\Delta G_{A \to REF} = -k_B T \ln \left\langle e^{-(H_A - H_{REF})/k_B T} \right\rangle_{REF} \tag{1.22}$$

In one-step perturbation (OSP) approaches [18], REF can also be a nonphysical state that is designed to sample relevant conformational space for multiple real compounds. This works best if the compounds are closely related, so it is useful in the lead optimization stage. Note that in OSP, one is not restricted in the number of physical states the reference state represents, and the effect of small neutral substituents on a common scaffold can easily be screened for, e.g. a halogen scan. In practice, the method is more challenging if the differences between compounds involve larger changes in the charge distribution or the flexibility of the molecules.

The reference state in OSP needs to be selected carefully and may involve soft-core interactions, such that relevant conformations can be sampled for smaller and larger ligands in a single simulation. Alternative applications involve promiscuous stereocenters that sample multiple chemical configurations of a single scaffold. In practice, it can be quite difficult to design the reference state such that it results in adequate phase space overlap for all ligands under investigation.

Another method that is based on the simulation of a single reference state is enveloping distribution sampling (EDS) [19] The main difference with OSP is that EDS has an automated way of combining the Hamiltonians of the n ligands under investigation into the reference state. The n end-state Hamiltonians are combined into the reference state by Boltzmann weighting

$$H_{REF} = -k_B T \ln \left(\sum_{i=1}^{n} e^{-(H_i - \Delta F_i^R)/k_B T} \right) \tag{1.23}$$

where ΔF_i^R are free energy offset parameters. These ΔF_i^R correspond to the relative free energies of the end states if all states are sampled with equal probability. This complex energy surface is subsequently further smoothened by the use of a smoothening parameter or acceleration factors. The method typically requires two stages: a (set of) simulations in which the optimal parameters are derived, followed by a production simulation to obtain the free-energy differences.

1.3.7 Challenges in Alchemical Free Energy Calculations

The simplest functional form for connecting molecules A and B through the coupling parameter λ would be a linear combination of the two Hamiltonians

$$H(\lambda) = (1 - \lambda)H_A + \lambda H_B \tag{1.24}$$

However, in simulations where atoms are removed or added, this will cause major problems with the van der Waals interactions. When other atoms come close to the disappearing atom, E^{vdW}, which is an important nonbonded contribution of the Hamiltonians, becomes infinitely large, even at very small λ values. This restricts the sampling of configurational space. The same happens for the electrostatic interactions between atoms of equal charge. In order to address this problem, soft-core potentials are applied. The exact definition of the soft-core potential depends on the simulation software used. They all have in common that the singularity in the energy when the interatomic distance r_{ij} between two atoms approaches 0 is removed and the interaction function is smoothened for $\lambda \neq 0, 1$.

When alchemical changes involve a change in the net charge of the ligands, additional care needs to be taken. MD simulations are currently typically performed in explicit water under periodic boundary conditions, which means that the simulation systems are of the order of nanometers. At this scale, the simulation methodology will always treat the electrostatic interactions in an approximate way. One possibility is the use of cutoffs with reaction-field contributions to account for a homogeneous medium outside the cutoff sphere, and a second approximation is to use lattice summation to compute the electrostatic interaction for a strictly periodic molecular system. When a nonzero net charge is present during the simulation, these finite-size effects can lead to large charge-dependent artifacts that are propagated into the binding free energies [20]. To obtain results that do not depend on the size of the simulation box, one can either apply post-simulation corrections [21, 22] or try to avoid the net charge changes. Under some conditions, the latter can, e.g. be done by simultaneously performing opposite charge changes on a counter ion [23, 24]. Addition of explicit ionic solution tends to further screen the remaining artifacts.

Generally, one will want to add or remove as few atoms as possible during an alchemical perturbation. However, this may not be the case if rings are involved. If one compound has a ring and the other is very similar, with just the difference that the ring is no longer closed, it may not always be a good idea to just break the bond. This perturbation will not converge since there will basically be no phase space overlap between the two compounds. Instead it is much better to remove all atoms involving the ring and let the other atoms appear, even if this means that many more atoms need to be added and removed. Several tools are available to determine the optimal perturbation pathway between two compounds [17, 25].

Water molecules can play an important part in binding as they are able to stabilize the interactions between the protein and ligand. When investigating several ligands in the same hydrated active site, it is possible that the optimal number of water molecules is different for different ligands [26]. This is important to keep in mind when relative binding free energy calculations are performed, especially when the active site is buried and the water molecules cannot easily move in or out of the active site. It might be necessary to alchemically make a water molecule disappear simultaneously with the changing ligand [27] or to perform the simulation in a grand-canonical ensemble [28], in which the number of water molecules in the active site may be adjusted during the simulations.

Alchemical methods can also be used to compute the full binding free energy of a ligand A, by defining the second molecule in the thermodynamic cycle as a noninteracting dummy molecule B. Effectively, the electrostatic and van der Waals interactions of the molecule of interest are scaled from fully interacting in state A to 0 in state B. For λ values close to 1, the lack of interactions will most likely lead to the molecule flying through the simulation box. In order to accelerate the convergence of the simulation of the (partly) decoupled ligand, restraints are applied to keep the ligand within the binding site. Several restraints can be used, i.e. a single distance restraint or additional angles and dihedrals can be used to restrain the ligand in its binding mode [29]. The obtained free energy difference subsequently needs to be corrected for the fact that a restraint was applied to the decoupled state, which can be done analytically.

Conformational changes of the protein (or ligand) during the perturbation can hamper the efficient convergence of the simulations. Longer simulation times might improve the situation if the free energy barrier between the conformations is not too high. Otherwise, enhanced sampling techniques can be applied to improve the convergence of the simulations. There are many techniques available; one of them is replica exchange molecular dynamics (REMD) [30]. Here, multiple noninteracting replicas are simulated simultaneously, each at a different temperature. The replicas at higher temperatures are more likely to overcome energy barriers, whereas replicas at a low temperature can get trapped in a local energy minimum. At certain intervals, an attempt is made to switch the configurations of two neighboring replicas.

The switch is then either accepted or rejected, according to the Metropolis criterion. This ensures that the copy with the lowest temperature (which is the original temperature of the system) also gets conformations for which the energetic barrier has been overcome. Instead of replicas with different temperatures, REMD can also be done with different Hamiltonians (HREMD), which can correspond to the end-states and the intermediate states along the free-energy calculation. When the source of the energetic barrier is known, precautions can be taken to ensure that the barrier is absent or easily crossed in at least one of the intermediate replicates.

Many different issues for alchemical methods remain, many of which are discussed in recent best-practices reviews [31, 32]. In addition, the performance of several free energy calculation methods has been evaluated in industrial drug design settings, offering an insight into real-life challenges [33–36].

1.4 Pathway Methods

The above methods give an insight into the (relative) binding free energy, the binding poses, and the interactions between the ligand and protein. However, there is no information on the binding process itself. Especially when the active site is buried, information on the binding path can also be very valuable, as it will give more insight into the binding kinetics, rather than just the binding thermodynamics. This is where pathway methods come into play [37, 38]. Probably the most intuitive way is to run an MD simulation of a solution with protein and ligand molecules present. Run this simulation long enough such that multiple binding and unbinding events are sampled and determine the binding free energy from the equilibrium binding constant

$$\Delta G_{bind} = -k_B T \ln K^\circ_{bind} = -k_B T \ln \left(\frac{[PL]C^\circ}{[P][L]} \right) \qquad (1.25)$$

where K°_{bind} is the equilibrium constant of the reversible binding process, $[PL]$, $[P]$, and $[L]$ are the concentrations of the complex, the protein, and the ligand, respectively, and C° is the standard-state concentration. In practice, this is not as simple as it sounds. Because simulations typically only involve a single protein and a single ligand, this is intrinsically different from an equilibrium of many bound and unbound molecules, requiring additional adjustments to the sampled volume, and the fraction in the equation above should only contain the number of observed bound vs. unbound configurations $P_{bound}/P_{unbound}$ [39].

But there are more challenges to observe the actual binding equilibrium in a straightforward simulation. First of all, a large simulation box is required, in order to be able to sample ligand configurations that are not interacting with the protein at all. Second, the ligand can spend a lot of time moving through the (large) simulation box before it finds the binding site of the protein. Third, after the ligand is finally bound to the protein, the unbinding still needs to be sampled to really observe the binding equilibrium. In the case of a strong binder, this often takes prohibitively long. All in all, the simulation time required to sample the reversible binding and unbinding events until equilibrium is reached is only very rarely feasible.

In order to speed up the binding and/or unbinding processes, additional restraining or pulling forces can be applied to the ligand. Starting from a bound configuration, one can define a reaction coordinate along which the ligand will move. This is usually the radial or linear distance between the centers of mass between the protein and the ligand, but much more elaborate coordinates can be designed. Unbinding can then be sampled by either a single simulation in which the ligand is gradually pulled toward a predefined free state (nonequilibrium simulation) or by multiple simulations with the ligand restrained to a slightly different part of the reaction coordinate (e.g. with umbrella sampling [US]). The nonequilibrium pulling simulations need a careful choice of pulling strength. It should be strong enough such that there is sufficient speed up in the simulation but not too strong, to avoid disruption of the protein structure. In order to obtain an equilibrium binding free energy estimate, the nonequilibrium simulations need to be repeated many times, and consequently, an exponential averaging according to the Jarzynski formalism needs to be performed.

In US, several intermediate states are generated in which the ligand is restrained to a different distance along the reaction coordinate. These biasing potentials make sure that unfavorable regions, as well as regions that require a conformational change, are properly sampled. When the individual umbrella simulations are converged and the phase space overlap between them is sufficient, the results are corrected for their biasing potentials, and the potential of mean force (PMF) can be constructed. Keeping the standard state correction in mind, the binding free energy can then be determined from the PMF. The efficiency of the US is very dependent on the system and the choice of the restraints. When neighboring umbrellas are too far apart, there is not enough phase space overlap, and the PMF is not properly converged. Similar problems occur when the restraining potential is too strong, such that only a very narrow range of distances is sampled during a simulation. Too weak restraints will cause the ligand to avoid regions with higher energy. Simulations at umbrellas that show conformational changes of the protein or the ligand can require very long simulations in order to reach convergence. This is especially the case with buried binding pockets where, i.e. amino acid side chains need to make space for the ligand to pass.

Pathway methods usually focus on the dissociation of the ligand from the protein. The advantage here is that it is not necessary to have knowledge about the binding path prior to the simulations. As mentioned above, the dissociation can be, e.g. enforced along a radial distance or a predefined path between the ligand and the (center of mass of the) active site. Enforcing the association process, however, is not so straightforward if the binding path is not known *a priori*. Gradually pulling the ligand to smaller radial distances is not very likely to result in binding. At a large radial distance, the ligand has a lot of space to move and can easily go to the other site of the protein. Pulling the ligand closer at this moment will just result in the ligand getting stuck at the surface of the protein, far away from the active site. Once a defined path is sampled and the simulations are converged, the binding free energy can be calculated. Although this value will be correct, it is possible that alternative paths are available, which are not sampled here, and thus the free energy profile along the path might be wrong.

1.5 Final Thoughts

We have outlined some of the commonly used methods above to compute the binding free energy in the context of drug discovery and drug design. In the last decade, the methods have typically become much more user-friendly and are partially incorporated into large drug discovery pipelines and software packages. This is a development that is extremely satisfying from the point of view of academic method development and furthermore crucial to ensure a broad application of such methods in efficient drug design. Recent examples of computational workflows (which include free energy calculations) driving drug discovery forward include the identification of novel allosteric binders for KRAS^{G12C} [40], the discovery of potent noncovalent inhibitors of the main protease of SARS-CoV-2 [41], and lead optimization of an inhibitor of phosphodiesterase 2A (PDE2A) [42], and others [43].

However, we also emphasize that all of these methods come with their own set of approximations and limitations, which we have tried to highlight wherever appropriate. Even if everyday use of the methods is becoming easier, we feel it is crucial for the user to understand the background of the methods that are being applied and to be aware of the intrinsic limitations they come with [36]. This is the only way to ensure that a free energy is predicted that is appropriate for the problem at hand and that can be used to guide new experiments and new designs.

The list of methods is far from exhaustive. Many alternative methods, modifications to the ones described, and further improvements have been described. The aim of this work was not to give a full overview of all methods available, but rather to offer a starting point to understand the key principles in free-energy calculations. The interested reader is encouraged to check out the work in the further reading section below.

References

1 Chipot, C. and Pohorille, A. (eds.) (2007). *Free Energy Calculations. Theory and Applications in Chemistry and Biology*. Berlin, New York: Springer Verlag.
2 Srinivasan, J., Cheatham, T.E., Cieplak, P. et al. (1998). Continuum solvent studies of the stability of DNA, RNA, and phosphoramidate–DNA helices. *J. Am. Chem. Soc.* 120 (37): 9401–9409. https://doi.org/10.1021/ja981844+.
3 Genheden, S. and Ryde, U. (2015). The MM/PBSA and MM/GBSA methods to estimate ligand-binding affinities. *Expert Opin. Drug Discov.* 10 (5): 449–461. https://doi.org/10.1517/17460441.2015.1032936.

4 Wang, E., Sun, H., Wang, J. et al. (2019). End-point binding free energy calculation with MM/PBSA and MM/GBSA: strategies and applications in drug design. *Chem. Rev.* https://doi.org/10.1021/acs.chemrev.9b00055.

5 King, E., Aitchison, E., Li, H., and Luo, R. (2021). Recent developments in free energy calculations for drug discovery. *Front. Mol. Biosci.* 8.

6 Lee, F.S., Chu, Z.-T., Bolger, M.B., and Warshel, A. (1992). Calculations of antibody-antigen interactions: microscopic and semi-microscopic evaluation of the free energies of binding of phosphorylcholine analogs to McPC603. *Protein Eng.* 5 (3): 215–228. https://doi.org/10.1093/protein/5.3.215.

7 Sham, Y.Y., Chu, Z.T., Tao, H., and Warshel, A. (2000). Examining methods for calculations of binding free energies: LRA, LIE, PDLD-LRA, and PDLD/S-LRA calculations of ligands binding to an HIV protease. *Proteins Struct. Funct. Bioinforma.* 39 (4): 393–407. https://doi.org/10.1002/(SICI)1097-0134(20000601)39:4<393::AID-PROT120>3.0.CO;2-H.

8 Åqvist, J., Medina, C., and Samuelsson, J.-E. (1994). A new method for predicting binding affinity in computer-aided drug design. *Protein Eng.* 7 (3): 385–391. https://doi.org/10.1093/protein/7.3.385.

9 de Ruiter, A. and Oostenbrink, C. (2012). Efficient and accurate free energy calculations on trypsin inhibitors. *J. Chem. Theory Comput.* 3686–3695. https://doi.org/10.1021/ct200750p.

10 Kirkwood, J.G. (1935). Statistical mechanics of fluid mixtures. *J. Chem. Phys.* 3 (5): 300. https://doi.org/10.1063/1.1749657.

11 Bennett, C.H. (1976). Efficient estimation of free energy differences from Monte Carlo data. *J. Comput. Phys.* 22 (2): 245–268.

12 Zwanzig, R.W. (1954). High temperature equation of state by a perturbation method. I. Nonpolar gases. *J. Chem. Phys.* 22: 1420.

13 de Ruiter, A. and Oostenbrink, C. (2016). Extended thermodynamic integration: efficient prediction of lambda derivatives at nonsimulated points. *J. Chem. Theory Comput.* 12 (9): 4476–4486. https://doi.org/10.1021/acs.jctc.6b00458.

14 Shirts, M.R. and Chodera, J.D. (2008). Statistically optimal analysis of samples from multiple equilibrium states. *J. Chem. Phys.* 129 (12): 124105. https://doi.org/10.1063/1.2978177.

15 Jarzynski, C. (1997). Equilibrium free-energy differences from nonequilibrium measurements: a master-equation approach. *Phys. Rev. E* 56 (5): 5018. https://doi.org/10.1103/PhysRevE.56.5018.

16 Goette, M. and Grubmüller, H. (2009). Accuracy and convergence of free energy differences calculated from nonequilibrium switching processes. *J. Comput. Chem.* 30 (3): 447–456.

17 Liu, S., Wu, Y., Lin, T. et al. (2013). Lead optimization mapper: automating free energy calculations for lead optimization. *J. Comput. Aided Mol. Des.* 27 (9): https://doi.org/10.1007/s10822-013-9678-y.

18 Mark, A.E., Xu, Y., Liu, H., and van Gunsteren, W.F. (1995). Rapid non-empirical approaches for estimating relative binding free energies. *Acta Biochim. Pol.* 42 (4): 525–535.

19 Christ, C.D. and van Gunsteren, W.F. (2007). Enveloping distribution sampling: a method to calculate free energy differences from a single simulation. *J. Chem. Phys.* 126 (18): 184110. https://doi.org/10.1063/1.2730508.

20 Hunenberger, P.; Reif, M. Single-Ion Solvation; 2011. https://doi.org/10.1039/9781849732222.

21 Rocklin, G.J., Mobley, D.L., Dill, K.A., and Hünenberger, P.H. (2013). Calculating the binding free energies of charged species based on explicit-solvent simulations employing lattice-sum methods: an accurate correction scheme for electrostatic finite-size effects. *J. Chem. Phys.* 139 (18): 184103. https://doi.org/10.1063/1.4826261.

22 Reif, M.M. and Oostenbrink, C. (2014). Net charge changes in the calculation of relative ligand-binding free energies via classical atomistic molecular dynamics simulation. *J. Comput. Chem.* 35 (3): 227–243. https://doi.org/10.1002/jcc.23490.

23 Chen, W., Deng, Y., Russell, E. et al. (2018). Accurate calculation of relative binding free energies between ligands with different net charges. *J. Chem. Theory Comput.* 14 (12): 6346–6358. https://doi.org/10.1021/acs.jctc.8b00825.

24 Clark, A.J., Negron, C., Hauser, K. et al. (2019). Relative binding affinity prediction of charge-changing sequence mutations with FEP in protein–protein interfaces. *J. Mol. Biol.* 431 (7): 1481–1493. https://doi.org/10.1016/j.jmb.2019.02.003.

25 Petrov, D. (2021). Perturbation free-energy toolkit: an automated alchemical topology builder. *J. Chem. Inf. Model.* 61 (9): 4382–4390. https://doi.org/10.1021/acs.jcim.1c00428.

26 Bodnarchuk, M.S. (2016). Water, water, everywhere … it's time to stop and think. *Drug Discov. Today* 21 (7): 1139–1146. https://doi.org/10.1016/j.drudis.2016.05.009.

27 Maurer, M., Hansen, N., and Oostenbrink, C. (2018). Comparison of free-energy methods using a tripeptide-water model system. *J. Comput. Chem.* 39 (26): 2226–2242. https://doi.org/10.1002/jcc.25537.

28 Bruce Macdonald, H.E., Cave-Ayland, C., Ross, G.A., and Essex, J.W. (2018). Ligand binding free energies with adaptive water networks: two-dimensional grand canonical alchemical perturbations. *J. Chem. Theory Comput.* 14 (12): 6586–6597. https://doi.org/10.1021/acs.jctc.8b00614.

29 Boresch, S., Tettinger, F., Leitgeb, M., and Karplus, M. (2003). Absolute binding free energies: a quantitative approach for their calculation. *J. Phys. Chem. B* 107 (35): 9535–9551. https://doi.org/10.1021/jp0217839.

30 Sugita, Y. and Okamoto, Y. (1999). Replica-exchange molecular dynamics method for protein folding. *Chem. Phys. Lett.* 314 (1–2): 141–151. https://doi.org/10.1016/S0009-2614(99)01123-9.

31 Lee, T.-S., Allen, B.K., Giese, T.J. et al. (2020). Alchemical binding free energy calculations in AMBER20: advances and best practices for drug discovery. *J. Chem. Inf. Model.* 60 (11): 5595–5623. https://doi.org/10.1021/acs.jcim.0c00613.

32 Mey, A.S.J.S., Allen, B.K., McDonald, H.E.B. et al. (2020). Best practices for alchemical free energy calculations [Article v1.0]. *Living J. Comput. Mol. Sci.* 2 (1): 18378–18378. https://doi.org/10.33011/livecoms.2.1.18378.

33 Homeyer, N., Stoll, F., Hillisch, A., and Gohlke, H. (2014). Binding free energy calculations for lead optimization: assessment of their accuracy in an industrial drug design context. *J. Chem. Theory Comput.* 10 (8): 3331–3344. https://doi.org/10.1021/ct5000296.

34 Breznik, M., Ge, Y., Bluck, J.P. et al. (2022). Prioritizing small sets of molecules for synthesis through in-silico tools: a comparison of common ranking methods. *ChemMedChem* e202200425. https://doi.org/10.1002/cmdc.202200425.

35 Schindler, C.E.M., Baumann, H., Blum, A. et al. (2020). Large-scale assessment of binding free energy calculations in active drug discovery projects. *J. Chem. Inf. Model.* 60 (11): 5457–5474. https://doi.org/10.1021/acs.jcim.0c00900.

36 Meier, K., Bluck, J.P., and Christ, C.D. (2021). Use of free energy methods in the drug discovery industry. In: *Free Energy Methods in Drug Discovery: Current State and Future Directions; ACS Symposium Series*, vol. 1397, 39–66. American Chemical Society https://doi.org/10.1021/bk-2021-1397.ch002.

37 Dickson, A., Tiwary, P., and Vashisth, H. (2017). Kinetics of ligand binding through advanced computational approaches: a review. *Curr. Top. Med. Chem.* 17 (23): 2626–2641.

38 Bruce, N.J., Ganotra, G.K., Kokh, D.B. et al. (2018). New approaches for computing ligand–receptor binding kinetics. *Curr. Opin. Struct. Biol.* 49: 1–10. https://doi.org/10.1016/j.sbi.2017.10.001.

39 De Jong, D.H., Schäfer, L.V., De Vries, A.H. et al. (2011). Determining equilibrium constants for dimerization reactions from molecular dynamics simulations. *J. Comput. Chem.* 32 (9): 1919–1928. https://doi.org/10.1002/jcc.21776.

40 Mortier, J., Friberg, A., Badock, V. et al. (2020). Computationally empowered workflow identifies novel covalent allosteric binders for KRASG12C. *ChemMedChem* 15 (10): 827–832. https://doi.org/10.1002/cmdc.201900727.

41 Zhang, C.-H., Stone, E.A., Deshmukh, M. et al. (2021). Potent noncovalent inhibitors of the main protease of SARS-CoV-2 from molecular sculpting of the drug perampanel guided by free energy perturbation calculations. *ACS Cent. Sci.* 7 (3): 467–475. https://doi.org/10.1021/acscentsci.1c00039.

42 Tresadern, G., Velter, I., Trabanco, A.A. et al. (2020). [1,2,4]Triazolo[1,5-*a*]pyrimidine phosphodiesterase 2A inhibitors: structure and free-energy perturbation-guided exploration. *J. Med. Chem.* 63 (21): 12887–12910. https://doi.org/10.1021/acs.jmedchem.0c01272.

43 Abel, R. (2022). Advanced computational modeling accelerating small-molecule drug discovery. In: *Contemporary Accounts in Drug Discovery and Development*, 9–25. Wiley. https://doi.org/10.1002/9781119627784.ch2.

2

Gaussian Accelerated Molecular Dynamics in Drug Discovery

Hung N. Do, Jinan Wang, Keya Joshi, Kushal Koirala, and Yinglong Miao

Center for Computational Biology and Department of Molecular Biosciences, University of Kansas, Lawrence, KS 66047, USA

2.1 Introduction

Molecular dynamics (MD) is a powerful computational technique for simulating biomolecular dynamics at an atomistic level [1]. Due to advancements in computing hardware and software, timescales accessible to MD simulations have increased while costs have decreased [2, 3]. However, conventional MD (cMD), which makes no use of any enhanced sampling schemes, is often limited to tens to hundreds of microseconds [3–10] for simulations of biomolecular systems and cannot reach the timescales required to observe many biological processes of interest, which typically occur over milliseconds or longer, due to high energy barriers (e.g. 8–12 kcal/mol) [3–10].

Many enhanced sampling techniques have been developed during the last several decades to overcome the challenges mentioned above [11–15]. One class of enhanced sampling techniques uses predefined collective variables (CVs) or reaction coordinates (RCs), including umbrella sampling (US) [16, 17], metadynamics [18, 19], adaptive biasing force [20, 21], and steered MD [22]. However, it can be challenging to define proper CVs prior to simulation [3], and predefined CVs might significantly limit the sampling of conformational space during simulations [3]. Another class of enhanced sampling techniques, including replica exchange MD (REMD) [23, 24] or parallel tempering [25], self-guided Langevin MD [26–28], and accelerated MD (aMD) [29, 30], do not require predefined CVs. The latter class of unconstrained enhanced sampling techniques remains attractive to improve the sampling of biomolecular dynamics and obtain sufficient accuracy in free energy calculations.

Gaussian accelerated molecular dynamics (GaMD) is an unconstrained enhanced sampling technique that works by applying a harmonic boost potential to smooth the biomolecular potential energy surface [31]. Since this boost potential usually exhibits a near Gaussian distribution, cumulant expansion to the second order

("Gaussian approximation") can be applied to achieve proper energy reweighting [32]. GaMD allows for simultaneous, unconstrained enhanced sampling and free energy calculations of large biomolecules [31]. GaMD has been successfully demonstrated on enhanced sampling of ligand binding [31, 33–36], protein folding [31, 35], protein conformational changes [34, 37–40], protein-membrane [41], protein–protein [42–44], and protein-nucleic acid [45, 46] interactions.

Furthermore, GaMD has been combined with other enhanced sampling methods, including the REMD [47, 48] and weighted ensemble [49], to further improve conformational sampling and free energy calculations [3]. Notably, a novel multi-level enhanced sampling strategy was developed by combining GaMD, US, and a new adaptive sampling technique (AS) with the "dual-water" mode and AMOEBA [50–52] polarizable force field in TINKER-HP [53, 54] to accelerate the convergence of GaMD simulations [49]. In addition, "selective GaMD" algorithms, including ligand GaMD (LiGaMD) [55], peptide GaMD (Pep-GaMD) [56], and protein–protein interaction GaMD (PPI-GaMD) [44], have been developed to enable repetitive binding and dissociation of small-molecule ligands, highly flexible peptides, and proteins within microsecond simulations, which allow for highly efficient and accurate calculations of ligand/peptide/protein binding free energy and kinetic rate constants [3]. Furthermore, GaMD has been combined with deep learning (DL) and free energy profiling in a workflow (GLOW) to predict molecular determinants and map free energy landscapes of biomolecules [37].

Here, we review the principles and selected applications of GaMD in drug discovery. In particular, GaMD has been applied for advanced simulations of various biomolecular systems, including G-protein-coupled receptors (GPCRs), nucleic acids, and human angiotensin-converting enzyme 2 (ACE2) receptors.

2.2 Methods

2.2.1 Gaussian Accelerated Molecular Dynamics

Consider a system comprised of N atoms with their coordinates $r \equiv \{\vec{r}_1, \cdots, \vec{r}_N\}$ and momenta $p \equiv \{\vec{p}_1, \cdots, \vec{p}_N\}$. The system Hamiltonian can be expressed as:

$$H(r,p) = K(p) + V(r), \tag{2.1}$$

where $K(p)$ and $V(r)$ are the system's kinetic and total potential energies, respectively. Next, we decompose the potential energy into the following terms:

$$V(r) = V_b(r) + V_{nb}(r) \tag{2.2}$$

where V_b and V_{nb} are the bonded and nonbonded potential energies, respectively. According to classic force fields, the bonded and nonbonded potential energies can be expressed as the following:

$$V_b = V_{bonds} + V_{angles} + V_{dih} + V_{improbers} + V_{UB} + V_{CMAP}, \tag{2.3}$$

$$V_{nb} = V_{elec} + V_{vdW}. \tag{2.4}$$

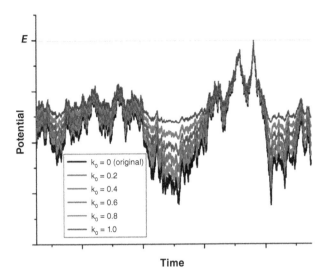

Figure 2.1 Schematic illustration of GaMD. When the threshold energy is set to the maximum potential ($E = V_{max}$), the system potential energy surface is smoothened by adding a harmonic boost potential that follows Gaussian distribution. The coefficient k_0 in the range of 0–1 determines the magnitude of the applied boost potential. With greater k_0, higher boost potential is added to the original energy surface in cMD, which provides enhanced sampling of biomolecules across decreased energy barriers. Source: Reproduced with permission of Miao et al. [31]/American Chemical Society / Public Domain CC BY 3.0.

To enhance biomolecular conformational sampling, we will add boost potential based on these potential energetic terms (*bonds, angles, dih, impropers, UB, CMAP, elec, VDW*).

GaMD works by adding a harmonic boost potential to smooth the potential energy surface when the system potential drops below a reference energy E [31] (Figure 2.1):

$$\Delta V(r) = \begin{cases} \frac{1}{2}k(E - V(\vec{r}))^2, & V(\vec{r}) < E \\ 0, & V(\vec{r}) \geq E, \end{cases} \quad (2.5)$$

where k is the harmonic force constant. The two adjustable parameters E and k can be determined based on three enhanced sampling principles. First, for any two arbitrary potential values $V_1(\vec{r})$ and $V_2(\vec{r})$ found on the original energy surface, if $V_1(\vec{r}) < V_2(\vec{r})$, ΔV should be a monotonic function that does not change the relative order of the biased potential values; i.e. $V_1^*(\vec{r}) < V_2^*(\vec{r})$. Second, if $V_1(\vec{r}) < V_2(\vec{r})$, the potential difference observed on the smoothed energy surface is smaller than that of the original, i.e. $V_2^*(\vec{r}) - V_1^*(\vec{r}) < V_2(\vec{r}) - V_1(\vec{r})$. The reference energy needs to be set in the following range:

$$V_{max} \leq E \leq V_{min} + \frac{1}{k}, \quad (2.6)$$

where V_{max} and V_{min} are the system's minimum and maximum potential energies. To ensure that Eq. (2.6) is valid, k must satisfy: $k \leq \frac{1}{V_{max} - V_{min}}$. Let us define $k \equiv k_0 \frac{1}{V_{max} - V_{min}}$, then $0 \leq k_0 \leq 1$. Third, the standard deviation of ΔV needs to be small enough (i.e. narrow distribution) to ensure proper energetic reweighting [32]: $\sigma_{\Delta V} = k(E - V_{avg})\sigma_V \leq \sigma_0$, where V_{avg} and σ_V are the average and standard deviation of the system's potential energies, $\sigma_{\Delta V}$ is the standard deviation of ΔV with σ_0 as a user-specified upper limit (e.g. $10k_BT$) for proper reweighting. When E is set to the lower bound $E = V_{max}$, k_0 can be calculated as:

$$k_0 = \min(1.0, k_0') = \min\left(1.0, \frac{\sigma_0}{\sigma_V} \frac{V_{max} - V_{min}}{V_{max} - V_{avg}}\right), \quad (2.7)$$

Alternatively, when the threshold energy E is set to its upper bound $E \leq V_{min} + \frac{1}{k}$, k_0 is set to:

$$k_0 = k_0'' \equiv \left(1.0 - \frac{\sigma_0}{\sigma_V}\right) \frac{V_{max} - V_{min}}{V_{avg} - V_{min}}, \quad (2.8)$$

if k_0'' is found to be between 0 and 1. Otherwise, k_0 is calculated using Eq. (2.7).

2.2.2 Ligand Gaussian Accelerated Molecular Dynamics

On the basis of GaMD, selective GaMD algorithms, including LiGaMD [55], have been developed for more efficient simulations and calculations of both free energy and kinetics of biological processes [3]. LiGaMD, in particular, allows for simulations of repetitive binding and dissociation of small-molecule ligands for calculating ligand binding free energies and kinetics [3, 55]. Given a simulation system consisting of ligand L, protein P and biological environment E, the system potential energy $V(r)$ from Eq. (2.1) could be decomposed into the following terms:

$$V(r) = V_{P,b}(r_P) + V_{L,b}(r_L) + V_{E,b}(r_E) + V_{PP,nb}(r_P) + V_{LL,nb}(r_L) + V_{EE,nb}(r_E)$$
$$+ V_{PL,nb}(r_{PL}) + V_{PE,nb}(r_{PE}) + V_{LE,nb}(r_{LE}) \quad (2.9)$$

where $V_{P,b}$, $V_{L,b}$, and $V_{E,b}$ are the bonded potential energies in protein P, ligand L, and environment E, respectively. $V_{PP,nb}$, $V_{LL,nb}$, and $V_{EE,nb}$ are the self-nonbonded potential energies in protein P, ligand L, and environment E, respectively. $V_{PL,nb}$, $V_{PE,nb}$, and $V_{LE,nb}$ are the nonbonded interaction energies between $P - L$, $P - E$, and $L - E$, respectively. On the other hand, the nonbonded potential energies are usually calculated as:

$$V_{nb} = V_{elec} + V_{vdW} \quad (2.10)$$

where V_{elec} and V_{vdW} are the system's electrostatic and van der Waals potential energies. In general, ligand binding involves mostly the nonbonded interaction energies of the ligand ($V_{LL,nb}$, $V_{PL,nb}$, and $V_{LE,nb}$). Therefore, a selective boost potential is added to the ligand nonbonded potential energy based on the GaMD algorithm [31]:

$$\Delta V_{L,nb}(r) = \begin{cases} \frac{1}{2}k_{L,nb}(E_{L,nb} - V_{L,nb}(r))^2, & V_{L,nb}(r) < E_{L,nb} \\ 0, & V_{L,nb}(r) \geq E_{L,nb} \end{cases} \quad (2.11)$$

where $E_{L,nb}$ is the threshold energy for applying boost potential and $k_{L,nb}$ is the harmonic constant.

In addition, multiple ligand molecules can be added to the solvent to facilitate ligand binding to proteins in MD simulations [3]. The higher the ligand concentration, the faster the ligand binds, as long as the ligand concentration is still within its solubility concentration [3]. Therefore, besides the selective boost added to the bound ligand, another boost potential could be applied to the unbound ligand molecules, protein, and environment to facilitate ligand dissociation and rebinding [3, 55]. The second boost potential is calculated using the total system potential energy excluding the nonbonded potential energy of the bound ligand as:

$$\Delta V_D(r) = \begin{cases} \frac{1}{2}k_D(E_D - V_D(r))^2, & V_D(r) < E_D \\ 0, & V_D(r) \geq E_D \end{cases} \tag{2.12}$$

where V_D is the total system potential energy excluding the nonbonded potential energy of the bound ligand, E_D is the threshold energy for applying the second boost potential, and k_D is the harmonic constant. Therefore, dual-boost LiGaMD has a total boost potential of $\Delta V(r) = \Delta V_{L,nb}(r) + \Delta V_D(r)$ [3, 55].

2.2.3 Energetic Reweighting of GaMD for Free Energy Calculations

For energetic reweighting of GaMD simulations, the probability distribution along a selected RC can be calculated from simulations as $p^*(A)$. Given the boost potential $\Delta V(r)$ of each frame in GaMD simulations, $p^*(A)$ can be reweighted to recover the canonical ensemble distribution, $p(A)$, as:

$$p(A_j) = p^*(A_j) \frac{\langle e^{\beta \Delta V(\bar{r})} \rangle_j}{\sum_{i=1}^{M} \langle p^*(A_i) e^{\beta \Delta V(\bar{r})} \rangle_i}, j = 1, \ldots, M \tag{2.13}$$

where M is the number of bins, $\beta = k_B T$ and $\langle e^{\beta \Delta V(\bar{r})} \rangle_j$ is the ensemble-averaged Boltzmann factor of $\Delta V(\bar{r})$ for simulation frames found in the jth bin. The ensemble-averaged reweighting factor can be approximated using cumulant expansion [31, 32]:

$$\langle e^{\beta \Delta V(\bar{r})} \rangle = \exp \left\{ \sum_{k=1}^{\infty} \frac{\beta^k}{k!} C_k \right\}, \tag{2.14}$$

where the first three cumulants are given by:

$$\begin{aligned} C_1 &= \langle \Delta V \rangle, \\ C_2 &= \langle \Delta V^2 \rangle - \langle \Delta V \rangle^2 = \sigma_{\Delta V}^2, \\ C_3 &= \langle \Delta V^3 \rangle - 3 \langle \Delta V \rangle^2 \langle \Delta V \rangle + 2 \langle \Delta V \rangle^3, \end{aligned} \tag{2.15}$$

The boost potential obtained from GaMD simulations usually shows near-Gaussian distribution [57]. Cumulant expansion to the second order thus provides a good approximation for computing the reweighting factor [31, 32]. The reweighted free energy $F(A) = -k_B T \ln p(A)$ is calculated as:

$$F(A) = F^*(A) - \sum_{k=1}^{2} \frac{\beta^k}{k!} C_k + F_c, \tag{2.16}$$

where $F^*(A) = -k_B T \ln p^*(A)$ is the modified free energy obtained from GaMD simulation and F_c is a constant.

To characterize the extent to which ΔV follows a Gaussian distribution, its distribution anharmonicity γ is calculated as [32]:

$$\gamma = S_{max} - S_{\Delta V} = \frac{1}{2}\ln(2\pi e\sigma_{\Delta V}^2) + \int_0^\infty p(\Delta V)\ln(p(\Delta V))d\Delta V \tag{2.17}$$

where ΔV is dimensionless as divided by $k_B T$ with k_B and T being the Boltzmann constant and system temperature, respectively, and $S_{max} = \frac{1}{2}\ln(2\pi e\sigma_{\Delta V}^2)$ is the maximum entropy of ΔV [32]. When γ is zero, ΔV follows the exact Gaussian distribution with sufficient sampling. Reweighting by approximating the exponential average term with cumulant expansion to the second order is able to accurately recover the original free energy landscape. As γ increases, the ΔV distribution becomes less harmonic, and the reweighted free energy profile obtained from cumulant expansion to the second order would deviate from the original. The anharmonicity of ΔV distribution serves as an indicator of the enhanced sampling convergence and accuracy of the reweighted free energy.

2.2.4 GLOW: A Workflow Integrating Gaussian Accelerated Molecular Dynamics and Deep Learning for Free Energy Profiling

Recently, GaMD and DL were integrated to develop GLOW–a workflow to identify important RCs and map free energy profiles of biomolecules [37]. First, dual-boost GaMD simulations are performed on the biomolecules of interest. Since simulation trajectories are collections of static PDB snapshots, the residue contact map of each simulation frame can be calculated and transformed into images. The specialized type of neural network for image classification, the two-dimensional (2D) convolutional neural network (CNN), is employed to classify the residue contact maps of target biomolecules. The default DL architecture of GLOW consists of four convolutional layers of 3 × 3 kernel size, with 32, 32, 64, and 64 filters, respectively, followed by three fully connected (dense) layers, the first two of which include 512 and 128 filters with a dropout rate of 0.5 each. The final fully connected layer is the classification layer. "ReLu" activation is used for all layers in the 2D CNN, except the classification layer, in which "softmax" activation is used. A maximum pooling layer of 2 × 2 kernel size is added after each convolutional layer [37]. The saliency (attention) map of residue contact gradients is calculated through backpropagation by vanilla gradient-based attribution [58] using the residue contact map of the most populated structural cluster of each system, from which important RCs can be identified. Finally, the free energy profiles of these RCs are calculated through reweighting of GaMD simulations to characterize the biomolecular systems of interest.

2.2.5 Binding Kinetics Obtained from Reweighting of GaMD Simulations

Provided sufficient sampling of repetitive protein dissociation and binding in the simulations, we recorded the time periods and calculated their averages for the ligand sampled in the bound (τ_B) and unbound (τ_U) states from the simulation trajectories. The τ_B corresponds to the protein residence time. Then, the ligand dissociation and binding rate constants (k_{off} and k_{on}) were calculated as:

$$k_{off} = \frac{1}{\tau_B} \tag{2.18}$$

$$k_{on} = \frac{1}{\tau_U \cdot [L]} \tag{2.19}$$

where [L] is the ligand concentration in the simulation system.

According to Kramers' rate theory, the rate of a chemical reaction in the large viscosity limit is calculated as [59]:

$$k_R \cong \frac{w_m w_b}{2\pi\xi} e^{-\Delta F/k_B T} \tag{2.20}$$

where w_m and w_b are frequencies of the approximated harmonic oscillators (also referred to as curvatures of free energy surface [60, 61]) near the energy minimum and barrier, respectively, ξ is the frictional rate constant and ΔF is the free energy barrier of transition. The friction constant ξ is related to the diffusion coefficient D with $\xi = k_B T/D$. The apparent diffusion coefficient D can be obtained by dividing the kinetic rate calculated directly using the transition time series collected directly from simulations by that using the probability density solution of the Smoluchowski equation [62]. In order to reweight protein kinetics from the GaMD simulations using the Kramers' rate theory, the free energy barriers of protein binding and dissociation are calculated from the original (reweighted, ΔF) and modified (no reweighting, ΔF^*) PMF profiles, similarly for curvatures of the reweighed (w) and modified (w^*, no reweighting) PMF profiles near the protein bound ("B") and unbound ("U") low-energy wells and the energy barrier ("Br"), and the ratio of apparent diffusion coefficients from simulations without reweighting (modified, D^*) and with reweighting (D). The resulting numbers are then plugged into Eq. (2.20) to estimate accelerations of the ligand binding and dissociation rates during GaMD simulations [59], which allows us to recover the original kinetic rate constants.

2.2.6 Gaussian Accelerated Molecular Dynamics Implementations and Software

GaMD has been implemented in widely used simulation packages including AMBER [31], NAMD [35], GENESIS [48], TINKER-HP [63], and OpenMM [64]. Overall, GaMD simulations consist of three main stages [55], including short cMD, GaMD equilibration, and GaMD production. During short cMD, a number of preparatory steps are first performed to equilibrate the system, and then the potential statistics (including V_{max}, V_{min}, V_{avg}, and σ_V) are collected [55]. During the GaMD pre-equilibration stage, boost potential is applied, but no boost parameters are updated [55]. In the GaMD equilibration stage, boost potential continues to be applied, and boost parameters are updated [55]. In the final stage of GaMD production, boost potential is applied, and the boost parameters are held fixed [55]. In a typical GaMD simulation, users need to specify the number of simulation steps for each stage as well as the number of simulation steps used to calculate the average and standard deviation of potential energies, a flag to apply boost potential, a flag to set the threshold energy E for applying boost potentials, a flag to restart GMD simulation, and the upper limits of the standard deviations of the first and second boost potentials. Additional resources related to GaMD can be found here: https://www.med.unc.edu/pharm/miaolab/resources/GaMD

2.3 Applications

2.3.1 G-Protein-Coupled Receptors

GPCRs are the largest family of human membrane proteins and the primary targets of ~34% of currently marketed drugs [65]. On the basis of sequence homology and functional similarity, GPCRs are classified into six different classes, including class A (Rhodopsin-like), class B (secretin receptors), which is further divided into subclasses of B1 (classical hormone receptors), B2 (adhesion GPCRs), B3 (methuselah-type receptors), class C (metabotropic glutamate receptors), class D (fungal mating pheromone receptors), class E (cyclic AMP receptors), and class F (frizzled/TAS2 receptors) [66, 67]. GPCRs share a characteristic structural fold of seven transmembrane (TM) α-helices (TM1-TM7), connected by three extracellular loops (ECL1-ECL3) and three intracellular loops (ICL1-ICL3). Next, we will review recent applications of GaMD in the studies of GPCRs (including adenosine (ADO) and muscarinic acetylcholine receptors).

2.3.1.1 Characterizing the Binding and Unbinding of Caffeine in Human Adenosine A_{2A} Receptor

Adenosine receptors (ARs) are a subfamily of class A GPCRs with ADO as the endogenous ligand [68], with four known subtypes: A_1AR, $A_{2A}AR$, $A_{2B}AR$, and A_3AR [69]. Despite their broad distribution in human tissues and functional differences, ARs share common antagonists of caffeine (CFF) and theophylline, both of which antagonize the receptors upon binding. Sequence alignment shows that the seven TM helix bundles of the A_1AR share high similarity with $A_{2A}AR$ by 71%, $A_{2B}AR$ by 70%, and A_3AR by 77%. Their sequence similarity is significantly reduced in the ECLs, being 43% for $A_{2A}AR$, 45% for $A_{2B}AR$, and 35% for A_3AR when compared with A_1AR [33].

All-atom GaMD simulations were performed to determine the binding and dissociation pathways of CFF to the human $A_{2A}AR$ [33], starting from the X-ray structure of $A_{2A}AR$ in complex with CFF (PDB: 5MZP) [70]. The T4-lysozyme, lipid molecules, CFF, water, and heteroatom molecules were removed. A total of 10 CFF ligand molecules were placed randomly at a distance >15Å from the extracellular surface of the $A_{2A}AR$. Spontaneous binding and dissociation of CFF in the receptor were successfully captured in the GaMD simulations [33]. A dominant binding pathway of CFF to the $A_{2A}AR$ was identified from the 63-ns GaMD equilibration (Figure 2.2a). CFF approached the $A_{2A}AR$ through interactions with ECL2, the extracellular mouth between ECL2, ECL3, and TM7, and finally, the receptor orthosteric site located deeply within the receptor TM bundle (Figure 2.2d). A slightly different binding pathway was observed when the orthosteric pocket of the $A_{2A}AR$ was already occupied by one CFF molecule in Sim2 (Figure 2.2b). In this pathway, the second CFF first explored a region between ECL3 and TM7 during the binding process (Figure 2.2e). The dissociation pathway of CFF was mostly the reverse of the dominant binding pathway (Figure 2.2c,f).

2.3.1.2 Unraveling the Allosteric Modulation of Human A_1 Adenosine Receptor

Recently, we investigated the allosteric effect of a newly discovered PAM MIPS521 in the A_1AR [71]. MIPS521 was identified to bind a novel lipid-facing pocket of the A1R and hardly changed the cryo-EM structure of the ADO-A1R-G_{i2} protein complex. Therefore, using the latest cryo-EM structures of A_1AR, we performed further GaMD simulations on four systems: ADO-A_1AR-G_{i2}-MIPS521, A_1AR-G_{i2}, ADO-A_1AR-MIPS521 and ADO-A_1AR (Figure 2.3) [71]. Three replicas of GaMD simulations lasting 500 ns or 1000 ns were performed. MIPS521 underwent high fluctuations in the ADO-A_1AR-MIPS521 system and could even dissociate in the GaMD simulations, while it remained bound to the ADO-A_1AR-G_{i2} complex during both the GaMD and cMD simulations. We then focused on the dynamics of the agonist, ADO. In the absence of G protein, ADO sampled a large conformational space in the orthosteric pocket and exhibited higher flexibilities in simulations with and without MIPS521 (Figure 2.3b,c). The presence of G protein decreased the conformational dynamics of ADO, consistent with a ternary complex model where the G protein allosterically stabilizes agonist binding in the orthosteric pocket (Figure 2.3d) [71]. In the presence of MIPS521, ADO was stabilized even further when G_{i2} was present in the A1R complex (Figure 2.3e). Next, we examined whether the effect of MIPS521 on ADO stability in the orthosteric pocket resulted from changes in receptor and G protein dynamics in the simulations. In the absence of G protein, the active receptor in the ADO-A_1AR GaMD simulations relaxed toward the inactive structure, for which the $R105^{3.50}$–$E229^{6.30}$ distance could decrease to ~8.3 Å (Figure 2.3f) [71]. Simulations in the presence of G protein, MIPS521, or both revealed TM3-6 distances consistent with the A_1AR in the active conformation, suggesting a direct stabilization of the A_1AR in a "G protein-bound-like" conformation by MIPS521 post-removal of G_{i2} (Figure 2.3g–i).

Finally, GLOW was applied to characterize GPCR activation and allosteric modulation, using the A_1AR as a model system [37]. GLOW characterization of GPCR activation was obtained through the classification of the A_1AR bound by "Antagonist" (PSB36), "Agonist" (adenosine), and "Agonist – Gi." GLOW achieved an overall accuracy of 99.34% and a loss of 1.85% on the validation data set after 15 epochs. GLOW revealed that the removal of the G_{i2} protein from the A_1AR bound by "Agonist-Gi" led to the deactivation of the A_1AR. The intracellular halves of TM3, TM5, TM6, and TM7 underwent significant conformational changes. In particular, TM6 drew closer to TM3, while TM7 moved away from TM3 and TM5 [37]. The receptor TM5 and TM6 intracellular domains and helix 8 also became more flexible in the absence of the G_{i2} protein. Replacement of "Agonist" with

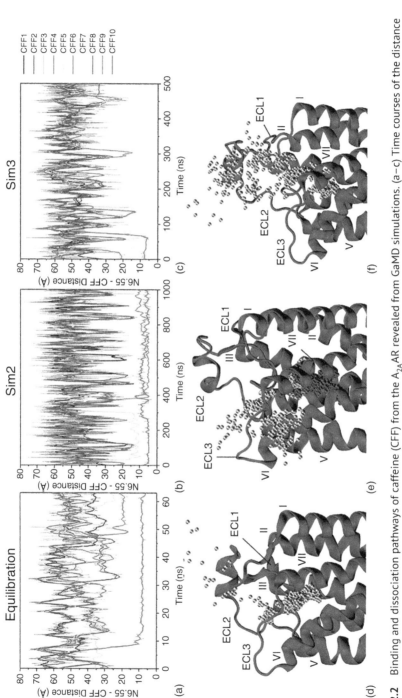

Figure 2.2 Binding and dissociation pathways of caffeine (CFF) from the $A_{2A}AR$ revealed from GaMD simulations. (a–c) Time courses of the distance between receptor residue N6.55 atom ND2 and CFF atom N1 calculated from GaMD equilibration, Sim2, and Sim3 GaMD production simulations. (d–f) Trace of CFF (orange and red) in the $A_{2A}AR$ observed in the GaMD equilibration, Sim2, and Sim3 GaMD production simulations. The seven transmembrane helices are labeled I–VII, and extracellular loops 1–3 are labeled ECL1-ECL3. Source: Adapted with permission from Do et al. [33]. Copyright 2021 Do, Akhter, and Miao. https://www.frontiersin.org/articles/10.3389/fmolb.2021.673170/full. Further permissions related to the material excerpted should be directed to Frontiers.

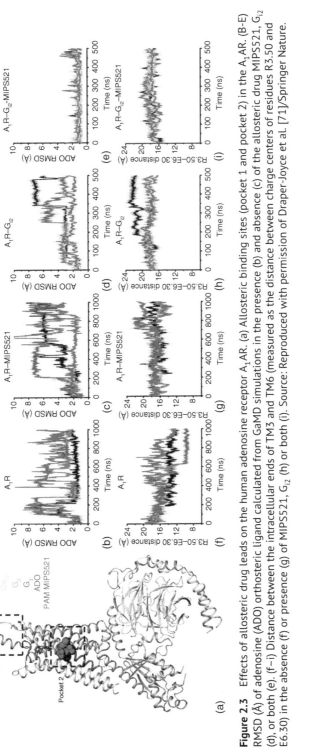

Figure 2.3 Effects of allosteric drug leads on the human adenosine receptor A_1AR. (a) Allosteric binding sites (pocket 1 and pocket 2) in the A_1AR. (B-E) RMSD (Å) of adenosine (ADO) orthosteric ligand calculated from GaMD simulations in the presence (b) and absence (c) of the allosteric drug MIPS521, G_{i2} (d), or both (e). (f–i) Distance between the intracellular ends of TM3 and TM6 (measured as the distance between charge centers of residues R3.50 and E6.30) in the absence (f) or presence (g) of MIPS521, G_{i2} (h) or both (i). Source: Reproduced with permission of Draper-Joyce et al. [71]/Springer Nature.

"Antagonist" in A_1AR led to the complete closure of the intracellular pocket, where TM3 formed intracellular residue contacts with TM6 and the NPxxY motif in TM7 moved away from TM3 and TM5. In addition, the ligand-binding extracellular domains and intracellular G-protein binding domains were found to be loosely coupled in GPCR activation [37, 72]. A_1AR sampled a more "open" conformational state of the extracellular mouth in the "Agonist-Gi" bound system due to mostly increased flexibility of the ECL1 and ECL2. Removal of the G_{i2} protein from the "Agonist-Gi"-A_1AR complex reduced the conformational space of the receptor ECL2, while replacement of "Agonist" by "Antagonist" confined the ECL2 significantly [37]. Characterization of GPCR allosteric modulation was done through the classification of the A_1AR bound by "Agonist-Gi" and "Agonist-Gi-PAM," with the PAM being MIPS521. GLOW achieved an overall accuracy of 99.27% and a loss of 1.78% on the validation data set after 15 epochs. Furthermore, GLOW showed that ECL2 played a critical role in the allosteric modulation of A_1AR [37], which is consistent with previous mutagenesis, structure, and molecule modeling studies [71, 73–76]. GLOW revealed that binding of a PAM (MIPS521) to the agonist-Gi-A_1AR complex biased the receptor conformational ensemble, especially in the ECL1 and ECL2 regions. PAM binding stabilized agonist binding within the orthosteric pocket of A_1AR, which confined the extracellular mouth of the receptor. Furthermore, PAM binding disrupted the $N148^{ECL2}$-$V152^{ECL2}$ α-helical hydrogen bond and distorted this portion of the ECL2 helix [37]. A small pocket was formed in the ECL2-TM5-ECL3 region because of this ECL2 helix distortion. Taken together, GaMD simulations provided mechanistic insight into the residue level of the positive cooperativity of MIPS521 in the A_1AR-G_{i2} complex.

2.3.1.3 Ensemble Based Virtual Screening of Allosteric Modulators of Human A_1 Adenosine Receptor

Virtual screening has been widely used for agonist/antagonist design targeting GPCRs [77]. However, it is rather challenging to apply virtual screening to identify allosteric modulators due to their low affinity compared with the agonist/antagonist. Recently, retrospective ensemble docking calculations of PAMs to the A_1AR combining GaMD simulations and Autodock [78] were performed [79]. Receptor structural ensembles obtained from GaMD simulations were used to increase the docking performance of known PAMs using the A_1AR as a model GPCR.

The GaMD simulations implemented in AMBER [31] and NAMD [35] were applied to generate receptor ensembles. The flexible docking and rigid body docking at different levels (i.e. short, medium, and long) were all evaluated. Docking scores corrected by the GaMD-reweighted free energy of the receptor structural cluster further improved the docking performances. The calculated docking enrichment factors and the area under the receiver operating characteristic curves are increased using ranking by the average binding energy in comparison with the minimum binding energy. Ensembles obtained from AMBER dual-boost GaMD simulations of the VCP171-bound ADO–A_1AR–Gi complex outperformed other ensembles for docking. Interactions between the PAM and receptor ECL2 in the VCP171-bound ADO–A_1AR–Gi complex might induce more suitable conformations for PAM binding, which were difficult to be sampled in the simulations of PAM-free (i.e. apo) A_1AR. Dual-boost GaMD with higher boost potential was observed to perform better than the dihedral-boost GaMD for ensemble docking. Overall, flexible docking performed significantly better than rigid-body docking at different levels with AutoDock, suggesting that the flexibility of protein side chains is also important in ensemble docking. In summary, docking performance has been highly improved by combining GaMD simulations with flexible docking, which effectively accounts for the flexibility of the backbone and side chains in receptors. Such an ensemble docking protocol will greatly facilitate future PAM design of the A_1AR and other GPCRs [3].

2.3.2 Nucleic Acids

2.3.2.1 Exploring the Binding of Risdiplam Splicing Drug Analog to Single-Stranded RNA

Risdiplam is the first approved small-molecule splicing drug for the treatment of spinal muscular atrophy (SMA). Five 500 ns GaMD simulations were performed to explore the binding interaction of a known risdiplam analog, SMN-C2, which binds to the target nucleic acids with GA-rich sequence [36]. The simulation structures of (Seq6) of DNA and RNA were built using NAB in the AmberTools [80] package. The ligand molecule was placed randomly at >15 Å away from the nucleic acid. The AMBER force field BSC1 [81] parameter sets were used for DNA, OL3 [82] for RNA, GAFF2 [83] for ligand, and TIP3P [84] for water in the system. Spontaneous binding of the SMN-C2 to both RNA and DNA Seq6 was observed in the GaMD simulations [36]. For the binding of SMN-C2 to the RNA Seq6, three low-energy conformational states were identified, including the "Unbound/Unfolded," "Bound/Intermediate" and "Bound/Folded" states (Figure 2.4a). SMN-C2 was able to interact with RNA Seq6 in two bound states (Figure 2.4b). When bound to RNA Seq6, the COM distance between RNA and ligand is reduced to ~6 Å and the RNA radius of gyration (Rg) reduced to ~8.0 Å. While for the binding of SMN-C2 to the DNA Seq6, the location of the bound small molecule was slightly different from that observed in RNA, as demonstrated by the COM distance between DNA and ligand being reduced to ~4 Å in the bound states [36] (Figure 2.4b). Three low-energy conformational states of the DNA-ligand system were also identified for DNA Seq6, including the "Bound/Unfolded," "Intermediate," and "Bound/Folded" states (Figure 2.4c). In the Bound/Folded state, a similar binding mode of SMN-C2 was observed in DNA as in RNA, with subtle differences [36] (Figure 2.4d). In both RNA and DNA, it appeared that the AAG trinucleotide in the GAAG motif is thus important for the binding of SMN-C2 through π-stacking interactions. Furthermore, the unfolded RNA Seq6 appeared to be more flexible than the unfolded DNA, and SMN-C2 did not spontaneously bind to the unfolded RNA [36].

2.3.2.2 Uncovering the Binding of RNA to a Musashi RNA-Binding Protein

The Musashi 1 (MSI1) protein serves as a therapeutic drug target for treating several cancers, such as colorectal, ovarian, bladder, and myeloid leukemia. It is known

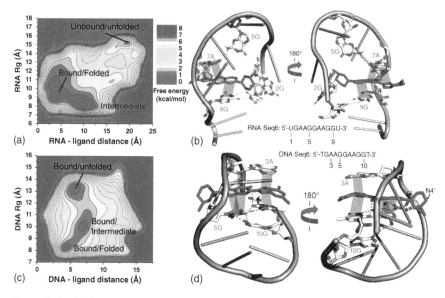

Figure 2.4 GaMD simulations revealed spontaneous binding of risdiplam to RNA and DNA Seq6. (a) 2D free energy profile of the center-of-mass (COM) distance between RNA-ligand and RNA radius of gyration (Rg). Three low-energy states were identified, namely the Unbound/Unfolded, Intermediate, and Bound/Folded. (b) Representative conformation of risdiplam-bound RNA Seq6 in the folded state. (c) 2D free energy profile of the COM distance between DNA-ligand and DNA radius of gyration (Rg). Three low-energy states were identified, namely the Unbound/Unfolded, Intermediate, and Bound/Folded. (d) Representative conformation of risdiplam-bound DNA Seq6 in the folded state. The color scheme is as follows: magenta = risdiplam, yellow = interacting nucleotides, cyan = other nucleotides, green dashed line = polar interaction, light red shade = $\pi-\pi$ or lone pair-π stacking. Source: Reproduced with permission of Tang et al. [36]/Oxford University Press/Public Domain CC BY 4.0.

to bind and suppress translation of the 3′-UTR of Numb mRNA [86], however, the molecular mechanism of this interaction remains elusive, which is important for effective drug design. GaMD simulations were performed to study the binding mechanism between Numb RNA and MSI1 [85]. For system setup, the Numb RNA was placed ~30 Å away from the MSI1. The AMBER force fields were used with ff14SBonlysc for protein, RNA.LJbb [82] for RNA, and TIP3P [84] model for water molecules. Spontaneous binding of Numb RNA to the MSI1 protein was successfully captured in 6 out of the 19 independent 1200 ns of GaMD simulations. In Sim1, RNA binding was observed at ~100 ns, where the RMSD of RNA relative to the NMR structure reduced to ~2.50 Å. In Sim2, RNA binding was observed at ~1010–1130 ns followed by RNA dissociation into the bulk solvent. The RNA bound to MSI1 after ~800 ns in Sim3, Sim4, and Sim5. In Sim6, spontaneous binding of RNA was observed after ~1000 ns. Five low-energy minima were characterized from GaMD simulations, including the "Bound," "Intermediate I1," "Intermediate I2," "Intermediate I3," and "Unbound" (Figure 2.5). These states were identified at the backbone RMSD of the RNA core and $N_{contacts}$ as (2.0 Å, 1500), (5.2 Å, 480), (9.5 Å, 200), (25.0 Å, 10), and (40 Å, 0), respectively. The "Unbound," "Intermediate I1,"

"Intermediate I2," and "Intermediate I3" states were identified at the backbone RMSD of the RNA core and R_g of Numb being (40.0 Å, 6.2 Å), (5.0 Å, 7.2 Å), and (6.9 Å, 6.2 Å), respectively (Figure 2.5a–d). From the GaMD simulations, the R_g of the Numb RNA in the "Bound" state was observed to have a wider range as compared to the "Unbound" and "Intermediate" conformations, which suggested an induced fit mechanism of Numb RNA binding to MSI1. The I1 state showed interactions of Numb RNA with the β2-β3 loop and C terminus of MSI1. Three hydrogen bonds were formed between MS1 C terminus residue R99 and the RNA nucleotide A106 (Figure 2.5e). The I2 state showed flipping of the MS1 residue R61 sidechain toward the solvent, leading to the formation of hydrogen bonds and salt-bridge interactions with the phosphate oxygen of the sidechain and backbone of RNA nucleotide A106, respectively (Figure 2.5f). The I3 state showed large conformational changes in the MS1 C terminus, where hydrogen bond and salt bridge interactions were observed between the C terminus residue R99 and the sidechain and backbone of RNA nucleotide A106, respectively (Figure 2.5g). These important understandings of the RNA binding mechanism to MSI1 provided by GaMD simulations can aid rational structure-based drug design against MSI1 and other related diseases.

2.3.3 Human Angiotensin-Converting Enzyme 2 Receptor

Angiotensin-converting enzyme 2 (ACE2), which helps to convert angiotensin II to angiotensin 1–7 and angiotensin to angiotensin 1–9 [87], has been identified as the function receptor for severe acute respiratory syndrome coronaviruses, including SARS-CoV and SARS-CoV-2 [88]. SARS-CoV-2 is responsible for the 2019 coronavirus pandemic (COVID-19). The entry of SARS-CoV-2 is mediated by the interaction of the receptor binding domain (RBD) in the virus spike protein S1 subunit with the host ACE2 receptor. Therefore, inhibiting the interaction between the viral RBD and host ACE2 presents a promising strategy for blocking SARS-CoV-2 entry into human cells [34].

LiGaMD simulations were performed to explore the binding mechanism of MLN-4760, a highly selective and potent inhibitor, in the human ACE2 receptor [34], starting from the 1R4L [89] PDB structure. Besides the bound ligand, nine additional ligands were placed randomly in the solvent. During the equilibration phase, the bound MLN-4760 was observed to dissociate from ACE2 active site to the bulk solvent after the conformational changes in the subdomain I of ACE2 [34]. LiGaMD simulations were further extended to ten 700–2000 ns independent production simulations with randomized initial atomic velocities. Repetitive binding and unbinding of a ligand molecule to ACE2 was observed in three out of ten simulations [34]. The receptor was found to undergo various large-scale conformational changes, with four primary low-energy conformational states observed, including "Open," "Partially Open," "Closed," and "Fully Closed" conformations (Figure 2.6a,b). Among the four low-energy conformations, three of them were already observed in the experimental structures, including "Open," "Partially Open," and "Closed" conformations, whereas the "Fully Closed" conformation

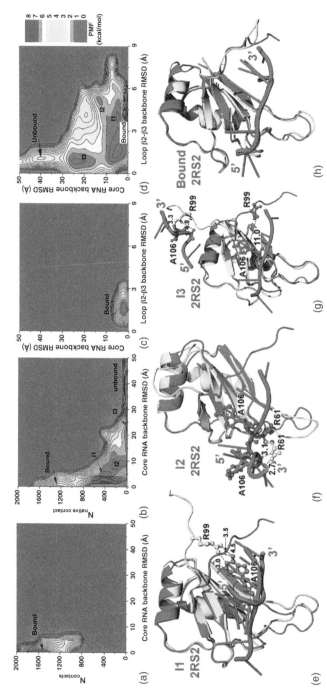

Figure 2.5 Binding of Musashi 1 (MSI1) RNA-binding protein to Numb mRNA. (a, b) 2D free energy profiles of the core RNA backbone RMSD relative to the first NMR conformation (PDB: 2RS2) and number of native contacts between MSI1 and Numb mRNA calculated from GaMD simulations starting from the (a) Bound and (b) Unbound states of the MSI1-Numb system. (c, d) 2D free energy profiles of the MSI1 β2-β3 loop backbone RMSD and core RNA backbone RMSD relative to the first NMR conformation (PDB: 2RS2) calculated from GaMD simulations starting from the (c) Bound and (d) Unbound states of the MSI1-Numb system. (E, H) Low-energy conformational states (I1, I2, and I3) and "Bound" state as identified from the 2D free energy profiles of the MSI1-RNA simulation system started from the Unbound state. The MSI1 protein and Numb RNA are shown in green and red, respectively. The NMR structure of the MSI1-Numb complex is shown in blue for comparison. Source: Reproduced with permission of Wang et al. [85]/Elsevier.

was uncovered from the LiGaMD simulations (Figure 2.6a,b) [34]. The presence of the polar and charged groups in a different part of the receptor was found to make favorable interactions with polar chloride and nitrogen atoms and the charged carboxylate group of the ligand molecules. Similarly, subdomain II of protein was relatively stable during LiGaMD simulations, adopting ~2–4 Å RMSD in relative to 1R4L PDB structure, whereas subdomain I showed higher flexibility with conformational changes ranging ~3–10 Å. Notably, two primary binding and dissociation pathways were observed in these production simulations (Figure 2.6c,d). The binding pathway involved the opening between α2 and α4 helices during the transition of subdomain I from the "Closed" to "Open" conformation [34] (Figure 2.6c), while the dissociation required interactions between MLN-4760 molecules and an interface formed between α5 helices and ACE2 $3_{10}H_4$ [34] (Figure 2.6d).

2.3.4 Discovery of Novel Small-Molecule Calcium Sensitizers for Cardiac Troponin C

Cardiac troponin C (TnC) is a calcium-dependent protein in the troponin complex responsible for the activation of muscle contraction. Disorders of TnC may trigger heart diseases and then cause death. One of the current therapies [90] requires the design of small molecules that can stabilize the open structure of the TnC and facilitate binding of the TnC switch peptide. To identify the potent small molecules, Coldren et al. [91] combined GaMD with high-throughput virtual screening to predict binding conformations and affinities of small molecules in TnC. GaMD simulations were compared with experiments for the TnC protein structures in complex with calcium sensitivity modulators. Independent 300ns GaMD simulations were performed on each system to obtain protein-ligand binding poses. GaMD trajectories were clustered to obtain the top 10 most populated conformations using the agglomerative hierarchical clustering algorithm, which were then used for virtual screening and docking. Their work identified a number of novel compounds that reduced the calcium dissociation rate and showed an overall calcium sensitization effect. One of the compounds exhibited high binding affinity in TnC and was further verified by the stopped-flow kinetic experiment.

2.3.5 Binding Kinetics Prediction from GaMD Simulations

Binding kinetics have recently been recognized to be more relevant for drug design. In particular, the dissociation rate constant that determines the drug residence time appears to better correlate with drug efficacy [92, 93]. However, binding kinetic rates are even more challenging to predict than binding free energies, largely due to slow dissociation processes [93]. In order to efficiently capture both the binding and unbinding processes of ligand/peptide/protein, we have recently developed selective GaMD methods, including LiGaMD, Pep-GaMD, and PPI-GaMD. The LiGaMD, Pep-GaMD, and Pep-GaMD simulations were able to capture multiple events of ligand/peptide/protein binding and unbinding within microsecond of simulation time. These highly efficient simulations thus allowed us to accurately

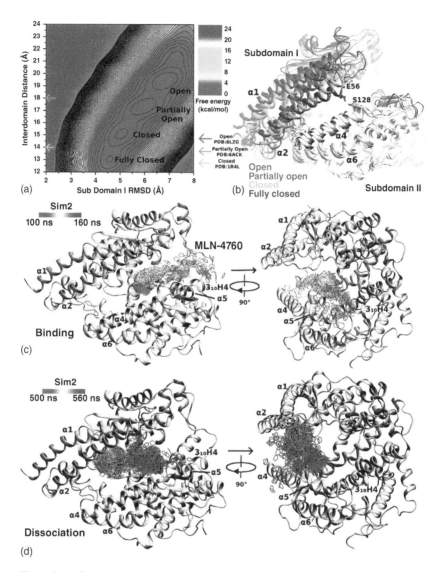

Figure 2.6 Binding and dissociation of the MLN-4760 inhibitor in the human ACE2 receptor. (a) 2D potential of mean force (PMF) of the subdomain I RMSD and interdomain distance calculated by combining the ten LiGaMD simulations. (b) Low-energy conformations of the ACE2 receptor with subdomain I found in the "Open" (red), "Partially Open" (blue), "Closed" (green), and "Fully Closed" (brown) states in the LiGaMD simulations. Subdomain II is stable and colored in white. (c) Two different views of the ligand binding pathways were observed in "Sim2," for which the center ring of MLN-4760 is represented by lines and colored by simulation time in a blue-white-red (BWR) color scale. (d) Two different views of the ligand dissociation pathway were observed in "Sim2," for which the center ring of MLN-4760 is represented by lines and colored by simulation time in a blue-white-red (BWR) color scale. Source: Reproduced with permission of Bhattarai et al. [34]/American Chemical Society.

characterize the ligand/peptide/protein binding thermodynamics and kinetics [55, 56].

LiGaMD has been proposed to quantitatively characterize ligand binding thermodynamics and kinetics [55]. Hundreds-of-nanosecond LiGaMD simulations captured repetitive guest binding and unbinding in the β-cyclodextrin host. The calculated guest binding free energies were in good agreement with experimental data with errors <1.0 kcal/mol in comparison with converged μs-timescale cMD simulations. Particularly, ligand kinetic rate constants estimated using Kramers' rate theory were accurately predicted. Furthermore, repetitive binding and unbinding of the benzamidine inhibitor in trypsin was observed in 1 μs LiGaMD simulations, allowing us to accurately calculate ligand binding free energy and kinetic rate constants. The predicted values were in excellent agreement with the experimental data [55].

Pep-GaMD [56] has been demonstrated on binding of three model peptides to the SH3 domains [94, 95], which include "PPPVPPRR" (PDB: 1CKB), "PPPALPPKK" (PDB: 1CKA) and "PAMPAR" (PDB: 1SSH). Repetitive peptide binding and unbinding processes were captured in independent 1 μs Pep-GaMD simulations, allowing us to calculate peptide binding thermodynamics and kinetics. The predicted values from Pep-GaMD were in good agreement with available experimental data. PPI-GaMD [56] has been demonstrated on a model system of the ribonuclease barnase binding to barstar. Six independent 2 μs PPI-GaMD simulations have successfully captured repetitive barstar dissociation and rebinding events. The barstar binding free energy predicted from PPI-GaMD was highly consistent with experimental data (-17.79 vs -18.90 kcal/mol) [96]. In addition, the predicted protein binding kinetic rates k_{on} and k_{off} were $21.7 \pm 13.8 \times 10^8$ M^{-1} s^{-1} and $7.32 \pm 4.95 \times 10^{-6}$ s^{-1}, being highly consistent with the corresponding experimental values of 6.0×10^8 M^{-1} s^{-1} and 8.0×10^{-6} s^{-1}, respectively. Furthermore, PPI-GaMD simulations have provided mechanistic insights into barstar binding to barnase, which involve long-range electrostatic interactions and multiple binding pathways, being consistent with previous experimental and computational findings of this model system.

2.4 Conclusions

In this work, we have reviewed the important developments and applications of GaMD in the field of drug discovery. GaMD is an unconstrained enhanced sampling technique that allows for the exploration of large biomolecular conformational spaces and complex biological interactions. Furthermore, the boost potential in GaMD exhibits a Gaussian distribution, enabling accurate reweighting of the simulations using cumulant expansion to the second order. Given its strengths, GaMD was applied to reveal the binding mechanisms of various ligands to GPCRs, nucleic acids, and human ACE2 receptors, as well as the effects of allosteric drug leads in GPCRs at the residue level. Additional applications of GaMD uncovered the mechanisms of protein-membrane [41] interactions, identified cryptic pockets within the SARS-CoV-2 main protease [97], explored drug binding to protease [98], and revealed the conformational landscape of drug binding to GPCRs [99]. Nevertheless, more efficient GaMD algorithms and enhanced sampling methods are still needed to characterize the thermodynamics and kinetics of important protein–protein/nucleic acid interactions and explore the structural dynamics in systems of increasing sizes, such as viruses and cells [3]. GaMD can be potentially applied to predict ADMET properties (e.g. membrane permeation), especially when combined with compatible enhanced sampling techniques such as replica exchange US [48]. Further developments in both supercomputing hardware and enhanced sampling methods should help tackle these challenges in the future.

References

1 Karplus, M. and McCammon, J.A. (2002). *Nat. Struct. Biol.* 9: 646–652, https://doi.org/10.1038/Nsb0902-646.
2 Hollingsworth, S. and Dror, R. (2018). *Neuron* 99: 1129–1143.
3 Wang, J. et al. (2021). *WIREs Comput. Mol. Sci.* e1521, https://doi.org/10.1002/wcms.1521.
4 Henzler-Wildman, K. and Kern, D. (2007). *Nature* 450: 964–972.
5 Harvey, M.J., Giupponi, G., and Fabritiis, G.D. (2009). *J. Chem. Theory Comput.* 5: 1632–1639.
6 Johnston, J.M. and Filizola, M. (2011). *Curr. Opin. Struct. Biol.* 21: 552–558.
7 Shaw, D.E. et al. (2010). *Science* 330: 341–346.
8 Lane, T.J., Shukla, D., Beauchamp, K.A., and Pande, V.S. (2013). *Curr. Opin. Struct. Biol.* 23: 58–65.
9 Vilardaga, J.-P., Bünemann, M., Krasel, C. et al. (2008). *Nat. Biotechnol.* 21: 807–812.
10 Miao, Y. and Ortoleva, P.J. (2006). *J. Chem. Phys.* 125: 214901.
11 Spiwok, V., Sucur, Z., and Hosek, P. (2015). *Biotechnol. Adv.* 33: 1130–1140.
12 Gao, Y.Q., Yang, L.J., Fan, Y.B., and Shao, Q. (2008). *Int. Rev. Phys. Chem.* 27: 201–227.
13 Liwo, A., Czaplewski, C., Oldziej, S., and Scheraga, H.A. (2008). *Curr. Opin. Struct. Biol.* 18: 134–139.
14 Christen, M. and van Gunstere, W. (2008). *J. Comput. Chem.* 29: 157–166.
15 Miao, Y. and McCammon, J.A. (2016). *Mol. Simul.* 42: 1046–1055.
16 Torrie, G. and Valleau, J. (1977). *J. Comput. Phys.* 23: 187–199.
17 Kumar, S., Rosenberg, J., Bouzida, D. et al. (1992). *J. Comput. Chem.* 13: 1011–1021.
18 Laio, A. and Gervasio, F. (2008). *Rep. Prog. Phys.* 71: 126601.
19 Besker, N. and Gervasio, F. (2012). *Computational Drug Discovery and Design*, 501–513. Berlin: Springer.
20 Darve, E., Rodriguez-Gomez, D., and Pohorille, A. (2008). *J. Chem. Phys.* 128: 144120.
21 Darve, E., Wilson, M., and Pohorille, A. (2002). *Mol. Simul.* 28: 113–144.

22 Isralewitz, B., Baudry, J., Gullingsrud, J. et al. (2001). *J. Mol. Graph. Model.* 19: 13–25.
23 Sugita, Y. and Okamoto, Y. (1999). *Chem. Phys. Lett.* 314: 141–151.
24 Okamoto, Y. (2004). *J. Mol. Graph. Model.* 22: 425–439.
25 Hansmann, U. (1997). *Chem. Phys. Lett.* 281: 140–150.
26 Wu, X. and Brooks, B. (2003). *Chem. Phys. Lett.* 381: 512–518.
27 Wu, X., Brooks, B., and Vanden-Eijnden, E. (2016). *J. Comput. Chem.* 37: 595–601.
28 Wu, X. and Wang, S. (1998). *J. Phys. Chem. B.* 102: 7238–7250.
29 Hamelberg, D., Mongan, J., and McCammon, J.A. (2004). *J. Chem. Phys.* 120: 11919–11929.
30 Voter, A. and Hyperdynamics, F. (1997). *Phys. Rev. Lett.* 78: 3908.
31 Miao, Y., Feher, V.A., and McCammon, J.A. (2015). *J. Chem. Theory Comput.* 11: 3584–3595.
32 Miao, Y. et al. (2014). *J. Chem. Theory Comput.* 10: 2677–2689.
33 Do, H., Akhter, S., and Miao, Y. (2021). *Front. Mol. Biosci.* 8: 242.
34 Bhattarai, A., Pawnikar, S., and Miao, Y. (2021). *J. Phys. Chem. Lett.* 12: 4814–4822.
35 Pang, Y., Miao, Y., and McCammon, J.A. (2017). *J. Chem. Theory Comput.* 13: 9–19.
36 Tang, Z. et al. (2021). *Nucleic Acids Res.* 49: 7870–7883.
37 Do, H., Wang, J., Bhattarai, A., and Miao, Y. (2022). *J. Chem. Theory Comput.* 18: 1423–1436.
38 Bhattarai, A., Devkota, S., Bhattarai, S. et al. (2020). *ACS Central Sci.* 6: 969–983.
39 Bhattarai, A. et al. (2022). *J. Am. Chem. Soc.* 144: 6215–6226.
40 Miao, Y. and McCammon, J.A. (2016). *Proc. Natl. Acad. Sci. U. S. A.* 113: 12162–12167.
41 Bhattarai, A., Wang, J., and Miao, Y. (2020). *J. Comput. Chem.* 41: 460–471.
42 Miao, Y. and McCammon, J.A. (2018). *Proc. Natl. Acad. Sci. U. S. A.* 115: 3036–3041.
43 Wang, J. and Miao, Y. (2019). *J. Phys. Chem. B.* 123: 6462–6473.
44 Wang, J. and Miao, Y. (2022). *J. Chem. Theory Comput.* 18: 1275–1285.
45 East, K.W. et al. (2020). *J. Am. Chem. Soc.* 142: 1348–1358.
46 Ricci, C.G. et al. (2019). *ACS Central Sci.* 5: 651–662.
47 Huang, Y.-M., McCammon, J.A., and Miao, Y. (2018). *J. Chem. Theory Comput.* 14: 1853–1864.
48 Oshima, H., Re, S., and Sugita, Y. (2019). *J. Chem. Theory Comput.* 15: 5199–5208.
49 Ahn, S.H., Ojha, A.A., Amaro, R.E., and McCammon, J.A. (2021). *J. Chem. Theory Comput.* 17: 7938–7951.
50 Ponder, J.W. et al. (2010). *J. Phys. Chem. B.* 8: 2549–2564.
51 Shi, Y. et al. (2013). *J. Chem. Theory Comput.* 9: 4046–4063.
52 Zhang, C. et al. (2018). *J. Chem. Theory Comput.* 14: 2084–2108.
53 Lagardere, L. et al. (2018). *Chem. Sci.* 9: 956–972.
54 Adjoua, O. et al. (2021). *J. Chem. Theory Comput.* 17: 2034–2053.

55 Miao, Y., Bhattarai, A., and Wang, J. (2020). *J. Chem. Theory Comput.* 16: 5526–5547.

56 Wang, J. and Miao, Y. (2020). *J. Chem. Phys.* 153: 154109.

57 Miao, Y. and McCammon, J.A. (2017). *Annu. Rep. Comp. Chem.* 13: 231–278, https://doi.org/10.1016/bs.arcc.2017.06.005.

58 *Keras-Vis* (GitHub, 2017).

59 Miao, Y. (2018). *J. Chem. Phys.* 149: 072308, https://doi.org/10.1063/1.5024217.

60 Doshi, U. and Hamelberg, D. (2011). *J. Chem. Theory Comput.* 7: 575–581, https://doi.org/10.1021/ct1005399.

61 Frank, A.T. and Andricioaei, I. (2016). *J. Phys. Chem. B* 120: 8600–8605, https://doi.org/10.1021/acs.jpcb.6b02654.

62 Hamelberg, D., Shen, T., and Andrew McCammon, J. (2005). *J. Chem. Phys.* 122: 241103, https://doi.org/10.1063/1.1942487.

63 Celerse, F. et al. (2022). *J. Chem. Theory Comput.* 18: 968–977.

64 Copeland, M.C. et al. (2022). *J. Phys. Chem. B.* 126: 5810–5820.

65 Hauser, A.S. et al. (2018). *Cell* 172: 41–54.

66 Stevens, R.C. et al. (2013). *Nat. Rev. Drug Discov.* 12: 25–34.

67 Isberg, V. et al. (2016). *Nucleic Acids Res.* 44: D365–D364.

68 Fredholm, B. et al. (1997). *Trends Pharmacol. Sci.* 18: 79–82.

69 Jacobson, K.A. and Gao, Z.-G. (2006). *Nat. Rev. Drug Discov.* 5: 247–264.

70 Cheng, R. et al. (2017). *Structure* 25: 1275–1285.

71 Draper-Joyce, C.J. et al. (2021). *Nature* 597: 571–576, https://doi.org/10.1038/s41586-021-03897-2.

72 Dror, R.O. et al. (2011). *Proc. Natl. Acad. Sci. U. S. A.* 108: 18684–18689.

73 Avlani, V. et al. (2007). *J. Biol. Chem.* 282: 25677–25686.

74 Peeters, M. et al. (2012). *Biochem. Pharmacol.* 84: 76–87.

75 Nguyen, A. et al. (2016). *Mol. Pharmacol.* 90: 715–725.

76 Miao, Y., Bhattarai, A., Nguyen, A. et al. (2018). *Sci. Rep.* 8: 16836.

77 Wang, J. et al. (2020). *GPCRs* (ed. B. Jastrzebska and P.S.H. Park), 283–293. Academic Press.

78 Morris, G.M. et al. (2009). *J. Comput. Chem.* 30: 2785–2791.

79 Bhattarai, A., Wang, J., and Miao, Y. (2020). *Biochim. Biophys. Acta Gen. Subj.* 1864: 129615.

80 Amber 2021 (University of California, San Francisco, 2021).

81 Ivani, I. et al. (2016). *Nat. Methods* 13: 55–58.

82 Zgarbova, M., Otyepka, M., Šponer, J. et al. (2011). *J. Chem. Theory Comput.* 7: 2866–2902.

83 Wang, J., Wolf, R., Caldwell, J. et al. (2004). *J. Comput. Chem.* 25: 1157–1174.

84 Mark, P. and Nilsson, L. (2001). *Chem. A Eur. J.* 105: 9954–9960.

85 Wang, J., Lan, L., Wu, X., Xu, L. & Miao, Y. bioRxiv, 2020.2010.2030.362756, https://doi.org/10.1101/2020.10.30.362756 (2021).

86 Kudinov, A.E., Karanicolas, J., Golemis, E.A., and Boumber, Y. (2017). *Clin. Cancer Res.* 23: 2143–2153, https://doi.org/10.1158/1078-0432.Ccr-16-2728.

87 Gross, L.Z.F. et al. (2020). *ChemMedChem* 15: 1682–1690, https://doi.org/10.1002/cmdc.202000368.

88 Hoffmann, M. et al. (2020). *Cell* 181: 271.
89 Towler, P. et al. (2004). *J. Biol. Chem.* 279: 17996–18007.
90 Remme, W.J. and Swedberg, K. (2001). *Eur. Heart J.* 22: 1527–1560, https://doi.org/10.1053/euhj.2001.2783.
91 Coldren, W.H., Tikunova, S.B., Davis, J.P., and Lindert, S. (2020). *J. Chem. Inf. Model.* 60: 3648–3661, https://doi.org/10.1021/acs.jcim.0c00452.
92 Schuetz, D.A. et al. (2017). *Drug Discov. Today* 22: 896–911, https://doi.org/10.1016/j.drudis.2017.02.002.
93 Tonge, P. and Drug-Target, J. (2018). *ACS Chem. Nerosci.* 9: 29–39, https://doi.org/10.1021/acschemneuro.7b00185.
94 Ahmad, M. and Helms, V. (2009). *Chem. Cent. J.* 3: O22.
95 Ball, L.J., Kuhne, R., Schneider-Mergener, J., and Oschkinat, H. (2005). *Angew. Chem. Int. Ed. Engl.* 44: 2852–2869, https://doi.org/10.1002/anie.200400618.
96 Schreiber, G. and Fersht, A.R. (1993). *Biochemistry* 32: 5145–5150.
97 Sztain, T., Amaro, R.E., and McCammon, J.A. (2021). *J. Chem. Inf. Model.* 61: 3495–3501.
98 Wang, Y.-T. et al. (2022). *Phys. Chem. Chem. Phys.* 24: 22898–22904.
99 Zhang, H. et al. (2021). *Nat. Commun.* 12: 4151.

3

MD Simulations for Drug-Target (Un)binding Kinetics

Steffen Wolf

Biomolecular Dynamics, Institute of Physics, University of Freiburg, Hermann-Herder-Strasse 3, Freiburg 79104, Germany

3.1 Introduction

3.1.1 Preface

Compared to the free energy calculations employed for drug design presented in the preceding chapters, determining protein-drug binding and unbinding kinetics from molecular dynamics (MD) simulations is a comparatively young field. Though a large body of methods has evolved within the last decade, these developments have yet to result in commonly accepted best practices as well as a gold standard of test systems at the time of writing this chapter.

Because of the still-fast-paced rate of development of the field, the author here can only attempt to give a general overview of the current state-of-the-art. Additionally, the versatility of approaches employed prohibits an exhaustive explanation of the underlying theory. Instead, this chapter aims at providing the interested computer-aided drug design (CADD) researcher with a basic overview of the problem of predicting kinetics, its formal basis, and a practical guideline on the strengths and shortcomings of selected methods. For the interested reader, links to the core literature and helpful reviews will be provided.

3.1.2 Motivation for Predicting (Un)binding Kinetics

Experimental approaches in drug design focus on determining a diverse set of constants, e.g. inhibition constants K_i, half maximal effective concentrations EC_{50}, or dissociation constants K_D. Especially the latter K_D can be reformulated in the form of a standard binding free energy

$$\Delta F^0 = -k_B T \ln K_D. \tag{3.1}$$

with an implicit reference concentration of 1 M. Accordingly, finding a compound with large ΔF^0 directly translates into finding a high-affinity compound, which is the basic principle behind free energy-based computational as well as early-stage experimental drug development.

Computational Drug Discovery: Methods and Applications, First Edition.
Edited by Vasanthanathan Poongavanam and Vijayan Ramaswamy.
© 2024 WILEY-VCH GmbH. Published 2024 by WILEY-VCH GmbH.

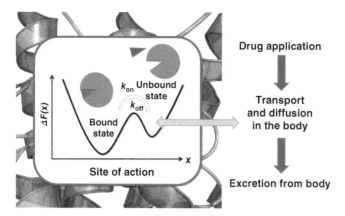

Figure 3.1 Schematic of nonequilibrium aspects of drug binding and pharmacokinetics. Red arrows represent irreversible steps and orange arrows reversible steps. The coordinate x represents the distance between the ligand and its binding site in a target protein.

Starting in the midst of the 2000s, this focus on high-affinity compounds was challenged by several works proposing that the efficacy of a compound, which is the finally relevant criterion for the successful applicability of a compound as a drug, may not result from high affinity but from long residence times of a drug at its target protein [1–4]. A prime example of this hypothesis is the tyrosine kinase inhibitor Gleevec (Imatinib) [2, 5, 6], which gains its selectivity for Abl over other kinases by such slow unbinding kinetics.

Where does this difference between affinity and residence time come from? In short, from the difference between equilibrium and nonequilibrium statistical mechanics. Free energies are thermodynamic variables describing closed systems in equilibrium in which no exchange of particles with the surrounding occurs. This is markedly different from the situation in the human body, which is a strictly nonequilibrium environment with particle uptake and release. Figure 3.1 depicts a simplified yet instructive representation of the process: if a dose of a drug is administered, one initially observes transport and diffusion to its active sites, followed by a local equilibration between the drug-bound and the unbound states. Concurrently, the liver and kidneys start to remove the drug from the body and thus deplete the population of unbound drugs. In case that the bound and unbound states exchange quickly, a drug that binds tightly to its target will nevertheless be quickly eliminated and thus be only shortly active. Consequently, the compound needs to be applied more often, raising the possibility for off-target binding and thus undesired side effects.

3.1.3 The Time Scale Problem of MD Simulations

Given the large interest in predicting residence times, it is natural to consider performing MD simulations for their calculation. Here, however, one faces a major challenge based on the fundamental physics underlying MD simulations: this method is based on numerical solutions of Newton's equations of motion, in which in a new position of an atom $q(t)$ can be calculated in its simplest form as

$$q(t + \Delta t) = q(t) + v(t)\Delta t + \frac{f(t)}{2m}\Delta t^2 \tag{3.2}$$

based on the atom's mass m, velocity $v(t)$, and the force $f(t)$ the atom experiences from its environment. Such solutions require the definition of a time increment Δt, which needs to be small enough to sample the fastest motion in a simulation system. In biomolecular systems, these motions are O–H stretch vibrations with a period of about ten femtoseconds. Accordingly, Δt needs to be on the order of 1–2 fs. This time range stands in stark contrast to pharmacologically relevant timescales, as especially unbinding events occur on the order of seconds to hours (see, e.g. [7, 8]). As the solution of Eq. (3.2) needs to be performed iteratively, parallelization only remedies the computational costs of calculating forces per time step but cannot solve this discrepancy. At the time of writing of chapter, the best-performing purpose-built simulation computer, Anton3 of D. E. Shaw research [9], was said to be able to calculate 100 μs per day. While this is an achievement of its own, even simulating a single second of real-time would still require 33 years. Therefore, there is no computational hardware in sight that may solve this conundrum. Instead, smart simulation approaches from the group of enhanced sampling techniques are required, which help to overcome this timescale problem. While this type of technique has been introduced in Chapter 2, we have to briefly review the theory underlying such calculations to better understand the way these approaches can help in calculating molecular process rates.

3.2 Theory of Molecular Kinetics Calculation

3.2.1 Nonequilibrium Statistical Mechanics in a Nutshell

To better understand the challenges encountered in the calculation of kinetics, we briefly need to review their formal basis and the differences to free energy calculations.[1] Let us assume that for investigating the (un)binding process of a ligand from its target protein, it is sufficient to regard the process only along the distance $x(t)$ of the center of mass (COM) between ligand and protein. This distance is formally a *collective variable* (CV), i.e. a linear combination of atomic mass-weighed atom positions $q(t)$. If a CV is capable of sufficiently describing a molecular process of interest, it qualifies as a *reaction coordinate*. Given an MD simulation system in an *NVT* ensemble,[2] one can calculate a free energy profile along x

$$\Delta F(x) = -k_B T \ln P(x) \tag{3.3}$$

with the Boltzmann constant k_B and the probability distribution $P(x)$. This distribution can simply be obtained, e.g. by histogramming $x(t)$ from an unbiased

---

1 For readers who want to dive deeper into the formal basis of nonequilibrium statistical mechanics, the author recommends the respective books by Zwanzig [10] and Pottier [11].
2 We use the ensemble here for formal reasons. The solutions in a *NPT* ensemble contain an additional volume work term, but are analogous in their form.

MD simulation along x. One can directly see that $P(x)$ does not contain the time dependence of $x(t)$ anymore. Instead, $\Delta F(x)$ constitutes the time-independent potential of mean force with

$$-\frac{dF(x)}{dx} = \langle f(x) \rangle. \qquad (3.4)$$

where $\langle \cdot \rangle$ denotes a mean over time (specifically here overall points in t that correspond to a specific x). To recover the formal time dependence of $x(t)$, it is possible to rationalize dynamics along the CV of choice, e.g. as a Markovian Langevin equation [10]

$$m\ddot{x} = -\frac{dF(x)}{dx} - \dot{x}\Gamma + \sqrt{2k_B T \Gamma}\, \xi(t), \qquad (3.5)$$

with a friction coefficient Γ. On the right side of Eq. (3.5), the first term is the mean force as given in Eq. (3.4) and the second term is a friction force. The third term, representing a fluctuating force, consists of a zero-mean, normally distributed stochastic term $\xi(t)$ with variance unity (i.e. a random number drawn from a standard Gaussian probability distribution) and an amplitude $\sqrt{2k_B T \Gamma}$ that is coupled to the friction due to the fluctuation-dissipation theorem [11]. Friction factors can be calculated, e.g. based on the force autocorrelation function (ACF) [12]

$$C(t') = \langle (f(t) - \langle f \rangle)(f(t+t') - \langle f \rangle) \rangle \qquad (3.6)$$

as integral over time

$$\Gamma = \frac{1}{k_B T} \int_{t_0}^{\infty} C(t')\, dt' \qquad (3.7)$$

If we assume that $C(t')$ follows an exponential decay $C(t') \approx \langle f_0^2 \rangle \exp(-t'/\tau)$, Eq. (3.7) becomes $\Gamma = \frac{\tau}{k_B T} \langle f_0^2 \rangle$. To be able to recover the dynamics along x, we need not only to know the time-independent mean force given by the free energy but also the variance of the force. On a microscopic level, the mean force represents the average interaction of a ligand with its surroundings, while the variance contains the effect of chaotic dynamics such as collisions of the ligand with water molecules or the motion of side chains. The latter two effects can both decelerate and accelerate the ligand, hence the appearance of friction factors in both friction and fluctuation forces.

3.2.2 Kramers Rate Theory

Based on Eq. (3.5), we require both $\Delta F(x)$ and Γ for being able to calculate rates. Indeed, Kramers derived an analytical expression for such rates based on these two values [13, 14] for the high friction regime: given a free energy surface as displayed in Figure 3.1, i.e. with a single free energy minimum and a single free energy barrier with a height $\Delta F^{\neq}$, a ligand will cross the barrier at a rate

$$k = \frac{\sqrt{|F''(x_{min})||F''(x_{barrier})|}}{2\pi \Gamma} \exp\left(-\frac{\Delta F^{\neq}}{k_B T}\right) \qquad (3.8)$$

Here, the absolute value of the second derivative at the minimum $|F''(x_{min})|$ and the barrier $|F''(x_{barrier})|$, and the friction Γ is hard to calculate. As a consequence, the fraction in Eq. (3.8) usually cannot be solved analytically. However, the structure of the equation helps us to understand the principles of calculating molecular rates: the fraction in Eq. (3.8) has units of inverse time and represents the *attempt frequency*, i.e. how often a ligand attempts to leave the bound state. The exponential term represents the Boltzmann probability to reach the least likely position along x from the bound state, which is the *transition state* barrier with a height $\Delta F^{\neq}$. All biased techniques employed for rate prediction modulate these two parts of Eq. (3.8) such that a sufficient number of transitions occurs in a reasonable time frame of simulations (preferably within a few nanoseconds), and that the underlying unbiased rate can be extracted from these calculations with a suitable reweighing scheme.

3.2.3 Biased MD Methods

We here briefly review the theory of biased MD simulations, which are of most importance for the calculation of molecular kinetics. For further information, the author would like to refer the reader to Chapter 2 of this book as well as to a recent review [15].

3.2.3.1 Temperature- and Barrier-Scaling

As can be seen directly in Eq. (3.8), the simplest approach to accelerate sampling is to include a tuning factor a into the Boltzmann probability, leading to accelerated kinetics

$$k_a = k \exp\left(-\frac{\Delta F^{\neq}}{k_B T}[a-1]\right) \tag{3.9}$$

There are two options available on how to implement factor a into simulations: on one hand, it can be applied as a scaling factor of temperature, leading to *temperature-accelerated MD* [16, 17]. While this approach is well applicable in simulations of protein folding [18], it is less suited for the problems discussed here, as protein structures are sensitive to high temperatures that would be necessary for sufficient acceleration. Therefore, given that $\Delta F^{\neq} = \Delta U^{\neq} - T\Delta S^{\neq}$ with potential (inner) energy U and entropy S, the factor is usually included in the free energy as a modification of the potential energies U, leading to the approach of *scaled* or *smoothed potential MD* [19]. An alternative development here is Gaussian accelerated MD [20], which uses a quadratic scaling of U below threshold energy that dampens the influence of deep minima in ΔF.

3.2.3.2 Bias Potential-Based Methods

Instead of accelerating dynamics per se, biased potential methods employ a potential $V_{bias}(x)$ that is added to the system Hamiltonian, leading to an adjusted probability $P_{bias}(x)$ in Eq. (3.3). The free energy of the unbiased system $\Delta F(x)$ can then be calculated from the biased simulation as [21]

$$\Delta F(x) = -k_B T \ln P_{bias}(x) - V_{bias}(x) \tag{3.10}$$

If $V_{bias}(x)$ is well-chosen, then $P_{bias}(x)$ is close to uniform enough to sample well the relevant range of x. Torrie and Valleau originally suggested the usage of harmonic potentials $V_{bias}(x) = \frac{1}{2}(x - x_0)^2$ [21]. Alternatively, in the approach of *metadynamics* [22, 23], $V_{bias}(x)$ is constructed from a sum of Gaussian functions

$$V_{bias}(x) = \sum_i A_i \exp\left(-[x - x_i]^2 / 2\sigma^2\right), \tag{3.11}$$

which are sequentially placed in i steps during simulation until $P_{bias}(x)$ is constant, i.e. a ligand diffuses freely in and out of its binding pocket. Under these conditions, $-k_B T \ln P_{bias}(x)$ becomes a constant that can be ignored, and $\Delta F(x) = -V_{bias}(x)$.

3.2.3.3 Bias Force-Based Methods

Instead of employing bias potentials to drive a ligand out from its binding site, it is feasible to pull it out actively, i.e. to apply an external force. This approach is reminiscent of atomic force microscopy, and indeed, the first such computational approaches were carried out on the streptavidin-biotin complex in combination with such experiments [24]. Employing bias forces

$$f_{bias}(x) = -\frac{d}{dx} V_{bias} \tag{3.12}$$

from a harmonic potential with a time-dependent center point

$$V_{bias}(x) = \frac{1}{2}(x - [x_0 + vt])^2 \tag{3.13}$$

gives rise to *steered MD* simulations [25, 26].

A theory-intrinsic alternative to these harmonic restraints is *targeted MD* [27], which employs a constant velocity constraint

$$\Phi(x) = x - [x_0 + vt] = 0, \tag{3.14}$$

which then gives rise to a constraint force

$$f_{bias}(x) = \lambda \frac{d}{dx} \Phi(x). \tag{3.15}$$

with a Lagrange multiplier λ. The constraint force is calculated in each time step to ensure that the ligand moves out of the binding site with a velocity v.

For both approaches listed above, $f_{bias}(x)$ is applied within MD simulations along a pre-defined vector between protein and ligand, e.g. the distance between the ligand's COM and the COM of some reference atoms within the protein. This setup requires a general idea in which direction an unbinding pathway can be found. If this knowledge is not given, it is possible to apply a small f_{bias} with a pre-set amplitude, but in a random direction, giving rise to *random acceleration MD* [28]. This approach allows for employing simpler reaction coordinates, e.g. root mean square distances of a ligand from its initial position, and therefore allows to explore possible unbinding pathways without prior knowledge about them.

3.2.3.4 Knowledge-Biased Methods

All three biasing approaches described above have the disadvantage that the introduced bias interferes with the dynamics of the system. Another option is to apply a knowledge-based bias: similar to the action of a "Maxwell's daemon," trajectories can be sorted based on whether they naturally evolve along a reaction coordinate of choice, resulting in such methods as *Weighted Ensembles* (WE) [29] or *Milestoning* [30]. Giving an appropriate short simulation time, the final structures from trajectories that have successfully crossed a pre-defined threshold along the reaction coordinate are used as seeds for new simulations. The cycle is repeated until the final state sought after has been reached. To find the correct $\Delta F^{\neq}$, *Transition Path Sampling* (TPS) [31, 32] can be carried out. Here, after an initial (enforced) transition from bound to unbound state, a redistribution of velocities is applied to a structure at or around the transition state, and the resulting state is numerically propagated forward and backward in time to check if the unbound and bone states are reached, respectively. If a successful transition is found, the procedure is repeated to generate an ensemble of transition trajectories.

3.2.3.5 Coarse-graining and Master Equation Approaches

All methods listed before focus on accelerating dynamics in biasing fully-atomistic MD simulations. However, alternative approaches exist that coarse-grain the overall dynamics, which reduces the necessary computational power by several orders of magnitude. One such approach is the integration of the Langevin Equation (3.5) [33]. Alternatively, it is possible to reduce the dynamics to the waiting times in the bound and unbound states and any intermediate states in between. If the transitions between all these states are Markovian, i.e. the transition between two states at time t does not depend on any information from a prior time, a *Markov State Model* (MSM) [34, 35] can be constructed. From the MSM, binding and unbinding times can then be determined. The advantage of this approach is that short trajectories, e.g. from knowledge-based simulation methods, can be used for the construction of an MSM rate matrix that encodes the transition times between the states.

3.3 Challenges and Caveats in Rate Prediction

3.3.1 Finding Reaction Coordinates and Pathways

At this position in the chapter, it is necessary to highlight the challenges a researcher faces when attempting to calculate (un)binding rates. The major and crucial difference to standard free energy calculations is the coordinate dependence of $\Delta F(x)$. While free energy calculations in principle are possible based on the single-step difference between bound and unbound states using Zwanzig's equation [36], determining $\Delta F(x)$ requires finding both the correct reaction coordinate x and the right transition state $\Delta F^{\neq}$ along it. In time-dependent processes, x describes the slowest process in the molecular system of interest. If this process is, say, a conformational change of the protein, such as in the Abl kinase mentioned above, then a simulation of unforced ligand unbinding ignoring this change will inevitably yield wrong results. Furthermore, a good reaction coordinate such as dihedral angles describing this conformational change may not be a suitable biasing coordinate,

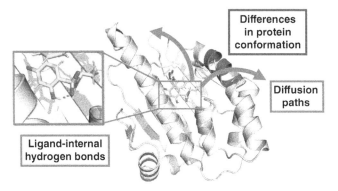

Figure 3.2 Possible reaction coordinates that need to be taken into account in the search for pathways for the example of the N-terminal domain of Hsp90. Protein as cartoon, ligand as sticks.

as a biasing coordinate requires the possibility to define forces that can serve as input for the integrator Eq. (3.2). Lastly, it may be that x is not one-dimensional but a multi-dimensional $\boldsymbol{x}$, e.g. if conformational change and ligand diffusion to and from the binding site are coupled with each other, if a ligand can take different routes out of a binding site (see Figure 3.2), or if ligand translation is coupled to a specific bond rotation. In such cases, ligands may take paths through $\boldsymbol{x}$ that need to be found to understand the unbinding mechanism and to determine the most relevant unbinding path. Accordingly, methods have been developed for finding or learning unbinding reaction coordinates on the fly [32, 37, 38], detecting pathways in trajectory ensembles a posteriori [33, 39–43], or performing outright brute-force pathway exploration [44–46]. The reader needs to note that pathways can change drastically between ligands despite only small chemical differences. For example, we found for two ligands bound to the N-terminal domain of Hsp90 that swapping an amide with a sulfonamide moiety results in a completely different unbinding behavior [43]. The reason here is the formation of a ligand-internal hydrogen bond, which is not formed with a carbonyl oxygen but exists with a sulfonyl oxygen atom. The presence of the hydrogen bond led to an additional rotation barrier within the ligand, causing the sulfonamide ligand to be less flexible than its amide counterpart. The corresponding hydrogen bond donor/acceptor distance indeed turned out to be a good additional reaction coordinate to discriminate unbinding pathways.

3.3.2 Error Ranges of Estimates

Another general issue is the error range in absolute (un)binding rate predictions. We see in Eq. (3.8) that this error is exponentially dependent on the accuracy of transition barrier height estimation. If we consider the mean error range currently achievable by free energy calculations [47] $\Delta\Delta F^{\neq} \approx 1.7$ kcal/mol $\approx 3\,k_\mathrm{B}T$, then the corresponding mean error of a rate is $\Delta k \approx k \cdot \exp(3)$, so about a factor of 17. Predicting a binding rate correctly within a magnitude is already considered a successful prediction and within the best range achievable.

3.3.3 A Need for Reliable Benchmarking Systems

Lastly, a significant challenge is the current absence of a gold standard system to benchmark theoretical prediction methods. Commonly employed systems that allow a computational comparison between rates predicted by biased methods and the true unbiased rates are a NaCl ion pair in water [33], the F506 binding protein (FBKP) with bound DMSO, DSS of 4-hydroxy-2-butanonte as ligands [48], or the L99A mutant of T4 lysozyme with bound benzene or xylene [49]. It is easy to note that these compounds lack the chemical complexity that constitutes a real drug. On the side of experimentally measured kinetics, the trypsin-benzamidine complex [50] has emerged as a generally investigated benchmarking system. The problem here is that the kinetics of only a single ligand is known and used for predictions, which does not allow filtering of true from false positives. A system that can serve as a "gold standard" for benchmarking drug kinetics has not been agreed on yet. The author would like to highlight the N-terminal domain of Hsp90 [8] as well as the A2 adenosine receptor [7] as a possible benchmarking system, for which diverse sets of protein–ligand complexes have been characterized both in their atomic structure and their binding kinetics within the framework of the kinetics for drug discovery consortium [51]. Especially for Hsp90, a library of chemically diverse ligands is available that cover a range in rates of about four orders of magnitude in time [8, 52, 53]. Furthermore, long ligand residence is believed to depend on a conformational change of a protein helix covering the binding site [8], which poses a challenge for reaction coordinate prediction methods.

3.3.4 Problems with Force Fields

Besides all these issues, a further problem might be found in the force fields used in MD simulations themselves: transition state barrier heights $\Delta F^{\neq}$ may be overestimated because of missing polarization effects. While this problem is well known from simulations of ion transfer rates over membrane channels [54–56], it has recently been proposed to exist in the case of the M2 muscarinic receptor, as well [57]. A possible remedy could be the utilization of polarizable force fields [58, 59] introduced in Chapter 6, for which reliable ligand force fields still have to be developed.

3.4 Methods for Rate Prediction

In the following, several high-accuracy (within the expected error range, see above) applications of theories from Section 3.2 will be briefly introduced, and their advantages and limitations will be discussed. Comprehensive and recent reviews of existing methods and their detailed performance can be found in Refs. [60–67].

3.4.1 Unbinding Rate Prediction

3.4.1.1 Empirical Predictions
The first class to consider are methods that fulfill a similar role like Docking in the prediction of high-affinity ligands. These approaches do not attempt to predict

absolute rate constants, but relative parameters with a proposed proportionality to rate constants such as mean biased unbinding times or the mean nonequilibrium work to enforce unbinding, which are then used to score ligands. Such simulations are comparatively inexpensive and require a few to several hundred nanoseconds of simulated time per ligand. The accuracy of predictions suffers from similar issues as Docking [68], e.g. a dependence on prediction accuracy on protein target and ligand chemical scaffold, but at least allows a qualitative to semi-quantitative classification of ligands that is fast.

Scaled MD and smoothed potential MD have been successfully employed for scoring ligands bound to the N-terminal domain of Hsp90 [40, 42] as well as glucokinases [69]. The advantage of these approaches is that no reaction coordinate needs to be employed to accelerate such simulations. The first limitation of scaled and smoothed potential MD is the comparatively high simulation time for slowly unbinding compounds of several microseconds. Furthermore, scaling potential energy barriers following Eq. (3.9) causes a destabilization of the protein's structure, requiring the application of soft restraints onto protein atoms, which causes problems when a protein conformational change is involved in the unbinding process.

From the side of biased simulations, fast adiabatic metadynamics [70] and steered MD [71] have been employed for several G protein-coupled receptor-ligand combinations. The currently fastest scoring method [63] is nonequilibrium targeted MD [72], which only requires a few nanoseconds of cumulative simulation time to score a ligand and is independent of the unbinding rate of a ligand. The major limitation of this approach is that only compounds with similar chemical scaffolds can be compared to each other. Furthermore, all three approaches require a relatively good intuition along which pathway unbinding occurs for a definition of the biasing coordinate.

The best compromise between computational cost, prediction accuracy, and versatility for empirical predictions is found in tauRAMD [53], based on random acceleration MD. This method has been successfully applied to a range of different protein–ligand complex classes, including G protein-coupled receptors (GPCRs) and kinases [49, 73, 74]. Computationally, investigating a single slowly unbinding compound may still require hundreds of nanoseconds of simulation time. However, it is particularly well suited in cases where unbinding pathways are not known and, therefore, can serve for pathway exploration as well.

3.4.1.2 Prediction of Absolute Unbinding Rates

In comparison to scoring ligands according to unbinding rates, the prediction of absolute unbinding rates is still a daunting task and requires significant computational effort. As transitions along x need to be sufficiently sampled, the calculation of explicit kinetical rates even when using biased simulations usually requires accumulated trajectory data on the order of several µs for a single ligand. Computation of explicit kinetics, therefore, is a precision tool that should only be used in the later stages of a drug development program.

The so far best-established method for unbinding rate prediction is infrequent metadynamics [39, 75]: filling up the initial free energy minimum of the bound state results in an increased transition probability in Eq. (3.8). The unbiased rate k can then be extracted from the biased results via

$$k = k_M \left\langle \exp\left(\frac{V_{bias,M}}{k_B T}\right) \right\rangle_M^{-1}, \qquad (3.16)$$

with the mean rate observed in the M bias-accelerated simulations k_M and the bias potential $V_{bias,M}$ that needed to be added in each simulation until an unbinding event happened. The method has already been successfully applied to a range of proteins, from test systems [39, 48] to GPCRs [37, 57] and kinases [6, 76]. Due to the flexible implementation of the MD simulation interface PLUMED [77], a range of different reaction coordinates can be used for biasing. Judged by the 5 µs of accumulated simulation data for the trypsin-benzamidine complex, the computational requirements are relatively low. As a downside, metadynamics requires a prior definition of a range of parameters in Eq. (3.11), especially amplitudes, width, and placement frequency, as well as the position of the added Gaussian functions, for which a suitable choice depends on the underlying free energy landscape and therefore is not known beforehand.

Concerning methods that do not require the introduction of a bias on dynamics, Milestoning approaches have been used in the form of the SEEKR algorithm [78, 79] on the example of the trypsin–benzamidine complex. This approach further reduces the computational cost for the generation of single trajectories via the implementation of Brownian Dynamics at distances larger than a suitable threshold from the binding site, where the details of near-ordering and dynamics of water molecules are not required to be taken into account anymore. Here, the unbinding and binding rates are calculated from the flux, i.e. the number of trajectories, crossing thresholds along x, and the respective time they require for doing so. The computational requirements are moderate: ca 20 µs of accumulated MD simulation time were needed for the trypsin-benzamidine complex [78]. A similar approach is *adaptive multisplitting* [80], which resulted in a comparable prediction accuracy, albeit with only requiring ca 2.5 µs simulation time for said complex.

A completely different approach is taken in *dissipation-corrected targeted MD* (dcTMD) [81]: calculating a bias work $W(x) = \int_{x_0}^{x} dx' f_{bias}(x')$ from the constant velocity constraint bias work of in Eq. (3.15). Based on a second-order cumulant expansion of Jarzynski's equality [82], free energies and friction profiles are then calculated as

$$\Delta F(x) = \langle W \rangle_N - \frac{1}{2k_B T} \langle \Delta W^2 \rangle_N \qquad (3.17)$$

$$\Gamma(x) = \frac{1}{2k_B T v} \frac{d}{dx} \langle \Delta W^2 \rangle_N \qquad (3.18)$$

with the mean and variance over a set of N-independent pulling simulations. The two profiles then serve as input for numerical integration of the Langevin equation (3.5), which already allows for the calculation of unbinding kinetics on the order of microseconds within a few minutes of wall clock time. To reach biomedically relevant time scales, a temperature acceleration similar to that in Eq. (3.9) can be employed: determining $\Delta F(x)$ and $\Gamma(x)$ at a temperature of interest T and setting them to be independent of temperature, Langevin equation simulations at higher temperatures T' yield rate constants k'. These constants then can be scaled back to predict k at the temperature T via the *temperature-boosting* equation [33]

$$k = k' \exp\left(-\frac{\Delta F^{\neq}}{k_B}\left[\frac{1}{T} - \frac{1}{T'}\right]\right), \tag{3.19}$$

which in turn is based on Kramers rate theory Eq. (3.8). dcTMD and temperature-boosted Langevin simulations have been applied to several proteins such as the trypsin-benzamidine complex and the N-terminal domain of Hsp90 [33] as well as to a library of fragments bound to SARS-CoV-2 amin protease [83]. The advantage of this approach is a small computational requirement for pulling data acquisition, which was only 0.4 µs for the case of the trypsin-benzamidine complex, while the cost of the Langevin equation simulations is negligible. In addition, $\Delta F(x)$ and $\Gamma(x)$ yield valuable information on the molecular effects determining the dynamics of protein–ligand unbinding. Recently, an implementation of dcTMD into the Galaxy platform [83] has become available, allowing for a close connection with chemoinformatic tools. The major limit of this approach is the second-order cumulant expansion mentioned above, which requires W to follow a normal distribution and only appears to hold for individual unbinding pathways [43, 56]. Before dcTMD analysis, pathways, therefore, have to be found based on inter-trajectory similarity.

3.4.2 Binding Rate Prediction

Contrary to the prediction of unbinding rates, few options exist to simulate binding events. A ligand finding the correct conformation and near-ordering of contacts with surrounding amino acids within the binding site simply takes time and cannot be biased. Conversely, enforcing unbinding simply requires breaking these contacts [72]. A few examples of binding simulations exist in the form of long-time simulations on the β_2-adrenergic receptor with artificially high ligand concentrations [84]. Alternatively, structural coarse-graining of a protein and ligand can be employed to speed up binding simulations and assess details of this process [85], but this approach comes at the cost of losing the connection between coarse-grained simulation time and real-world (fine-grained) binding time. MSMs are able to capture the binding kinetics of molecular fragments [86] and of benzamidine to trypsin [87, 88], and it should be noted that they have been successfully applied for the prediction of protein-protein association rates [89], but these methods require a significant computational effort up to several hundred microseconds of total simulated time.

An exception here is the combination of dcTMD and temperature-boosted Langevin equation simulations [33]. As the free energy profiles generated with this approach cover the full range of x, both binding and unbinding events are sampled at the same time, i.e. binding and unbinding rates are obtained at the same time. Recently, a similar capability of capturing both dissociation and association kinetics has been reported for *ligand Gaussian accelerated molecular dynamics* (LiGaMD) for the trypsin-benzamidine complex [90] and SARS-CoV2 protease [91]. Furthermore, SEEKR and adaptive multi-splitting provide information on both binding and unbinding rates, as well, as the respective calculated trajectory flux.

3.5 State-of-the-Art in Understanding Kinetics

With all these methods having been established, we now take a look at what knowledge has been gained on the molecular discriminants determining such kinetics and how this knowledge may be exploited for the design of drugs with tailored binding and unbinding rates. From Eq. (3.8), we see that both $\Delta F^{\neq}$ and Γ have an impact on kinetics, with the transition state being the major contributor. On a general basis, the ligand mass, hydrophobic effect, and protein conformational changes have been implied to impact residence times [7, 92]. A special role has been found for electrostatic interactions in the N-terminal domain of Hsp90 [72]: while a salt bridge between a ligand and the protein in the bound state may prolong the ligand's residence time, transient charge interactions between a ligand and an amino acid along the unbinding path can accelerate unbinding of bulkier ligands by facilitating ligand-internal conformational changes as well.

Major contributors to (un)binding dynamics seem to be protein- and ligand-bound water molecules: in trypsin [93] as well as in Hsp90 [52], binding site and ligand desolvation appear to steer the ligand binding rate, which is in agreement with experimental investigations [94]. An intermediate ligand desolvation site has been observed in the before-mentioned simulations of ligand binding to the β_2-adrenergic receptor [84]. Conversely, ligand and binding site hydration shell formation or rearrangements cause a rise in friction and the relaxation of the free energy directly after the transition state in trypsin [33]. Therefore, both changing ligand charge and hydration characteristics appear as feasible start points for an optimization of small organic molecules kinetics.

3.6 Conclusion

While the field of calculating binding and unbinding kinetics is a comparatively new field, much progress has been made over the last decade. It is therefore feasible to assume that the fine-tuning of a drug's kinetic profile will become a standard option in the CADD toolbox in the coming decade. Currently, the largest barrier for this to happen is the high computational cost of such an undertaking. As an experimentalist once pointed out to the author of this chapter, computational predictions of kinetics are only helpful if they are either faster than synthesizing a library of small compounds or if they provide a significant (monetary or informational) benefit over such a brute-force experimental approach. The coming years will show if MD simulation-based predictions of kinetics, possibly with the help of machine learning models trained by MD input data [37, 38, 43, 95–97], can fulfill the promise of providing such benefits.

References

1 Copeland, R.A., Pompliano, D.L., and Meek, T.D. (2006). Drug–target residence time and its implications for lead optimization. *Nat. Rev. Drug Discov.* 5 (9): 730–739.

2 Swinney, D.C. (2006). Biochemical mechanisms of new molecular entities (NMEs) approved by United States FDA during 2001-2004: Mechanisms leading to optimal efficacy and safety. *Curr. Top. Med. Chem.* 6 (5): 461–478.

3 Swinney, D.C. (2012). Applications of Binding Kinetics to Drug Discovery. *Pharm. Med.* 22 (1): 23–34.

4 Copeland, R.A. (2016). The drug-target residence time model: a 10-year retrospective. *Nat. Rev. Drug Discov.* 15 (2): 87–95.

5 Agafonov, R.V., Wilson, C., Otten, R. et al. (2014). Energetic dissection of Gleevec's selectivity toward human tyrosine kinases. *Nat. Struct. Mol. Biol.* 21 (10): 848–853.

6 Shekhar, M., Smith, Z., Seeliger, M.A., and Tiwary, P. (2022). Protein flexibility and dissociation pathway differentiation can explain onset of resistance mutations in kinases. *Angew. Chem. Int. Ed. Engl.* 61 (28): e202200983.

7 Segala, E., Guo, D., Cheng, R.K.Y. et al. (2016). Controlling the dissociation of ligands from the adenosine A2A receptor through modulation of salt bridge strength. *J. Med. Chem.* 59 (13): 6470–6479.

8 Amaral, M., Kokh, D.B., Bomke, J. et al. (2017). Protein conformational flexibility modulates kinetics and thermodynamics of drug binding. *Nat. Commun.* 8 (1): 2276.

9 Shaw, D.E., Adams, P.J., Azaria, A. et al. (2021). Anton 3: twenty microseconds of molecular dynamics simulation before lunch. In: SC, https://doi.org/10.1145/3458817.3487397.

10 Zwanzig, R.W. (2001). *Nonequilibrium Statistical Mechanics*. New York, NY: Oxford Univ. Press.

11 Pottier, N. (2010). *Nonequilibrium Statistical Physics: Linear Irreversible Processes*. New York, NY: Oxford Univ. Press.

12 Vogelsang, R. and Hoheisel, C. (1987). Determination of the friction coefficient via the force autocorrelation function. A molecular dynamics investigation for a dense Lennard-Jones fluid. *J. Stat. Phys.* 47 (1–2): 193–207.

13 Kramers, H.A. (1940). Brownian motion in a field of force and the diffusion model of chemical reactions. *Physica* 7 (4): 284–304.

14 Hänggi, P., Talkner, P., and Borkovec, M. (1990). Reaction-rate theory: fifty years after Kramers. *Rev. Mod. Phys.* 62 (2): 251–341.

15 Hénin, J., Lelievre, T., Shirts, M.R. et al. (2022). Enhanced sampling methods for molecular dynamics simulations [Article v1.0]. *Living J. Comp. Mol. Sci.* 4 (1): 1583–1583. https://doi.org/10.33011/livecoms.4.1.1583.

16 Rensen, M.R.S. and Voter, A.F. (2000). Temperature-accelerated dynamics for simulation of infrequent events. *J. Chem. Phys.* 112 (21): 9599–9606.

17 Maragliano, L. and Vanden-Eijnden, E. (2006). A temperature accelerated method for sampling free energy and determining reaction pathways in rare events simulations. *Chem. Phys. Lett.* 426 (1–3): 168–175.

18 Abrams, C.F. and Vanden-Eijnden, E. (2010). Large-scale conformational sampling of proteins using temperature-accelerated molecular dynamics. *Proc. Natl. Acad. Sci. U. S. A.* 107 (11): 4961–4966.

19 Mollica, L., Decherchi, S., Zia, S.R. et al. (2015). Kinetics of protein-ligand unbinding via smoothed potential molecular dynamics simulations. *Sci. Rep.* 5: 11539.

20 Miao, Y., Feher, V.A., and McCammon, J.A. (2015). Gaussian accelerated molecular dynamics: unconstrained enhanced sampling and free energy calculation. *J. Chem. Theory Comput.* 11: 3584–3595.

21 Torrie, G.M. and Valleau, J.P. (1977). Nonphysical sampling distributions in Monte Carlo free-energy estimation: Umbrella sampling. *J. Comput. Phys.* 23 (2): 187–199.

22 Laio, A. and Parrinello, M. (2002). Escaping free-energy minima. *Proc. Natl. Acad. Sci. U. S. A.* 99 (20): 12562–12566.

23 Bussi, G. and Laio, A. (2020). Using metadynamics to explore complex free-energy landscapes. *Nat. Rev. Phys.* 23 (4): 1–13.

24 Grubmüller, H., Heymann, B., and Tavan, P. (1996). Ligand binding: molecular mechanics calculation of the streptavidin-biotin rupture force. *Science* 271 (5251): 997–999.

25 Izrailev, S., Stepaniants, S., Isralewitz, B. et al. (1999). Steered molecular dynamics, Computational Molecular Dynamics: Challenges, Methods, Ideas, Heidelberg. In: 39–65.

26 Isralewitz, B., Gao, M., and Schulten, K. (2001). Steered molecular dynamics and mechanical functions of proteins. *Curr. Opin. Struct. Biol.* 11 (2): 224–230.

27 Schlitter, J., Engels, M., and Krüger, P. (1994). Targeted molecular dynamics: a new approach for searching pathways of conformational transitions. *J. Mol. Graph.* 12 (2): 84–89.

28 Lüdemann, S.K., Lounnas, V., and Wade, R.C. (2000). How do substrates enter and products exit the buried active site of cytochrome P450cam? 1. Random expulsion molecular dynamics investigation of ligand access channels and mechanisms. *J. Mol. Biol.* 303 (5): 797–811.

29 Zuckerman, D.M. and Chong, L.T. (2017). Weighted ensemble simulation: review of methodology, applications, and software. *Annu. Rev. Biophys.* 46 (1): 43–57.

30 Faradjian, A.K. and Elber, R. (2004). Computing time scales from reaction coordinates by milestoning. *J. Chem. Phys.* 120 (23): 10880.

31 Dellago, C., Bolhuis, P.G., Csajka, F.S., and Chandler, D. (1998). Transition path sampling and the calculation of rate constants. *J. Chem. Phys.* 108 (5): 1964–1977.

32 Bolhuis, P.G., Chandler, D., Dellago, C., and Geissler, P.L. (2002). Transition path sampling: throwing ropes over rough mountain passes, in the dark. *Annu. Rev. Phys. Chem.* 53: 291–318.

33 Wolf, S., Lickert, B., Bray, S., and Stock, G. (2020). Multisecond ligand dissociation dynamics from atomistic simulations. *Nat. Commun.* 11 (1): 2918.

34 Bowman, G.R., Pande, V.S., and Noé, F. (2013). *An Introduction to Markov State Models and Their Application to Long Timescale Molecular Simulation*. Springer Science & Business Media.

35 Thayer, K.M., Lakhani, B., and Beveridge, D.L. (2017). Molecular dynamics-Markov state model of protein ligand binding and allostery in CRIB-PDZ: conformational selection and induced fit. *J. Phys. Chem. B* 121 (22): 5509–5514.

36 Zwanzig, R.W. (1954). High-temperature equation of state by a perturbation method. I. Nonpolar gases. *J. Chem. Phys.* 22: 1420–1426.

37 Ribeiro, J.M.L., Provasi, D., and Filizola, M. (2020). A combination of machine learning and infrequent metadynamics to efficiently predict kinetic rates, transition states, and molecular determinants of drug dissociation from G protein-coupled receptors. *J. Chem. Phys.* 153 (12): 124105.

38 Badaoui, M., Buigues, P.J., Berta, D. et al. (2022). Combined free-energy calculation and machine learning methods for understanding ligand unbinding kinetics. *J. Chem. Theory Comput.* 18 (4): 2543–2555.

39 Tiwary, P., Limongelli, V., Salvalaglio, M., and Parrinello, M. (2015). Kinetics of protein–ligand unbinding: predicting pathways, rates, and rate-limiting steps. *Proc. Natl. Acad. Sci. U. S. A.* 112 (5): E386–E391.

40 Schuetz, D.A., Bernetti, M., Bertazzo, M. et al. (2019). Predicting residence time and drug unbinding pathway through scaled molecular dynamics. *J. Chem. Inf. Model.* 59 (1): 535–549.

41 Kokh, D.B., Doser, B., Richter, S. et al. (2020). A workflow for exploring ligand dissociation from a macromolecule: efficient random acceleration molecular dynamics simulation and interaction fingerprint analysis of ligand trajectories. *J. Chem. Phys.* 153 (12): 125102.

42 Bianciotto, M., Gkeka, P., Kokh, D.B. et al. (2021). Contact map fingerprints of protein-ligand unbinding trajectories reveal mechanisms determining residence times computed from scaled molecular dynamics. *J. Chem. Theory Comput.* 17 (10): 6522–6535.

43 Bray, S., Tänzel, V., and Wolf, S. (2022). Ligand unbinding pathway and mechanism analysis assisted by machine learning and graph methods. *J. Chem. Inf. Model.* 62 (19): 4591–4604.

44 Capelli, R., Carloni, P., and Parrinello, M. (2019). Exhaustive search of ligand binding pathways via volume-based metadynamics. *J. Phys. Chem. Lett.* 3495–3499.

45 Capelli, R., Bochicchio, A., Piccini, G.M. et al. (2019). Chasing the full free energy landscape of neuroreceptor/ligand unbinding by metadynamics simulations. *J. Chem. Theory Comput.* 15 (5): 3354–3361.

46 Rydzewski, J. and Valsson, O. (2019). Finding multiple reaction pathways of ligand unbinding. *J. Chem. Phys.* 150 (22): 221101.

47 Fu, H., Zhou, Y., Jing, X. et al. (2022). Meta-analysis reveals that absolute binding free-energy calculations approach chemical accuracy. *J. Med. Chem.* 65 (19): 12970–12978.

48 Pramanik, D., Smith, Z., Kells, A., and Tiwary, P. (2019). Can one trust kinetic and thermodynamic observables from biased metadynamics simulations?: Detailed quantitative benchmarks on millimolar drug fragment dissociation. *J. Phys. Chem. B* 123 (17): 3672–3678.

49 Nunes-Alves, A., Kokh, D.B., and Wade, R.C. (2021). Ligand unbinding mechanisms and kinetics for T4 lysozyme mutants from tauRAMD simulations. *Curr. Res. Struct. Biol.* 3: 106–111.

50 Guillain, F. and Thusius, D. (1970). Use of proflavine as an indicator in temperature-jump studies of the binding of a competitive inhibitor to trypsin. *J. Am. Chem. Soc.* 92 (18): 5534–5536.

51 Schuetz, D.A., de Witte, W., Arnout, E. et al. (2017). Kinetics for drug discovery: an industry-driven effort to target drug residence time. *Drug Discov. Today* 22 (6): 896–911.

52 Schuetz, D.A., Richter, L., Amaral, M. et al. (2018). Ligand desolvation steers on-rate and impacts drug residence time of heat shock protein 90 (Hsp90) inhibitors. *J. Med. Chem.* 61 (10): 4397–4411.

53 Kokh, D.B., Amaral, M., Bomke, J. et al. (2018). Estimation of drug-target residence times by τ-random acceleration molecular dynamics simulations. *J. Chem. Theory Comput.* 14 (7): 3859–3869.

54 Peng, X., Zhang, Y., Chu, H. et al. (2016). Accurate evaluation of ion conductivity of the Gramicidin A channel using a polarizable force field without any corrections. *J. Chem. Theory Comput.* 12 (6): 2973–2982.

55 Ngo, V., Li, H., Mackerell, A.D. et al. (2021). Polarization effects in water-mediated selective cation transport across a narrow transmembrane channel. *J. Chem. Theory Comput.* 17 (3): 1726–1741.

56 Jäger, M., Koslowski, T., and Wolf, S. (2022). Predicting ion channel conductance via dissipation-corrected targeted molecular dynamics and Langevin equation simulations. *J. Chem. Theory Comput.* 18 (1): 494–502.

57 Capelli, R., Lyu, W., Bolnykh, V. et al. (2020). Accuracy of molecular simulation-based predictions of K_{off} values: a metadynamics study. *J. Phys. Chem. Lett.* 6373–6381.

58 Lopes, P.E.M., Huang, J., Shim, J. et al. (2013). Force field for peptides and proteins based on the classical Drude oscillator. *J. Chem. Theory Comput.* 9 (12): 5430–5449.

59 Shi, Y., Xia, Z., Zhang, J. et al. (2013). Polarizable atomic multipole-based AMOEBA force field for proteins. *J. Chem. Theory Comput.* 9 (9): 4046–4063.

60 Bruce, N.J., Ganotra, G.K., Kokh, D.B. et al. (2018). New approaches for computing ligand-receptor binding kinetics. *Curr. Opin. Struct. Biol.* 49: 1–10.

61 Ribeiro, J.M.L., Tsai, S.-T., Pramanik, D. et al. (2019). Kinetics of ligand-protein dissociation from all-atom simulations: Are we there yet? *Biochemistry* 58 (3): 156–165.

62 Bernetti, M., Masetti, M., Rocchia, W., and Cavalli, A. (2019). Kinetics of drug binding and residence time. *Annu. Rev. Phys. Chem.* 70: 143–171.

63 Nunes-Alves, A., Kokh, D.B., and Wade, R.C. (2020). Recent progress in molecular simulation methods for drug binding kinetics. *Curr. Opin. Struct. Biol.* 64: 126–133.

64 Limongelli, V. (2020). Ligand binding free energy and kinetics calculation in 2020. *WIREs Comput. Mol. Sci.* 8 (93): e1358.

65 Ahmad, K., Rizzi, A., Capelli, R. et al. (2022). Enhanced-sampling simulations for the estimation of ligand binding kinetics: current status and perspective. *Front. Mol. Biosci.* 9: 899805.

66 Wang, J., Do, H.N., Koirala, K., and Miao, Y. (2023). Predicting biomolecular binding kinetics: a review. *J. Chem. Theory Comput* 19 (8): 2135–2148. https://doi.org/10.1021/acs.jctc.2c01085.

67 Sohraby, F. and Nunes-Alves, A. (2023). Advances in computational methods for ligand binding kinetics. *Trends Biochem. Sci.* 48 (5): 437–449. https://doi.org/10.1016/j.tibs.2022.11.003.

68 Chen, Y.-C. (2015). Beware of docking! *Trends Pharmacol. Sci.* 36 (2): 78–95.

69 Mollica, L., Theret, I., Antoine, M. et al. (2016). Molecular dynamics simulations and kinetic measurements to estimate and predict protein-ligand residence times. *J. Med. Chem.* 59 (15): 7167–7176.

70 Bortolato, A., Deflorian, F., Weiss, D.R., and Mason, J.S. (2015). Decoding the role of water dynamics in ligand-protein unbinding: CRF1R as a test case. *J. Chem. Inf. Model.* 55 (9): 1857–1866.

71 Potterton, A., Husseini, F.S., Southey, M.W.Y. et al. (2019). Ensemble-based steered molecular dynamics predicts relative residence time of A2A receptor binders. *J. Chem. Theory Comput.* 15 (5): 3316–3330.

72 Wolf, S., Amaral, M., Lowinski, M. et al. (2019). Estimation of protein-ligand unbinding kinetics using non-equilibrium targeted molecular dynamics simulations. *J. Chem. Inf. Model.* 59 (12): 5135–5147.

73 Kokh, D.B. and Wade, R.C. (2021). G protein-coupled receptor-ligand dissociation rates and mechanisms from τRAMD simulations. *J. Chem. Theory Comput.* 17 (10): 6610–6623.

74 Berger, B.-T., Amaral, M., Kokh, D.B. et al. (2021). Structure-kinetic relationship reveals the mechanism of selectivity of FAK inhibitors over PYK2. *Cell Chem. Biol.* 28 (5): 686–698.e7.

75 Tiwary, P. and Parrinello, M. (2013). From metadynamics to dynamics. *Phys. Rev. Lett.* 111 (23): 230602.

76 Casasnovas, R., Limongelli, V., Tiwary, P. et al. (2017). Unbinding kinetics of a p38 MAP kinase type II inhibitor from metadynamics simulations. *J. Am. Chem. Soc.* 139 (13): 4780–4788.

77 The PLUMED Consortium (2019). Promoting transparency and reproducibility in enhanced molecular simulations. *Nat. Methods* 16 (8): 670–673.

78 Votapka, L.W., Jagger, B.R., Heyneman, A.L., and Amaro, R.E. (2017). SEEKR: simulation enabled estimation of kinetic rates, a computational tool to estimate molecular kinetics and its application to trypsin-benzamidine binding. *J. Phys. Chem. B* 121 (15): 3597–3606.

79 Jagger, B.R., Ojha, A.A., and Amaro, R.E. (2020). Predicting ligand binding kinetics using a Markovian milestoning with Voronoi tessellations multiscale approach. *J. Chem. Theory Comput.* 16 (8): 5348–5357.

80 Teo, I., Mayne, C.G., Schulten, K., and Lelievre, T. (2016). Adaptive multilevel splitting method for molecular dynamics calculation of benzamidine-trypsin dissociation time. *J. Chem. Theory Comput.* 12 (6): 2983–2989.

81 Wolf, S. and Stock, G. (2018). Targeted molecular dynamics calculations of free energy profiles using a nonequilibrium friction correction. *J. Chem. Theory Comput.* 14: 6175–6182.

82 Jarzynski, C. (1997). Equilibrium free-energy differences from nonequilibrium measurements: a master-equation approach. *Phys. Rev. E* 56 (5): 5018–5035.

83 Bray, S., Dudgeon, T., Skyner, R. et al. (2022). Galaxy workflows for fragment-based virtual screening: a case study on the SARS-CoV-2 main protease. *J. Chem.* 14 (1): 1–13.

84 Dror, R.O., Pan, A.C., Arlow, D.H. et al. (2011). Pathway and mechanism of drug binding to G-protein-coupled receptors. *Proc. Natl. Acad. Sci. U. S. A.* 108 (32): 13118–13123.

85 Souza, P.C.T., Thallmair, S., Conflitti, P. et al. (2020). Protein–ligand binding with the coarse-grained Martini model. *Nat. Commun.* 11 (1): 1–11.

86 Linker, S.M., Magarkar, A., Köfinger, J. et al. (2019). Fragment binding pose predictions using unbiased simulations and Markov-state models. *J. Chem. Theory Comput.* 15 (9): 4974–4981.

87 Buch, I., Giorgino, T., and De Fabritiis, G. (2011). Complete reconstruction of an enzyme-inhibitor binding process by molecular dynamics simulations. *Proc. Natl. Acad. Sci. U. S. A.* 108 (25): 10184–10189.

88 Plattner, N. and Noé, F. (2015). Protein conformational plasticity and complex ligand-binding kinetics explored by atomistic simulations and Markov models. *Nat. Commun.* 6 (1): 7653.

89 Plattner, N., Doerr, S., De Fabritiis, G., and Noé, F. (2017). Complete protein–protein association kinetics in atomic detail revealed by molecular dynamics simulations and Markov modelling. *Nat. Chem.* 9: 1005–1011.

90 Miao, Y., Bhattarai, A., and Wang, J. (2020). Ligand Gaussian accelerated molecular dynamics (LiGaMD): characterization of ligand binding thermodynamics and kinetics. *J. Chem. Theory Comput.* 16: 5526–5547.

91 Wang, Y.-T., Liao, J.-M., Lin, W.-W. et al. (2022). Structural insights into Nirmatrelvir (PF-07321332)-3C-like SARS-CoV-2 protease complexation: a ligand Gaussian accelerated molecular dynamics study. *Phys. Chem. Chem. Phys.* 24 (37): 22898–22904.

92 Pan, A.C., Borhani, D.W., Dror, R.O., and Shaw, D.E. (2013). Molecular determinants of drug–receptor binding kinetics. *Drug Discov. Today* 18 (13–14): 667–673.

93 Ansari, N., Rizzi, V., and Parrinello, M. (2022). Water regulates the residence time of Benzamidine in Trypsin. *Nat. Commun.* 13 (1): 1–9.

94 Schiebel, J., Gaspari, R., Wulsdorf, T. et al. (2018). Intriguing role of water in protein-ligand binding studied by neutron crystallography on trypsin complexes. *Nat. Commun.* 9 (1): 166.

95 Ribeiro, J.M.L. and Tiwary, P. (2018). Towards achieving efficient and accurate ligand-protein unbinding with deep learning and molecular dynamics through RAVE. *J. Chem. Theory Comput.* 15 (1): 708–719.

96 Brandt, S., Sittel, F., Ernst, M., and Stock, G. (2018). Machine learning of biomolecular reaction coordinates. *J. Phys. Chem. Lett.* 2144–2150.

97 Komp, E., Janulaitis, N., and Valleau, S. (2022). Progress towards machine learning reaction rate constants. *Phys. Chem. Chem. Phys.* 24 (5): 2692–2705.

4

Solvation Thermodynamics and its Applications in Drug Discovery

Kuzhanthaivelan Saravanan and Ramesh K. Sistla

thinkMolecular Technologies Pvt. Ltd., #03, 1st Cross, Reliaable Tranquil Layout, Bengaluru, Karnataka 560102, India

> yo'pām āyatanaṃ veda | āyatanavān bhavati
>
> – taittirīya aruṇa praśna – 1.72
>
> He who knows the position of water, secures his position.

4.1 Introduction

Water is the major constituent in all organisms. All life processes play out in the medium of water. Therefore, it is essential to understand the role of the aqueous medium in various life processes. The role of the solvent in the processes of protein folding [1, 2] and molecular recognition [3] is well documented in the literature. The hydrophobic effect [4] has been proposed to be the key reason for protein folding. A simplistic demonstration of the hydrophobic effect lies in the immiscibility of oil and water and the coming together of oil particles on a water surface. Extending this visualization further, the hydrophobic amino acids in a protein move away from the solvent and come together to form the hydrophobic core of the protein. This was demonstrated in a graphical manner through hydropathy plots proposed by Kyte and Doolittle [5]. The partitioning of the nonpolar amino acids into the lipid membranes and the formation of the hydrophobic core in globular proteins unambiguously suggest the role of the aqueous medium in which the proteins are present.

4.1.1 Protein Folding

Proteins are synthesized in the ribosomes inside the cells. Upon synthesis, the proteins immediately fold into three-dimensional structures and perform precise functions ranging from scaffolding to being carriers to performing catalysis. The protein folding phenomenon has been the subject of extensive research over the last several decades.

Computational Drug Discovery: Methods and Applications, First Edition.
Edited by Vasanthanathan Poongavanam and Vijayan Ramaswamy.
© 2024 WILEY-VCH GmbH. Published 2024 by WILEY-VCH GmbH.

The "thermodynamic hypothesis" for protein folding was suggested by Anfinsen [6], who postulated that, out of the numerous conformations that a protein can adopt, it chooses the conformation in the physiological milieu whose Gibbs-free energy is the lowest. The Gibbs free energy of a system has two components and is given as follows:

$$\Delta G = \Delta H - T\Delta S \tag{4.1}$$

where ΔG is the change in Gibbs free energy between two states, and the corresponding changes in enthalpy and entropy are denoted by ΔH and ΔS, respectively. T represents the temperature in Kelvin.

Enthalpy is the total energy of the system and comprises electrostatic and dispersion terms. These components of enthalpy manifest in a variety of interatomic interactions, such as hydrogen bonds, van der Waals interactions, aromatic π-stacking, and hydrophobic interactions, to name a few. For the folded conformation of the protein, ΔH is negative compared to the unfolded conformation. This essentially results from the favorable interactions between hydrophobic amino acids present in the core of a folded globular protein and also from the key hydrogen bonds and salt bridges that are formed in the secondary structures of the protein upon folding. The enthalpy is further boosted by the interactions of the exposed polar amino acids of the protein with the solvent. However, it must be borne in mind that the unfolded polypeptide chain makes several polar interactions with the solvent. When this chain folds, these waters are removed out into what is known as the bulk solvent, and the polar interactions are formed among the amino acid residues. Therefore, a net change in enthalpy during the process of protein folding could be modest [7].

The entropy of a system is a measure of its randomness. It has its basis in the second law of thermodynamics. In simple terms, it means the number of degrees of freedom that are available to a system. In the case of the polypeptide chain in the aqueous medium, it means the number of degrees of freedom that each atom has. Water molecules near hydrophobic amino acids are highly constrained and hence have lower entropy. When the protein folds, these waters are removed to the bulk solvent, thereby enhancing the entropy of the solvent. The gain in entropy adds favorably to the Gibbs free energy of the system (protein + solvent).

Thus, even if the gain of enthalpy during the folding is modest, the contribution of the solvent entropy to Gibbs free energy is significant. Thus, it is the solvent that funds the energy needed for protein folding through the maximization of its entropy [7]. The entropy term is the sum of the entropy of the solvent, the conformational entropy of the protein, and the rotational/translational entropy. While the conformational and rotational/translational entropy of the protein might reduce during the folding, the enhancement in solvent entropy outdoes the reduction in the other two.

4.1.2 Protein–Ligand Interactions

The recognition of various proteins by their cognate ligands, cofactors, and other molecules and their complex formation are also influenced immensely by the solvent in which the molecules are present. The protein–ligand complexes are formed only when the energy exchange between the protein, ligand, solvent, and any other essential component like ions present there is favorable. In other words, just as in the case of protein folding, the complex formation is favored when the Gibbs free energy of the whole system is negative [8]. The association and dissociation of a protein (P) and ligand (L) can be written in the following manner:

$$P + L \rightleftharpoons P.L$$

This reaction has an on rate of K_{on}, which represents the association of P and L, and an off rate of k_{off} signifying the dissociation of the ligand from the protein. The dissociation constant of the ligand K_d is given as

$$K_d = \frac{k_{off}}{K_{on}} = \frac{[P].[L]}{[P.L]} \qquad (4.2)$$

This representation and estimation of K_d looks as if this phenomenon of the association of a protein and the ligand is purely binary and the solvent has little or no role to play. However, in reality, the role of the solvent is in fact rolled into both k_{off} and K_{on} and thus the solvent dictates the dissociation or association of a complex.

In physical terms, prior to complex formation, the protein and the ligand are individually solvated. Upon the formation of a collision complex, the ligand and the binding site undergo desolvation, and subsequently the protein and the ligand form a complex. The considerations of entropy and enthalpy between the protein and ligand are again on the lines similar to what we have seen in the protein folding in the preceding section. The solvent that interacted with the protein binding site and the ligand earlier, upon desolvation, is removed to the bulk, and its entropy increases. The association between the protein and the ligand, on the other hand, contributes to favorable interactions between complementary groups and causes an enthalpic gain. The gain in enthalpy of the protein–ligand system and the increase in the entropy of the solvent ensure an overall negative Gibbs free energy, which favors complex formation.

The process of a protein and ligand forming a complex is graphically and conceptually shown by the Born Haber cycle [9] shown below in Figure 4.1.

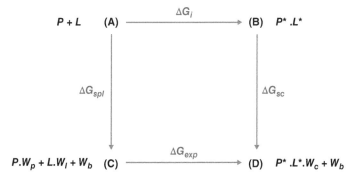

Figure 4.1 Born Haber cycle.

The Born Haber cycle provides a rigorous treatment of protein–ligand complex formation. In Figure 4.1, state **A** represents the protein and ligand, state **B** represents the unsolvated protein–ligand complex, state **C** represents the solvated states of the protein and ligand (W stands for water, its subscripts p, l, c, and b stand for protein, ligand, complex, and bulk, respectively) and state **D** represents the solvated protein–ligand complex and the bulk water. The transition from one state to another in the above process is associated with a change in Gibbs free energy. ΔG_i represents the intrinsic change in the free energy upon the complex formation (A → B). ΔG_{exp} represents the change in free energy that would be measured experimentally by techniques such as isothermal titration calorimetry (ITC). The other two terms ΔG_{spl} (A → C) and ΔG_{sc} (B → D) represent the solvation-free energies (SFEs) of the individual components and the complex, respectively.

The process A → C represents the solvation of the individual components – the protein and the ligand. The process B → D represents the solvation of the protein–ligand complex. The energies estimated in these processes are solvation energies, and those values with the opposite sign represent the desolvation energies. In the process of a protein–ligand complex formation, **A** is the initial state and **D** is the final state. Free energy being dependent only on the state and not on the path the system takes, or, in other words, being a state function, the energy equation can be written as

$$\Delta G_i + \Delta G_{sc} = \Delta G_{spl} + \Delta G_{exp} \tag{4.3}$$

Rearranging,

$$\Delta G_{exp} = \Delta G_i + (\Delta G_{sc} - \Delta G_{spl}) \tag{4.4}$$

The term in the bracket in Eq. (4.4) directly relates to the free energy of solvation of the complex and that of the individual components. Thus, the observed free energy change upon complex formation is a sum of the intrinsic change in the free energy and the change in free energy of solvation.

The free energy is in turn composed of enthalpy (H) and entropy (S). As referenced in the earlier section, the enthalpy is a result of various intermolecular interactions, and the entropy signifies the order or lack thereof of a system. The enthalpic change of solvation/desolvation may not always be favorable (negative). For example, when the solvent molecules are near hydrophobic groups of the protein/ligand and upon complexation move into the bulk, there is a net positive change in the enthalpy. However, the movement of the solvent to the bulk results in enhanced entropy. This change in entropy contributes favorably to the Gibbs free energy of the system and makes complex formation favorable. It is well known that the free energy change is related to the affinity (rate constant).

Thus far, we have seen a qualitative treatment of the process of solvation and the role played by solvent in protein folding as well as protein–ligand complex formation. The quantitative treatment of the contribution of water molecules to the entropy of the system through experimental techniques can be very limiting. NMR techniques provide average properties of the solvent that rapidly exchange between the binding site and the bulk solvent. Knowledge of the positions of water molecules by X-ray crystallography is highly resolution-dependent, and even in the best cases, only the structurally conserved water molecules can be assessed with confidence.

Docking the ligands in the binding sites of protein, and designing compounds based on the various interactions that are made by the ligand with the protein is a routine exercise in drug discovery. The ligands that dock in the binding site are ranked by the docking score, which is estimated using various scoring functions [10]. Ideally, the docking score computed from physics-based scoring functions should reflect the binding energy of the ligand. However, due to the limited accuracy of the forcefields employed, not considering the protein flexibility, and most importantly, not taking the solvent effect into account, the docking scores are not representative of the binding energy.

The effect of solvent is estimated in molecular modeling in two ways – (i) implicit solvent considerations and (ii) explicit solvent considerations. In either case, the goal is to incorporate the solvent effect into the estimation of the protein–ligand free energy.

In the case of the implicit solvent models, a dielectric continuum is considered in which the interactions between atoms of the protein and ligand (solute) take place. These interactions are influenced by the dielectric constant of the medium. Molecular mechanics Poisson–Boltzmann surface area (MM/PBSA) and molecular mechanics generalized Born surface area (MM/GBSA) [11] are popular endpoint free energy calculation methods that are more accurate compared to corresponding docking scores. Since it was first proposed [12] as an electrostatic interaction of a solute with the continuum, the implementations of the implicit solvent method have evolved and shown in several examples to result in better prediction of the binding affinity of the ligands [13]. As the name suggests, the medium in which the solute is present is represented by the dielectric, and there is no explicit consideration of solvent molecules in this approach. It is fast and can be employed in virtual screening campaigns. More recently, a machine-learning polarized continuum solvation model has been proposed [14]. The linear interaction energy is another method for estimating the ligand binding free energies [15]. This is based on forcefields and is a simplified method with several approximations, requiring only the intermolecular interactions between the ligand and the receptor to estimate the energy.

None of these methods treat the entropy associated with the solvent at all, and their limitations can be as such ascribed to that. In spite of this, these methods are computationally efficient and more accurate than docking scores and hence have great utility in large-scale virtual screening campaigns [16]. However, it has also been pointed out [17] that both the PBSA and GBSA performed poorly in estimating the solvtion free energies (SFE) compared to explicit solvent methods. The main reason for this is the poor estimation of apolar contributions, which stem from solvent entropy.

High-resolution crystal structures reveal the presence of only strongly interacting water molecules. Other water molecules that have a transient presence at certain sites do not show up in the structures. This being the case, treating the two regions as a homogenous dielectric medium can introduce errors in the results of the application of continuum theory. There are several examples in the literature where conserved water molecules play a key role in the binding of ligands. The directionality of the hydrogen bonds formed by the water molecules is an important determinant [18] in the potency of several molecules, which are not adequately treated by the continuum methods [19]. In this backdrop, it is imperative that the solvent be studied in an explicit manner rather than as a dielectric continuum.

Molecular dynamics (MD) [20] is a handy tool to treat water molecules explicitly and study their interactions with proteins and ligands. While the stability of the various protein–ligand interactions and conformational change of proteins is routinely studied by MD simulations, generally not much attention is placed on the estimations of the energetics of the water molecules, which are present in the protein pockets.

The enthalpic and entropic contributions of the water molecules are studied in a few different approaches – thermodynamic integration (TI), inhomogeneous solvation theory (IST), and reference interaction site models (RISM). The TI method computes the energy needed for the extraction of a water molecule from a particular position in the pocket.

The IST method draws information from the phase space of the solute–solvent complex generated from MD simulations. It considers the solute molecule to be central, evaluates the fluctuation of the density of water molecules, and estimates the enthalpies and entropies of these waters. Consequently, the implementation of IST is computationally intensive, but the reward is that the results are as accurate as the force field that is employed in these calculations. This approach has been implemented in some of the modern tools, such as Grid inhomogeneous solvation theory (GIST) [19, 21], Watermap [22], solvation thermodynamics of ordered water (STOW) [23], and solvation structure and thermodynamic mapping (SSTMap) [24]. Adequate sampling is needed to get accurate results with this method.

RISM, on the other hand, is a statistical mechanical integral approach [25]. This does not necessitate an extensive MD simulation and utilizes the optimized configuration of a solute–solvent complex. It then estimates for each solvent site a susceptibility function that is dependent on the positions and interactions of the solvent molecules with the solute and its polarizability. It solves the classic molecular Ornstein–Zernike equation that relates the correlation function between two atoms to the total correlation function of the solute and the solvent. Thus, it seeks to construct the molecular density distribution through atomic densities. The RISM approach has been implemented in several tools, viz. 3D-RISM, GCT, SZMAP, and WATsite.

A detailed description of each tool is beyond the scope of this chapter. In order to provide an insight into the contrasting techniques, in the upcoming section, a brief overview of the principles followed in the tools Watermap, GIST, and 3D-RISM will be presented, followed by case studies from the literature in Section 4.3.

4.2 Tools to Assess the Solvation Thermodynamics

To recap the earlier section, the Gibbs free energy has two components: enthalpy and entropy. The increase in solvent entropy has been pointed out to have a favorable contribution to the overall Gibbs free energy of the system. The enthalpy component is computed by summing up various interatomic interactions using the parameters provided in the force fields. We have also discussed the inadequacies of continuum solvent models. Therefore, a reasonable treatment of the entropy of the system, and particularly the solvent entropy, is in order. We have touched upon the general methods that are employed to compute solvent entropy. Now in this section, we will deal briefly with the principles behind the tools Watermap [22], GIST [19], and 3D-RISM [26].

4.2.1 Watermap

Watermap [22] seeks to characterize the waters in the binding site as per their energies as happy (low energy) and unhappy (high energy) waters. It does so by estimating the enthalpy and the entropy associated with individual waters. MD simulation of a rigid solute in a solvent box is performed to study the behavior of individual waters in the binding site under the influence of the interatomic forces of the amino acids in the binding pocket. The protein molecule is constrained throughout the simulation, and the hydration of the binding sites is studied. The water molecules that hydrate the binding site are of three different types. The first kind are those that make a full complement of hydrogen bonds in the binding site. The second kind are those that are fully satisfied enthalpically but do not form hydrogen bonds optimally. The third are those that are held by weak forces in the binding site and thus have their degrees of freedom severely restricted. The second and third kinds of waters, when they are replaced by ligand groups that make complementary interactions, boost the binding energy by moving the constrained water into the bulk solvent and contribute substantially to the binding energy. There is no advantage in displacing waters that are optimally bound in the cavities, as there will be a significant loss of enthalpy upon displacement, which may not be compensated by an increase in entropy or may just be a zero-sum game. The first type of water is termed "happy water" and the second and third types are termed "unhappy water."

Watermap divides the cavities in proteins into subvolumes. Each subvolume is termed a hydration site. A clustering algorithm scans through all the subvolumes and calculates the solvent density and solvent exposure at each point. The average number of water neighbors of each subvolume determines the solvent exposure at that point, and the degree of exposure is determined with respect to the solvent exposure of the bulk solvent. After identifying the hydration sites, the IST is used to determine the entropic cost of solvent ordering. Using this entropy calculation, the interaction energy of a water molecule at each hydration site with the rest of the system is computed.

The partial excess entropy of a given hydration site is computed through numerical integration, considering the orientational and spatial correlations of the water molecule at the given hydration site as per the following equation.

$$S^e = -\frac{k_b \rho_w}{\Omega} \int g_{sw}(r, \omega) \ln(g_{sw}(r, \omega)) dr d\omega \qquad (4.5)$$

$$\approx k_b \rho_w \int g_{sw}(r) \ln(g_{sw}(r)) dr - \frac{k_b N_W^V}{\Omega} \int g_{sw}(\omega) \ln(g_{sw}(\omega)) d\omega \quad (4.6)$$

In this equation, r signifies the cartesian positions, while ω is the Euler angle orientation of the water molecules. The distribution functions (g) and the density of the bulk water (ρ) are considered. Other enhancements of this entropy estimation are possible by including higher-order terms, but they increase the computational cost.

The system interaction energy computed for each hydration site characterizes the water as high-energy (unhappy) or low-energy (happy) water. Displacement of an unhappy water by a favorable group on the ligand has been shown to explain the structure-activity relationships observed in a congeneric series of ligands [27]. An example of a prospective use of Watermap in a computational triage will be presented in the next section.

Some of the limitations of Watermap may stem from the consideration of the protein as rigid and not considering higher-order entropy terms. However, neither of these limitations is general in nature. In fact, Watermap reveals a lot more waters [22] than most high-resolution crystal structures and also provides the energetics associated with those waters thus providing a very important qualitative insight into the hydration of the binding site.

4.2.2 GIST

GIST is another implementation of the IST [19]. As the name suggests, it solves the equations of IST on a grid made of discrete cells called voxels in a region of interest in the protein. The values of the solvation entropies, enthalpies, and free energies in each voxel are computed. The summation of these parameters is then done on a trajectory obtained from the MD simulations of a protein in a chosen solvent box, which yields the solvation thermodynamics. Unlike Watermap, GIST does not limit the calculations to the high-density water hydration sites but rather estimates the thermodynamic parameters in every voxel and thus provides a smooth variation of the character of water at every position in the region of interest on the protein. Thus, the hydration thermodynamic information is independent of the density of water in a given region relative to the bulk solvent. This is especially useful to identify the sites that are partially occupied by water molecules. Mapping each voxel with the calculated parameters, GIST is able to compute free energy from the states.

The solvation entropy of a flexible solute is given as

$$\int p(q) \Delta S_{solv}(q) dq \quad (4.7)$$

where $p(q)$ is the Boltzmann probability and ΔS_{solv} is solvation entropy. q defines the coordinates of the system. The solvation entropy is a sum of the solute–water entropy and the water–water entropy.

$$\Delta S_{solv} = \Delta S_{sw} + \Delta S_{ww} \quad (4.8)$$

Ignoring the water–water entropy term, which deals with the bulk water and does not concern us in the region of the binding sites,

4.2 Tools to Assess the Solvation Thermodynamics

$$\Delta S_{solv} \approx \Delta S_{sw} \equiv k_B \frac{\rho^o}{8\pi^2} \int g_{sw}(r,\omega) dr\, d\omega \qquad (4.9)$$

where k_B is the Boltzmann constant, g_{sw} is the pair correlation function between the solute–water and is a function of the cartesian coordinates (r) and the Euler angles (ω). The free energy would be a combination of the system interaction energy (enthalpy solute–water and water–water), solute–water translational entropy, and the solute–water rotational entropy (together the entropy of solvation). The above equation is discretized and the entropies are estimated for every voxel k and summed over the voxels in a region R.

The translational entropy for the voxel k is given as

$$\Delta S_{sw}^{trans}(r_k) \equiv k_B \rho^o \int_k g(r) \ln g(r) dr\, d\omega \approx k_B \rho^o\, V_k g(r_k) \ln g(r_k) \qquad (4.10)$$

$$g(r_k) \equiv \frac{N_k}{\rho^o V_k N_f} \qquad (4.11)$$

The translational entropy over a region R is the sum of those over the voxels in the region and is given as

$$\Delta S_{sw}^{R,trans} \approx \sum_{k \in R} \Delta S_{sw}^{trans}(r_k) \qquad (4.12)$$

The normalized entropy (per water in the region R containing n^R number of waters) would then become

$$\Delta S_{sw}^{R,trans,norm} \equiv \frac{\Delta S_{sw}^{R,trans}}{n^R} \qquad (4.13)$$

The orientational entropy is also estimated similarly for each voxel, summed over all the voxels in a region R, and is given as

$$\Delta S_{sw}^{R,orient,norm} \equiv \frac{\Delta S_{sw}^{R,orient}}{n^R} \qquad (4.14)$$

The solute–water interaction energy as well as the water–water interaction energy are also discretized for each voxel k, summed over a region R and finally normalized as given below.

$$\Delta E_{sw}(r_k) \equiv \int_k \Delta E_{sw}(r) dr \qquad (4.15)$$

$$\Delta E_{sw}^R = \sum_{k \in R} \Delta E_{sw}(r_k) \qquad (4.16)$$

$$\Delta E_{sw}^{R,norm} = \frac{\Delta E_{sw}^R}{n^R} \qquad (4.17)$$

The water–water interactions are defined between voxels k and l

$$E_{ww}^{R,corr} = \sum_{k \in R} E_{ww}(r_k) - \frac{1}{2} \sum_{\substack{k \in R \\ l \in R}}^{l \neq k} E_{ww}(r_k, r_l) \qquad (4.18)$$

$$E_{ww}^{R,norm} = \frac{\sum_{k \in R} E_{ww}(r_k)}{n^R} \qquad (4.19)$$

From the above terms, the total energy and entropy for each voxel are calculated as follows.

$$\Delta E_{total}(r_k) \equiv \Delta E_{sw}(r_k) + \Delta E_{ww}(r_k), \quad (4.20)$$

$$\Delta S_{sw}^{total}(r_k) = \Delta S_{sw}^{trans}(r_k) + \Delta S_{sw}^{orient}(r_k), \quad (4.21)$$

These two terms yield the free energy of every voxel

$$\Delta G(r_k) = \Delta E_{total}(r_k) - T\Delta S_{sw}^{total}(r_k) \quad (4.22)$$

Thus, GIST is an elegant implementation of the IST, which splits the region into a grid made of voxels and discretizes the energy and entropy of each voxel. Hence, it maps out the regions of favorable and unfavorable free energies in the protein with respect to the solvent, providing insight into the design of ligands. More recent work [21] has extended the GIST framework to use chloroform as a solvent to simulate the membrane permeability of molecules, and further, the method has been generalized to include any rigid molecule as a solvent. This important enhancement extends the utility of the tool to important parameters other than protein–ligand potency optimization.

4.2.3 3D-RISM

The RISM in the context of a solute–water interaction was proposed by Kovalenko [25]. Subsequently, it has been implemented in several software packages as an approach to study solvation through the three-dimensional RISM (3D-RISM) technique [26]. As rigorous as the formulation of the IST method and its implementation are, the results are influenced by the amount of sampling that the solvent undergoes. The adequacy of sampling in a given simulation is always a matter of contention.

The 3D-RISM method attempts to circumvent this by taking a purely statistical mechanical approach to the problem and applying the integral equation theory. In this method, first, the rigid solute is subjected to a standard 3D-RISM calculation. Subsequently, solvent molecules are placed at different sites, and the solvent distribution function $g(r)$, total correlation function $h(r)$, direct correlation $c(r)$, and the local interaction potential $u(r)$ at every solvent site are calculated and iterated until they are self-consistent.

Both the positions and the orientations of the solvent molecules are optimized until a preset cutoff is reached. A local population function and the location of maximum probability are computed through iterations to identify the solvent distribution. After identifying the locations of the solvents through this, the orientational distribution is identified.

Based on the thorough characterization of the solvent sites and estimation of the distributions of solvent density, the energy and the entropy of each site are then calculated using the following equations.

$$E_n^1 = \rho_0 \sum_{\gamma} \int_{V_n} g_{\gamma}(r) u_{\gamma}(r) dr \quad (4.23)$$

$$S_n^{1,trans} = -\frac{\rho_0 k_B}{N_{rot}} \int_{V_n} g_{anchor}(r) dr \int_\omega g(\omega r) \ln g(\omega r) d\omega \qquad (4.24)$$

Having briefly touched upon the methods employed in tools like Watermap, GIST, and 3D-RISM, case studies involving these tools from the literature are presented in the upcoming section. They are not meant to demonstrate the exhaustive use of these tools but rather to provide a flavor of their utilities.

4.3 Case Studies

4.3.1 Watermap

4.3.1.1 Background and Approach

Group-I p21-activated kinase 1 (PAK1) is essential for various cellular functions such as cytoskeletal organization, motility, mitosis, and angiogenesis. These roles make it an attractive therapeutic target for cancer [28], infectious diseases, and neurological disorders [29]. However, designing a highly selective PAK1 inhibitor is quite challenging because of the high homology of the kinase domain with other kinases.

In this work [30], a synergistic computational approach was used to repurpose the FDA-approved drugs from the Drugbank database [31] against PAK1. This synergistic approach includes molecular docking to understand the binding modes and potential affinity of the drug molecules revealed through noncovalent interaction energies. Since routine docking does not take into account the solvation effects, the authors utilized Watermap, which is a useful tool to predict the effects of explicit water molecules in the binding sites as described earlier. In this work, a short 2 ns MD simulation is performed using the Grand Canonical Monte Carlo (GCMC) sampling to predict structurally weak water clusters in the protein binding pocket.

The crystal structure of PAK1 kinase domain in complex with FRAX597 inhibitor (PDB id: 4EQC) with a resolution of 2.01 Å was selected for this study. The virtual screening process of the curated Drugbank molecules against PAK1 was performed using GLIDE molecular docking software in the Schrodinger suite. Out of 2162 FDA-approved drugs from the Drugbank database, 27 compounds were shortlisted based on the interactions that the docked poses made with the protein. These 27 compounds were then assessed with respect to the hydration site displacements predicted by Watermap in determining the binding affinity gains likely to be made by these compounds.

4.3.1.2 Results and Discussion

The Watermap calculations facilitated the identification of localized hydration sites around the binding cavity of PAK1, breaking down their thermodynamic energy profiles, viz. enthalpy (ΔH), entropy ($T\Delta S$), and differential binding energy ($\Delta\Delta G$). Based on $\Delta\Delta G$, the overlapping hydration sites on ligand functional groups could be categorized into displaceable ($\Delta\Delta G \gg 0$ and $\Delta H \gg 0$), replaceable ($\Delta H \ll 0$ and yet $\Delta\Delta G \gg 0$ or $\cong 0$), and stable ($\Delta\Delta G \ll 0$) water molecules. Out of the 27 drug molecules shortlisted through virtual screening by docking, 20 drug molecules were observed to displace a smaller number of high-energy hydration sites and hence may not offer better binding affinity to PAK1. Thus, the remaining seven molecules, namely, Mitoxantrone, Labetalol, Acalabrutinib, Sacubitril, Flubendazole, Trazodone, and Niraparib were shortlisted based on displacing more unfavorable hydration sites. The molecules were considered for exploring the implications of hydration site displacements in determining the binding affinity gains. The Watermap results indicated that the hinge region of PAK1 had two high-energy hydration sites (5.27 and 2.59 kcal/mol), and the remaining four high-energy hydration sites were concentrated in the back pocket of PAK1 (lining the gatekeeper residue) with hydration energies ranging from 3.93 to 6.44 kcal/mol. Except for Mitoxantrone, all other molecules were found positioned toward the hydrophobic back pocket of PAK1. It was observed that molecules Acalabrutinib, Flubendazole, and Trazodone had thermodynamically greater unfavorable hydration sites that were displaced, and therefore are anticipated to have stronger binding affinity gains to PAK1.

These shortlisted molecules were subjected to stability analysis using MD simulations, followed by molecular orbital and electrostatic surface potential analyses. The drug molecules Flubendazole, Niraparib, and Acalabrutinib scored better than others and are expected to have better binding affinity for PAK1. The three molecules that were identified thus from a large collection of >2000 molecules from the Drugbank would need further experimental testing, as noted by the authors.

The importance of this study stems from the fact that they have used computational triaging, including solvation thermodynamics, in a prospective manner. Additionally, the study also highlights the importance of drug repurposing. It is worth noting here that pandemics like Covid19 for which medication does not exist have benefited from drug repurposing efforts [32]. Therefore, this kind of computational triage can inform a drug repurposing effort.

A lot of other examples are present in the literature [27, 33], where Watermap has been used to explain the activity differences in a congeneric series. These have established the importance of solvation thermodynamics considerations in drug design.

4.3.2 Grid Inhomogeneous Solvation Theory (GIST)

4.3.2.1 Objective and Approach

In a structure-based drug discovery (SBDD) campaign, it is crucial to know the precise location of water molecules in order to predict the next candidate ligand molecule with optimized binding properties. For that, it is necessary to characterize the water structure of all end-states during the ligand-binding reaction. The configuration space of solvent molecules is strongly coupled to the configuration of solute molecules. In this study [34], GIST was used for the construction of a rational structure-activity relationship (SAR) based solely on solvent contributions. These solvent contributions are rationalized by building different physically motivated models that use data from MD simulations and associated GIST calculations in order to predict SFEs. The functional form of all these models was based on the previously described displaced-solvent functional [22, 35]. A highly congeneric series of 53 and 12 ligands were selected for binding to thrombin and trypsin, respectively. For all protein–ligand complexes, high-resolution crystal structures are available along with free energies measured by ITC or surface plasmon resonance (SPR) [36]. The ligands were sorted into matched pairs such that the affinity difference between ligands within a given pair can predominantly be attributed to a difference in solvation. The resulting 186 pairs for thrombin are used for further parameterization and testing of the solvent functionals.

MD simulations were carried out for the *apo* protein, protein–ligand complexes, and individual ligands in solution. Properties like the solvent energy, entropy, density, and entropy of protein–ligand association processes were calculated using the GIST method from these simulations. The solute atoms in each MD simulation were restrained to a reference structure.

In order to address the inaccuracies of the solvation calculations arising out of a fixed protein consideration, it was considered that conformational flexibility played a major role in the apo protein than in the more stabilized protein–ligand complexes. Thus, the apo structure was simulated in an unrestrained MD, and the trajectory was clustered into unique conformations. The cluster representatives were used as input for GIST calculations. For the protein–ligand complexes as well as the unbound ligand molecules, only fully restrained MD simulations were carried out, keeping the complex spatially fixed to the conformation found in the crystal structure.

In this work, three different basic solvent functionals, viz. F4, F5, and F6, for the ligand in solution (L/F4, L/F5 and L/F6), protein (P/F4, P/F5 and P/F6), and the protein–ligand complex (PL/F4, PL/F5 and PL/F6), have been used. These solvent functionals use the raw solvent entropy, energy, and density data from the GIST calculations and differ in the weighting parameters [22]. High-resolution X-ray structures of thrombin and trypsin were utilized to estimate the energy and entropy distributions in the binding site, and the high energy/high entropy regions in both enzymes were identified that would contribute to an enhanced binding energy of the ligand. The functionals were trained based on this experimental data, and they were then used to estimate the binding free energy, which was then compared with the experimental values.

4.3.2.2 Results and Discussion

The models based on the ligand alone in aqueous solution resulted in the best agreement with the experimental data. The solvent functionals L/F4, L/F5, and L/F6 gave the highest correlation ($r = 0.80$, 0.83, and 0.79, respectively) and performed better than similar functionals that were trained using shuffled data ($r = 0.20$, 0.25, and 0.24). The solvent functionals based on GIST data from the protein binding pocket, P/F4 to P/F6, performed well ($r = 0.77$, 0.79, and 0.77) but were worse than the ones based on the ligands alone. Almost similar performance was observed for the ones based on the protein binding pocket. In particular, the F5 basic solvent functional performed the best. The F5 basic functional led to a slightly increased mean unsigned error range compared to the other basic functionals F4 and F6. In the second part of this work, contributions from both the protein–ligand complex and the ligand molecule were considered in the same calculation. The performance of functionals PL-L/F4, PL-L/F5, and PL-L/F6 ($r = 0.42$, 0.61, and 0.65) was considerably worse than the corresponding functionals based on the individual displacement treatments. Interestingly, the predictive power of these functionals increased ($r = 0.40$, 0.85, and 0.76) when considering grid voxels up to 3.5 Å away from the surface of the ligand instead of only using grid voxels up to 3.0 Å. When an additional layer of grid voxels up to 4.0 Å is taken, no increase in performance was observed ($r = 0.37$, 0.84, and 0.73). For the other functionals, using only GIST data from the ligand molecule, no such performance increase is observed. This grid size variation is insightful as it captures the importance of the first hydration shell and possibly highlights the importance of considering all the waters that participate in it.

The work also compared their results with those obtained from MM-GBSA and 3D-RISM techniques, both in terms of computational efficiency and accuracy of predictions (correlation). While in terms of computational efficiency, the method was equally good, it outperformed the other two in a achieving better correlation with experimental binding free energy data for this set of compounds against thrombin and trypsin.

GIST provides useful insights into the protein binding pockets in terms of the continuous distribution of the hydrophobicity and polarity, which offers excellent insights into high energy and high entropy regions. This information can help in both drug design as well as estimating the druggability of a certain binding site and a target. However, the results obtained from this method may depend on the extent of training of the functionals, which could be highly dependent on the quality of the structure under consideration. The functionals are not transferable across targets.

4.3.3 Three-Dimensional Reference Interaction-Site Model (3D-RISM)

4.3.3.1 Objective and Background

Coronaviruses initiate cell entry via the binding of the receptor binding domain (RBD) of the viral spike protein to the receptor protein, angiotensin-converting enzyme 2 (ACE2), on the surface of cells [37]. This process triggers infection and proliferation, making the binding of spike proteins to ACE2 the most significant initiator of the SARS-CoV-2 infection. Various studies on the binding of SARS-CoV-2 have been reported and have pointed out the importance of hydrogen bonding between amino acids on the RBD–ACE2 binding interface, as well as the importance of the hydration structure changes and hydrogen-bond bridging afforded by water molecules [38]. In this work, the binding process of the SARS-CoV-2 spike protein RBD to ACE2 was investigated using MD and the three-dimensional reference interaction-site model (3D-RISM) theory to highlight the effect of solvation in the binding process.

The trajectory data set DESRES-ANTON- [10 857 295,10 895 671], available on the D. E. Shaw research website [39], was obtained via the accelerated weighted ensemble MD simulation of the RBD–ACE2 complex formation process (PDB: 6VW1).

Taking one of the trajectory outputs as the initial structure, MD simulations were performed for 10 ns in each window. To this output, 3D-RISM theory was employed coupled with the Kovalenko Hirata closure [25] to evaluate the correlation functions and the SFE every 500 ps. The number of grid points in the 3DRISM-KH calculations was 512 with a spacing of 0.5 Å.

4.3.3.2 Results and Discussion

The distance change between the ACE2 and the RBD during the simulation was monitored, and it was found that the binding initially reduced the distance between the two proteins, and subsequently a structural rearrangement of the complex took place. The protein structure energy and the SFE exhibited dramatic changes through the binding process. It was observed that a gradual decrease in the conformational energy as the distance between the proteins decreased was accompanied by an increase in the SFE. The reason behind this is stated to be the penalty of free energy caused by desolvation. The complex is stabilized by 9 hydrogen bonds and resulted in a stabilization energy of about −400 kcal/mol. Obviously, this stabilization energy is far more than the contribution of the hydrogen bonds and confirms the presence of a variety of other interactions, including the van der Waal interactions and electrostatic interactions between the two.

The SFE, which gives insight into the process of desolvation, had large fluctuations through the binding process and was investigated by monitoring the solvent-accessible surface area (SASA) and the partial molar volume (PMV). While the SASA decreased during the process, the PMV increased. The binding process results in the exclusion of the solvent from the first hydration shell of the individual proteins and thus contributes to the reduction in SASA. Water molecules present in the interface of RBD and ACE2 were identified using the placement algorithm. Water molecules bound to ACE2/LYS31, GLU35, and LYS353 and RBD/TYR489 and GLN498 were observed. Principal component analysis of protein structural change revealed that the ACE2 protein underwent a conformational change upon binding, while the RBD conformation was less varied.

The enterprising use of the solvent functional energy in the study of the protein–protein binding process and the useful observations therein certainly expand the scope of solvation thermodynamics considerations in drug discovery.

The advent of machine learning, deep learning, and artificial intelligence has opened up new ways of analysis by identifying interesting patterns of molecular behavior influenced by the changes in structure and interactions. A recent study [40] extended the traditional 3D-RISM method to develop an AI-based model using the SFE data of ∼4000 complexes estimated from this approach. The SFE predictions made from this model have resulted in a good correlation with experimental values.

Another study [41] applied deep learning to the output of MD simulations of thousands of protein structures and analyzed networks formed by the waters around the complexes. Their study affirmed that desolvation and water-mediated interactions are important. Additionally, the enthalpically favorable networks of first shell waters around the solvent-exposed portions of the ligands also play an important role in protein–ligand binding.

4.4 Conclusion

The developments in the theoretical framework for considering explicit water-based solvation thermodynamics with methods such as IST and 3D-RISM, coupled with giant leaps in computing power with high-performance clusters and graphics processing units (GPUs), have brought solvation thermodynamics into the realm of regular practice in drug design campaigns. While the methods discussed in the previous sections might have some shortcomings, it is undeniable that they bring immense value to the understanding of SAR in drug discovery programs and use the learnings in the prospective design of compounds. The solvation considerations impact not only the on-target and off-target potency of the compounds but also their disposition in terms of permeability and solubility. These are extremely important in the lead optimization phase of a drug discovery program, as they have a huge impact on driving the pharmacokinetic exposures needed for the compounds to show efficacy. As noted at the end of the last section, deep learning and machine learning techniques can augment and enhance the utility of physics-based methods in accurate prediction of solvation effects.

Thus, it can be visualized that solvation thermodynamics will soon mature into a mainstay of drug design processes, impacting the discovery process all the way from target identification to lead optimization.

References

1 Yu, Y., Wang, J., Shao, Q. et al. (2016). The effects of organic solvents on the folding pathway and associated thermodynamics of protcins: a microscopic view. *Sci. Rep.* 6: 19500. https://doi.org/10.1038/srep19500.
2 Lucent, D., Vishal, V., and Pande, V.S. (2007). Protein folding under confinement: a role for solvent. *PNAS* 104 (25): 10430–10434.
3 Yoshida, N. (2017). *J. Chem. Inf. Model.* 57 (11): 2646–2656. https://doi.org/10.1021/acs.jcim.7b00389.
4 Kyte, J. (2003). *Biophys. Chem.* 100: 193–203.
5 Kyte, J. and Doolittle, R.F. (1982). *J. Mol. Biol.* 157 (1): 105–132.
6 Anfinsen, C.B. (1973). *Science* 181: 4096.
7 Rose, G.D. (2021). *Biochemistry* 60 (49): 3753–3761.
8 Xing, D., Li, Y., Xia, Y.L. et al. (2016). Insights into protein–ligand interactions: mechanisms, models, and methods. *Int. J. Mol. Sci.* 17 (2): 144.
9 Homans, S.W. (2007). *Top. Curr. Chem.* 272: 51–82.
10 Li, J., Fu, A., and Zhang, L. (2019). An overview of scoring functions used for protein–ligand interactions in molecular docking. *Interdiscip. Sci. Comput. Life Sci.* 11: 320–328.
11 Wang, E., Sun, H., Wang, J. et al. (2019). End-point binding free energy calculation with MM/PBSA and MM/GBSA: strategies and applications in drug design. *Chem. Rev.* 119 (16): 9478-9508.

12 Miertuš, S., Scrocco, E., and Tomasi, J. (1981). Electrostatic interaction of a solute with a continuum. A direct utilizaion of AB initio molecular potentials for the prevision of solvent effects. *Chem. Phys.* 55 (1): 117–129.

13 Hou, T., Wang, J., Li, Y. et al. (2011). Assessing the performance of the MM/PBSA and MM/GBSA methods. 1. The accuracy of binding free energy calculations based on molecular dynamics simulations. *J. Chem. Inf. Model.* 51 (1): 69–82.

14 Alibrakshi, A. and Hartke, B. (2021). *Nat. Commun.* 12: 3584.

15 Aqvist, J. and Marelius, J. (2001). The linear interaction energy method for predicting ligand binding free energies. *Comb. Chem. High Throughput Screen* 4 (8): 613–626.

16 King, E., Aitchison, E., Li, H. et al. (2021). Recent developments in free energy calculations for drug discovery. *Front. Mol. Biosci.* 8: 712085.

17 Zhang, J., Zhang, H., Wu, T. et al. (2017). Comparison of implicit and explicit solvent models for the calculation of solvation free energy in organic solvents. *J. Chem. Theory Comput.* 13 (3): 1034–1043.

18 Nair, S., Kumar, S.R., Paidi, V.R. et al. (2020). Optimization of nicotinamides as potent and selective IRAK4 inhibitors with efficacy in a murine model of psoriasis. *ACS Med. Chem. Lett.* 11 (7): 1402–1409.

19 Nguyen, C.N., Young, T.K., and Gilson, M.K. (2012). Erratum:"Grid inhomogeneous solvation theory: hydration structure and thermodynamics of the miniature receptor cucurbit [7] uril". *J. Chem. Phys.* 137 (14): 044101.

20 Karplus and Kuriyan (2005). *PNAS* 102 (19): 6679–6685.

21 Waibl, F., Kraml, J., Hoerschinger, V.J. et al. (2022). Grid inhomogeneous solvation theory for cross-solvation in rigid solvents. *J. Chem. Phys.* 156 (20): 204101.

22 Abel, R., Young, T., Farid, R. et al. (2008). Role of the active-site solvent in the thermodynamics of factor Xa ligand binding. *J. Am. Chem. Soc.* 130 (9): 2817–2831.

23 Li, Z. and Lazaridis, T. (2012). *Methods Mol. Biol.* 819: 393–404.

24 Haider, K., Cruz, A., Ramsey, S. et al. (2018). Solvation structure and thermodynamic mapping (SSTMap): an open-source, flexible package for the analysis of water in molecular dynamics trajectories. *J. Chem. Theory Comput.* 14 (1): 418–425.

25 Kovalenko, A. and Hirata, F. (1998). *Chem. Phys. Lett.* 290: 237–244.

26 Imai, T., Kovalenko, A., and Hirata, F. (2004). Solvation thermodynamics of protein studied by the 3D-RISM theory. *Chem. Phys. Lett.* 395 (1–3): 1–6.

27 Wang, L., Berne, B.J., and Friesner, R.A. (2011). Ligand binding to protein-binding pockets with wet and dry regions. *PNAS* 108 (4): 1326–1330.

28 Vadlamudi, R.K. and Kumar, R. (2003). *Cancer Metastasis Rev.* 22: 385–393.

29 Meng, J., Meng, Y., Hanna, A. et al. (2005). Abnormal long-lasting synaptic plasticity and cognition in mice lacking the mental retardation gene Pak3. *J. Neurosci.* 25 (28): 6641–6650.

30 Biswal, J., Jayaprakash, P., Rayala, S.K. et al. (2021). Watermap and molecular dynamic simulation-guided discovery of potential PAK1 inhibitors using repurposing approaches. *ACS Omega* 6 (41): 26829–26845.

31 Wishart, D.S., Feunang, Y.D., Guo, A.C. et al. (2018). DrugBank 5.0: a major update to the DrugBank database for 2018. *Nucleic Acids Res.* 46 (D1): D1074–D1082.

32 Smith, D.P., Oechsle, O., Rawling, M.J. et al. (2021). Expert-augmented computational drug repurposing identified baricitinib as a treatment for COVID-19. *Front. Pharmacol.* 12: 709856.

33 Haider, K. and Huggins, D.J. (2013). *J. Chem. Inf. Model.* 53: 2571–2586.

34 Hufner-Wulsdorf, T. and Klebe, G. (2020). *J. Chem. Inf. Model* 60: 1409–1423.

35 Nguyen, C.N., Cruz, A., Gilson, M.K. et al. (2014). Thermodynamics of water in an enzyme active site: grid-based hydration analysis of coagulation factor Xa. *J. Chem. Theory Comput.* 10 (7): 2769–2780.

36 Sander, A., Hüfner-Wulsdorf, T., Heine, A. et al. (2019). Strategies for late-stage optimization: Profiling thermodynamics by preorganization and salt bridge shielding. *J. Med. Chem.* 62 (21): 9753–9771.

37 V'kovski, P., Kratzel, A., Steiner, S. et al. (2021). Coronavirus biology and replication: implications for SARS-CoV-2. *Nat. Rev. Microbiol.* 19 (3): 155–170.

38 Kobryn, A.E., Maruyama, Y., Velazquez-Martinez, C.A. et al. (2021). Modeling the interaction of SARS-CoV-2 binding to the ACE2 receptor via molecular theory of solvation. *New. J. Chem.* 45 (34): 15448–15457.

39 http://www.deshawresearch.com/resources_sarscov2.html

40 Osaki, K., Ekimoto, T., Yamane, T. et al. (2022). 3D-RISM-AI: a machine learning approach to predict protein–ligand binding affinity using 3D-RISM. *J. Phys. Chem. B* 126 (33): 6148–6158.

41 Mahmoud, A.H., Masters, M.R., Yang, Y. et al. (2020). Elucidating the multiple roles of hydration for accurate protein-ligand binding prediction via deep learning. *Commun. Chem.* 3 (1): 19.

5

Site-Identification by Ligand Competitive Saturation as a Paradigm of Co-solvent MD Methods

Asuka A. Orr and Alexander D. MacKerell Jr

University of Maryland, School of Pharmacy, Department of Pharmaceutical Sciences, 20 Penn Street, Baltimore, MD 21201, USA

5.1 Introduction

Computer-aided drug design (CADD) is an essential component of the repertoire of tools used in drug discovery. Ligand-based drug design (LBDD) approaches, most notably quantitative structure-activity relationships, have made important contributions to CADD from the 60's onward [1–9]. Alternatively, more widely used are structure-based drug design (SBDD) approaches, which rely on the availability of the 3D structure of the target macromolecule. SBDD approaches play roles in the full range of the steps to which CADD contributes to the drug design process. This ranges from database screening methods [10–13] for hit identification to lead optimization, using various approaches such as free energy perturbation [14–16], and addressing pharmacokinetic issues in drug development including contributions to improve absorption, disposition, metabolism, excretion, and toxicological (ADMET) considerations [17–20]. Multiple recent reviews of various CADD methods have been presented [21–27].

CADD methods are inherently limited due to the use of mathematical models to treat the complexity of ligand-protein interactions, including the extremely challenging problem of accounting for contributions from the environment. Two major challenges that counterbalance each other exist; the accuracy of the underlying Hamiltonian or potential energy function/model chemistry used to estimate the system energy and the ability to sample the full range of conformations that contribute to ligand–target interactions, including the configurational space required to calculate entropic contributions. Accordingly, molecular docking methods commonly used for virtual screening of larger databases of molecules in the hit identification stage typically use simplified energy functions that significantly approximate the contribution of the environment to ligand binding [21]. More rigorous treatment of the aqueous environment using continuum models, such as the Poisson–Boltzmann (PB) [28–30] or Generalized Born (GB) [31, 32] approaches, offers improved treatment of the solvent contributions in conjunction with MD

Computational Drug Discovery: Methods and Applications, First Edition.
Edited by Vasanthanathan Poongavanam and Vijayan Ramaswamy.
© 2024 WILEY-VCH GmbH. Published 2024 by WILEY-VCH GmbH.

simulations to predict binding affinities of ligands [33–36]. Free-energy perturbation (FEP) [37, 38] methods offer a formally rigorous approach to calculate the absolute or relative free energy of binding of ligands to a target in the presence of an explicit solvent [16, 39–43]. However, PB, GB, and FEP methods require extensive MD simulations to obtain the needed ensemble of conformations, such that empirical potential energy functions in conjunction with force fields, which are inherently limited in their accurate treatment of molecular interactions for the wide range of drug-like molecules, are typically applied [16, 34]. Quantum mechanical (QM) models in principle could overcome the limitation of potential energy functions, but they are computationally demanding as well as limited in their ability to treat long-range dispersion contributions to ligand–target interactions, limiting the wide-spread utilization of QM models in CADD to date [44–46].

Co-solvent MD methods can overcome the above limitations to various extents [47]. In the so-called co-solvent or mixed MD methods, which build upon experimental methods such as the multiple solvent crystal structures (MSCS) method [48] and early computational approaches such as Goodford's GRID [49] and the multiple copy simultaneous search (MCSS) [50] methods, a small solute is included in an explicit-solvent simulation system along with the target macromolecule. MD simulations are then run, from which the probability distribution of the solute molecule around the target is calculated. The resulting probability distribution, which is based on true free energy as stated above, is then used to inform compound design. This precomputation of the ensemble of conformations [51] required to calculate the probability distribution comes with a computational cost, typically on the order of days depending on the approach and system size. However, once the probability distribution is available, it may be used rapidly in ligand design, overcoming the need to recalculate an ensemble for each drug-like molecule as required in PB, GB, and FEP methods and thereby circumventing the need for extensive computational resources. In addition, while co-solvent MD methods use empirical potential energy functions, as the simulations only include the target molecule, water, and small solutes, their accuracy may be considered to be acceptable as such molecules are generally the most carefully optimized aspects of a force field. This partially overcomes the need to have highly refined parameters for the wide range of chemical spaces present in drug-like molecules.

The remainder of this chapter presents various details of different co-solvent MD methods. These methods include a variety of approaches such as MDmix [52–55], MixMD [56–63], probe-based MD [64–66], and the works of Yang and Wang [67–70]. In addition to the details presented below, we refer the reader to the review by Ghantoka and Carlson [58] on co-solvent MD methods. Details are then presented on the Site Identification by Ligand Competitive Saturation (SILCS) method developed in our laboratory and commercialized by SilcsBio LLC. This includes methodological details, including how they contrast with other co-solvent methods, and the various ways in which the SILCS methods may be used in the drug development process. Specific case studies are then presented to illustrate the SILCS approach.

Several co-solvent methods have been developed with probability distributions of the solute binding pattern used for various types of analysis. A summary of

Table 5.1 Co-solvent technologies and their applications in CADD.

Authors/method	Targeted applications
Privat et al. (Fragment dissolved MD)	Binding site identification [71].
Fabritiis et al.	Protein cryptic site prediction [72, 73].
Favia et al.	Cryptic ligand binding pockets prediction [74].
Zariquiey et al. (Cosolvent Analysis Toolkit)	Hotspot identification [75].
Yang et al.	Binding hotspots identification [67–70] and druggable protein conformations [76].
Bahar et al.	Determination of allosteric protein druggability [77]; pharmacophore modeling [78].
Barril et al. (MDmix)	Binding sites determination, predicting druggability [52], water displaceability and functional group mapping of protein binding sites [53], protein–ligand docking and binding affinity prediction [54], and pharmacophore modeling [55].
Carlson et al. (MixMD)	Hotspots mapping [57, 59], binding site identification [60, 63], kinetics of co-solvent on/off rates and identification of allosteric binding sites [62], free energies and entropies of binding sites [61], and displaceable water site identification [60].
Gorfe et al. (pMD)	Active and allosteric binding sites identification in proteins [64] including membrane-bound drug targets [64, 65].
Caflisch et al.	Co-solvent on-off rates and binding affinities [79] and water displaceability by co-solvents [80].
Tan and Verma (LMMD)	Hydrophobic and halogen binding site identification [81–83] and use of MD to map multiple ligand binding sites [84].
Lill et al.	Rank-ordering of ligand binding and identification of binding locations of functional groups [85, 86].
Yanagisawa et al. (EXPROPER)	Evaluation of hotspots and binding sites on proteins [87].
Takemura et al. (ColDock)	Prediction of protein–ligand docking structures [88].

published co-solvent methods and their applications is presented in Table 5.1. Co-solvent methods are typically based on MD simulations of a single solute molecule in aqueous solution that encompasses the entire target macromolecule, which is a protein in the majority of cases (Table 5.2). The solutes, or co-solvents, that are used are typically of a molecular weight of 150 daltons or less. The small size of the solute is important as it allows for adequate diffusion of the solute to sample the full region around the protein within the time scale of the MD simulations. Such sampling is further facilitated by the use of multiple solutes in the simulation systems (Table 5.2). In addition, enhanced sampling techniques, including accelerated MD [93] and lambda dynamics [86], have been implemented in co-solvent methods to facilitate solute sampling. However, the majority of efforts use standard MD simulations, typically spanning 20 to 200 ns, with multiple replicas performed to facilitate convergence. The importance of obtaining adequate convergence of the solute distributions cannot be understated. This is often assessed by comparing the calculated probability distributions of different portions of the MD simulations. This could be the first or second half of an extended MD simulation or, when multiple simulations are performed, calculating two sets of probability distributions from sets of simulations. Notably, the majority of co-solvent MD simulations can be performed with standard MD simulation packages such as AMBER [94], OpenMM [95], GROMACS [96], NAMD [97], and CHARMM [98], with available tools such as CPPTRAJ [99] used for calculation of the probability distributions and other analyses.

The solutes included in co-solvent simulations range from charged molecules, including acetate and methylammonium, to neutral polar molecules and to apolar molecules (Table 5.2). Polar neutral molecules include methanol, ethanol, acetamide, isopropyamine, acetic acid, isopropanol, and acetonitrile. Isopropanol was used in some early studies as it contains both polar alcohol and apolar aliphatic carbons. Apolar molecules include benzene, isobutane, and propane, as well as heterocycles such as pyridine and imidazole. In general, the majority of studies perform individual sets of simulations with a single solute molecule at various concentrations, with the probability distributions from the different sets of simulations combined for the various types of analysis listed in Table 5.1.

A critical consideration in co-solvent simulations is the concentration of the solutes included in the explicit aqueous solution. Typically, solute molecules are included in concentrations ranging from 0.1 to 1.5 M, though some workers use up to 3 and 12 M (Table 5.2). High concentrations of solute molecules may result in the denaturation of the target macromolecule [100]; however, low concentrations of solute molecules may result in slow convergence in the sampling of solute distribution around the full 3D space of the macromolecule. The undesired artifacts resulting from high concentrations of solute molecules can be circumvented through restraints to the targe macromolecule. In such cases, restraints to the macromolecule can be balanced to maintain the structural integrity of the macromolecule while simultaneously allowing sufficient flexibility for potential binding sites to open. An additional concern potentially exasperated by the choice of solute concentration includes the potential for aggregation of hydrophobic species, ion pairing between solutes, and proper convergence of the probability distributions. Apolar solutes can aggregate with themselves when used in co-solvent simulations. This issue was specifically addressed by Guvench and coworkers with respect to the potential for solutes to cause protein denaturation during co-solvent simulations [100]. To avoid hydrophobic aggregation as well as ion pairing, repulsive potentials may be introduced between selected solutes. This allows for an effective "ideal solution" behavior, thereby facilitating the sampling of the solutes in the full 3D space of the protein. However, such repulsive potentials also limit the sampling of, for example, two benzene solutes directly adjacent to each other in a binding pocket, as previously discussed [84].

Table 5.2 Overview of solutes used and their concentrations in co-solvent technologies.

Authors/method	Solutes investigated	Solute concentration
Privat et al. (Fragment dissolved MD)	Single solute per simulated system [71]. Investigated solute ligands include ethyl 3-amino-4-methylbenzoate, 2-methyl-2-(4-morpholinyl)-1-butanamin, (3r)-piperidin-3-yl(piperidin-1-yl)methanone, n-methyl-1-(1-methyl-1h-imidazol-2-yl) methanamine, 5-hydroxy-2-aminobenzimidazole, 2-aminopyrimidine, 1-aminoisoquinoline, 3-chloro-1-benzothiophene-2-carboxylate, 3-(4-chloro-3,5-dimethylphenoxy)propanoic acid, dimethyl-sulfoxide-methyl-(methylsulfinyl) methyl sulfide, and 6-azaniumylhexanoate in water [71].	Solute concentration dependent on size of solute molecule [71]: ~0.01 to ~0.09 M
Fabritiis et al.	Single solute per simulated system [72, 73]. Investigated 129 solute fragments [72, 73].	Single solute molecule per protein system [72, 73].
Favia et al.	Single solute per simulated system [74]. Investigated solutes include acetic acid, isopropanol, and resorcinol in water [74].	5% (m/m) [74]: ~3 M.
Zariquiey et al. (Cosolvent Analysis Toolkit)	Single solute per simulated system [75]. Investigated solutes include acetamide, benzene, acetanilide, imidazole, and isopropanol in water [75].	10% (m/m) [75]: ~6 M.
Yang et al.	Single solute per simulated system [68, 69, 76]. Investigated solutes include isopropanol in water [68, 69, 76].	20% (v/v) [68, 69, 76]: ~3.5 M
Bahar et al.	Single solute per simulated system [77] or multiple solutes per simulated system [78]. Investigated solutes include isopropanol, acetamide, imidazole, acetate, isopropylamine, and isobutane in water [77, 78].	20 : 1 water: solute ratio [77, 78]: ~3 M.
Barril et al. (MDmix)	Single solute per simulated system [52–55]. Investigated solutes include isopropanol, ethanol, acetamide, methylammonium, acetate, and acetonitrile in water [52–55].	20% (v/v) [52–55]: ~3 to 5 M.
Carlson et al. (MixMD)	Single solute per simulated system [57, 59, 60, 63] or multiple solutes per simulated system [60–63]. Investigated solutes include isopropanol, acetonitrile, pyrimidine, imidazole, n-methylacetamide, methylammonium, and acetate in water [57, 59–63].	50% (w/w) [57, 59]: ~8.5 to ~12 M 2.5% (v/v) [61, 62]: ~0.4 M 5% (v/v) [60–63]: ~0.6 M to ~1.5 M

(continued)

Table 5.2 (Continued)

Authors/method	Solutes investigated	Solute concentration
Gorfe et al. (pMD)	Single solute per simulated system [64] or multiple solutes per simulated system [65]. Investigated solutes include isopropanol, isobutane, acetamide, acetate, urea, dimethylsulfoxide, and acetone in water [64, 65].	~20 : 1 water : solute ratio [64]: ~3 M. 140 : 1 water : solute ratio [65]: ~0.4 M.
Caflisch et al.	Single solute per simulated system [79, 80]. Investigated solutes include 4-hydroxy-2-butanone, dimethylsulfoxide, 5-diethylamino-2-pentanone, methyl sulphinyl-methyl sulfoxide, 5-hydroxy-2-pentanone, tetrahydrothiophene 1-oxide, methanol, and ethanol in water [79, 80].	Single solute molecule per protein system [79]. 0.44 M [80]
Tan and Verma (LMMD)	Single solute per simulated system [82, 83] or multiple solutes per simulated system [84]. Investigated solutes include benzene, chlorobenzene, methanol, acetaldehyde, methylammonium, and acetate in water [82–84].	0.1 to 0.4 M [82–84]
Lill et al.	Multiple solutes per simulated system [85, 86]. Investigated solutes include propane, formamide, acetaldehyde, benzene, fluoro-benzene, chloro-benzene, bromo-benzene, and iodo-benzene in water [85, 86].	0.25 M [85, 86]
Yanagisawa et al. (EXPROPER)	Single solute per simulated system [87]. Investigated solutes include 138 molecules [87].	0.25 M [87]
Takemura et al. (ColDock)	Single solute per simulated system [88]. Investigated solutes include dimethylsulfoxide, methylsulfinyl-methylsulfoxinide, ε-aminocaproic acid, 4-(4-bromo-1H-pyrazol-1-yl) piperidinium, transaminomethyl-cyclohexanoic acid, and FK506 (Tacrolimus) [88].	~0.065 to ~0.15 M [88]
MacKerell et al. (SILCS)	Multiple solutes per simulated system [89–92]. Investigated solutes include benzene, propane, methanol, imidazole, acetaldehyde, formamide, methylammonium, acetate, dimethylether, fluoroethane, trifluoroethane carbon, fluorobenzene fluorine, chloroethane, chlorobenzene, and bromobenzene in water [89–92].	1 M[a][89, 90] 0.25 M [90–92]

a) The ~1 M concentrations were used in SILCS simulations containing only benzene and propane as co-solvents in early studies. All SILCS simulations thenceforth generally consist of 8 co-solvent molecules, each at ~0.25 M.

A strength of the co-solvent MD methods is their ability to convert the probability distributions into free energies. As the probability distributions are based on occupancies on a 3D grid, the conversion yields a 3D free energy grid associated with the binding of a solute molecule to the protein. In this procedure, as described by Alvarez and Barril and also performed early on by Bahar and coworkers [53, 77], the number of counts/MD snapshot in each 3D voxel is normalized by the expected number of counts based on the concentration of the solute in the simulation system. They then performed a standard state correction associated with a 1 M concentration. In that study, to estimate the binding free energy of individual hotspots, the free energies were calculated over all voxels within 2 Å of the hotspot, and then the Boltzmann weighted average over those voxels was calculated. It should be noted that defining concentration in a finite simulation system is not unique, given the inaccessible volume occupied by the target macromolecule. For example, calculating the concentration of solutes based on the total simulation volume is typically not appropriate as a significant portion of the system is not accessible to solutes or water. Accordingly, it is necessary to calculate solute concentrations relative to the amount of water in the system [53, 101]. An exception to this occurs with the application of the SILCS technology for the calculation of functional group-free energy distributions in bilayers, where the use of the total simulation system is appropriate [102].

Beyond the free energy functional group affinity information in the form of 3D maps generated from co-solvent MD methods, there is the comprehensive nature of that information for the entire 3D space of the protein or other target molecule being studied. This allows for a variety of different types of analyses to be performed based on affinity information, with those analyses being able to be performed rapidly given the pre-computed nature of the 3D maps. Listed in Table 5.1 are various types of analyses performed to date. The widest use of the technology has been for the identification of novel binding or hotspots that offer the potential to be novel allosteric sites for the discovery of novel positive and negative allosteric modulators. However, the successful identification of novel sites is impacted by the treatment of protein conformational flexibility in the MD simulations. The use of rigid protein structures can limit the extent to which putative binding sites not present in crystal structures may open, while the total lack of restraints may potentially lead to denaturation events, especially when high solute concentrations are used [100].

Once binding sites are known, co-solvent methods can be used for various aspects of ligand discovery and design. They have been used for the determination of pharmacophore models, which may subsequently be used for large-scale in silico database screening. Additionally, such methods can also be used to identify displaceable water sites and determine free energies and entropies of binding sites, facilitating drug design. An interesting variation of co-solvent methods has been the extraction of kinetics associated with the on- and off-rates of the co-solvents [79]. While the many variations of the co-solvent MD technology have seen wide use, it is safe to say that their full potential has yet to be achieved. To highlight this, in the remainder of the chapter, we will provide an overview of the various physical phenomena to which the SILCS technology has been implemented, along with case studies to offer explicit examples of those approaches.

5.2 SILCS: Site Identification by Ligand Competitive Saturation

SILCS was originally developed in the year 2008 by our lab through discussions involving Olgun Guvench and Alex MacKerell. The method was first implemented in the program CHARMM [98], and associated workflows were written to extract and calculate functional group probability distributions, which could readily be visualized using VMD [103], Pymol [104], or any visualization package that accepts grid density maps, such as those from X-ray crystallography. SILCS and other co-solvent approaches are simple: simulate the distribution of solutes presenting specific functional groups around a protein or other macromolecule to identify where such groups would, or would not, interact. Barril and coworkers published the approach on a single solute on several protein targets at approximately the same time as the initial SILCS publication in 2009 that examined the BCL-6 oncoprotein [52, 89]. Notably, SILCS included multiple solutes, initially propane and benzene [89], while the other earlier efforts included only a single solute [52, 56, 57, 77, 82]. This difference represented a unique capability that, along with the use of a repulsive potential between solutes in the simulation system, allowed for the technology to receive a US patent in 2018 (US Patent Number 10,002,228). Additionally, the SILCS technology has been commercialized in the form of SilcsBio LLC, while the development of the technology continues in the MacKerell laboratory. This combination has allowed the SILCS technology to be applied to a range of problems, as listed in Table 5.3 and described below.

SILCS, as with the other co-solvent MD approaches [52, 56], was motivated by the need to identify the location of different types of functional groups on the full surface of a protein. In contrast to other methods, we wanted to map multiple types of functional groups from a single set of simulations through "competitive saturation" between the different solutes representing the functional groups and with water. This contrasts with other methods that initially only included a single solute competing with water [52, 56, 57, 77, 82]. The original embodiment simply included benzene, propane, and water, representing aromatic, aliphatic, and hydrogen bond donors and acceptors, respectively [89]. The analysis focused on the probability distributions of the solute and water functional groups and how they recapitulated the known location of peptides binding to the BCL-6 protein lateral groove [89]. However, given the known importance of specific treatment of water versus other hydrogen bond donors and acceptors, the collection of solutes was expanded to a total of 8 solute molecules, each at a concentration of ~0.25 M, in work by Raman et al. [90, 91]. These 8 solute molecules included benzene, propane, methanol, imidazole, acetaldehyde, formamide, methylammonium, and acetate, with acetaldehyde subsequently replaced with dimethylether given the presence of an aldehyde moiety on formamide and the role of ether oxygens in drug-like molecules [92]. In addition, also motivated by their presence in drug-like molecules, was the implementation of a collection of halogen-containing solutes, termed SILCS-X [106], as performed by other groups [82, 86]. In SILCS, these halogen-containing solutes include fluoroethane, trifluoroethane, fluorobenzene,

Table 5.3 SILCS methods and tools.

SILCS Method	Description
SILCS GCMC/MD Simulations	Combined oscillating excess chemical potential Grand Canonical Monte Carlo/Molecular Dynamics Simulations [105] for generation of SILCS FragMaps using either the standard [90, 91] or halogen (SILCS-X) [92] solutes [90, 91].
SILCS-MC Docking	Monte Carlo docking of ligands into the SILCS FragMaps [92, 106].
SILCS-MC Pose Refinement	Local Monte Carlo docking of ligands from known orientations into the SILCS FragMaps.
SILCS-Hotspots	Identification of fragment binding sites via comprehensive SILCS-MC docking on entire 3D space of target macromolecule [107]. Subsequent SILCS-MC docking of a subset of FDA-approved compounds to identify binding sites for drug-like molecules including allosteric sites.
SILCS-Pharmacophore	Generation of target-based pharmacophore features for large-scale database screening [108, 109].
SILCS-PPI	Evaluation of protein–protein interaction distribution [110].
SILCS-RNA	Optimized SILCS simulation approach for RNA [111].
SILCS-Biologics	Rational selection of excipients for biologic (e.g. mAbs) formulation [112, 113].
SILCS-BML	Optimization of SILCS FragMaps targeting experimental binding affinities using a Bayesian Markov Chain Monte Carlo Machine Learning approach [106].

chloroethane, chlorobenzene, and bromobenzene [106]. Thus, the SILCS method, through two sets of GCMC/MD simulations, can generate a comprehensive set of functional group affinity patterns representing the majority of those commonly seen in drug-like molecules [92].

While the initial motivation for the SILCS and the other cosolvent approaches was primarily site identification, the power of the approach is more fully utilized when the probability distributions from the simulations are normalized and converted to free energies as described above. This approach is also applied in the SILCS approach with additional normalization performed for the number of functional group atoms in each solute [106], yielding grid free energy (GFE) FragMaps. The GFE FragMaps represent the binding free energies of each functional group on the 3D grid relative to being in an aqueous solution, where the GFE values equal zero. Thus, simply overlaying a ligand on the collection of GFE FragMaps and assigning atoms in the ligand to functional group types allows a GFE score to be assigned to the classified atoms, which may then be summed to yield the ligand grid free energy (LGFE). LGFE scores are an approximation of the experimental ligand binding affinities as discussed by Ustach et al. [106] Given that the SILCS FragMaps are precomputed, calculation of the LGFE score is virtually instantaneous; in the context of the SILCS-MC approach, MC sampling of translational, rotational, and dihedral rotational degrees of freedom of the ligands may be readily performed in minutes for ligand docking in a given binding site that includes a near exhaustive search of the binding site. Alternatively, if the bound orientation of a ligand or its parent is known, pose-refinement SILCS-MC may be performed from which the LGFE score of the ligand is obtained. This approach has been performed in a number of studies [12, 92, 106, 114–116], and in a comprehensive comparison, the use of SILCS-MC exhaustive docking has been shown to be comparable to highly computationally demanding FEP calculations for the prediction of relative affinities of ligand binding, with the SILCS-MC calculation being hundreds of times faster than the FEP approach [92]. Notably, the SILCS-MC docking approach does not require a known bound starting orientation of the ligand, as required by FEP. In addition, when experimental data is available on even a small set of the ligands (e.g. down to 10), the "best" SILCS-MC scoring regimen may be selected to improve the predictability of the LGFE scores. Moreover, the application of a Bayesian Markov Chain Monte Carlo Machine Learning approach (BML) [106] allows for optimization of the contribution of the different types of FragMaps to the LGFE scores, yielding predictive models that are systematically better than those from FEP methods [92].

A central theme in a medicinal chemistry campaign is developing an accurate and detailed structure-activity relationship (SAR) to lead the ligand design process. With SILCS, the use of the atomic GFE scores to yield the LGFE scores allows for the contribution of individual atoms as well as different chemical moieties to the binding affinity to be determined. This is powerful as, for example, the addition of a chemical moiety to a molecule may lead to a relatively small change in the overall binding affinity due to energetic gains of the added moiety being offset by less favorable contributions from other parts of the molecule. This was shown to occur in a series of allosteric inhibitors of heme oxygenase [117]. The use of GFE contributions has facilitated ligand design in a number of studies targeting proteins such as Bcl-6 [118], Mcl-1, and Bcl-xl [119].

A key development in the evolution of the SILCS technology was the implementation of the oscillating excess chemical potential, μ_{ex}, GCMC approach [105]. This was motivated by the need for adequate solute sampling in deep and fully occluded, cryptic binding pockets, such as those commonly seen in GPCRs and in nuclear receptors, respectively, in a thermodynamically correct fashion. As GCMC allows for the insertion and deletion, as well as rotations and translation, of solutes from an external bath into the simulation, system a proper equilibrium of the solutes and water in deep and occluded binding pockets may be achieved. From this, GFE and LGFEs can be obtained for these types of binding sites as compared to co-solvent MD methods, where the diffusion times and/or the need for the protein to partially unfold to adequately sample such sites are prohibitive. However, the use of standard GCMC, where μ_{ex} for water and the solutes are set to the experimental hydration free energy leads to very low acceptance rates for insertions and deletions, thereby requiring prohibitive amounts of GCMC sampling. The use of an oscillating μ_{ex}, where μ_{ex} is varied based on the target concentration of each solute or water, effectively acting as an umbrella potential in chemical potential space rather than conformational space, overcomes this limitation, yielding adequate acceptance rates. The approach was initially implemented and shown to yield accurate relative binding affinities of ligands to the T4 lysozyme pocket mutant, which contains a totally occluded binding pocket [105]. The method was subsequently applied to nuclear receptors and the β2 adrenergic receptor, a GPCR [120], and is now the standard in SILCS simulations in conjunction with the MD portion of the method, which is needed to further facilitate both solute and water sampling as well as conformational sampling of the protein or other target macromolecule. Recent efforts have ported the oscillating μ_{ex} GCMC method to GPUs, yielding significant speed enhancements, especially in larger simulation systems [121].

In the context of drug design and development, the SILCS technology represents an end-to-end resource. Qualitatively, visualization of the SILCS FragMaps may be used to facilitate the identification of possible binding sites, including cryptic and allosteric sites [117], facilitate decisions on what types of scaffolds can occupy a site, and then help the medicinal chemist determine the types of functional groups that may be added to a lead compound to improve the binding affinity while simultaneously considering synthetic accessibility. Quantitatively, the various applications are diverse. In the absence of known binding sites on a target macromolecule, the SILCS-Hotspots approach is of utility [107]. In SILCS-Hotspots, a library of fragment molecules common to drug-like compounds [122, 123] is comprehensively docked in the full 3D space of the target macromolecule and then subjected to 2 rounds of clustering to identify and rank putative fragment binding sites. This goes beyond simply using the solute binding locations typically performed in co-solvent methods, as the fragments are of a larger MW and more chemically diverse. Additionally, while other methods often assess the hotspots on macromolecules based on rigid fragments, during the SILCS-Hotspots docking, fragment conformation, as well as orientation are sampled. Due to the inclusion of macromolecule flexibility in the computation of SILCS FragMaps, SILCS-Hotspots are able to explore and potentially identify cryptic pockets [107]. The number of identified hotspots throughout the macromolecule with low LGFE scores can also indicate the propensity of the macromolecule to bind several classes of ligands at different sites. Once fragment binding sites are identified, the identification of binding sites for larger drug-like molecules can be performed through the identification of two or more adjacent hotspots followed by SILCS-MC docking of FDA-approved compounds into those sites. The average LGFE scores of the top 20 or 25 compounds are then obtained along with the relative solvent accessibility (rSASA) of those compounds in the presence and absence of the target macromolecule with putative binding sites typically having average LGFE scores <−10 kcal/mol and rSASA values >60%, where 100% indicates full exclusion of the ligands from the solvent by the protein. The SILCS-Hotspots approach has been used for the identification of cryptic, allosteric sites on β-Glucosidase A [124] and on the β2 adrenergic receptor, a GPCR [125, 126], and the identification of a site to block protein–protein interactions on the Ski8 complex [127].

When binding sites are known or have been identified as described in the preceding paragraph, SILCS offers effective tools for large-scale database screening. Utilizing the SILCS FragMaps, target-based pharmacophores may be generated

[108, 109]. The method typically generates 10 to 12 pharmacophore features for a binding site from which multiple 4 or 5-feature pharmacophore hypotheses are generated. These typically contain 2 to 3 apolar/hydrophobic features along with various polar or charged features. The pharmacophore hypotheses are used to screen pre-prepared virtual databases that contain multiple conformations and classified functional groups using programs such as Pharmer [128]. From this process up to 10 000 compounds for each hypothesis are selected and then subjected to SILCS-MC pose refinement based on the pharmacophore-docked orientations. The resulting LGFE scores are then used to rank the compounds with the top 1000 to 2000 selected. Bioavailability may then be predicted using various metrics, such as the 4D Bioavailability (4DBA) metric, which takes into account the terms in Lipinski's rule of 5 in a single scalar number [129]. The compounds may then be clustered based on chemical fingerprints from which final compounds from individual clusters may be selected for purchase and experimental assay. Once active hit compounds are identified, compounds similar to the hit compounds may be obtained from a virtual database and experimentally assayed to determine true lead compounds as well as initial SAR data to jump-start a ligand design effort [130]. A similar approach has been successfully applied to identify both agonists and antagonists of a number of proteins, including the β2 adrenergic receptor [120], the A18 ribonucleotide binding protein [131], and the WDR61 protein in the Ski8 complex [127]. In the β2 adrenergic receptor study, FragMaps for both the active and inactive forms of the GPCR were used to specifically perform database searching for agonists; 7 out of 15 compounds were shown to be active agonists [120].

When active compounds are known, the SILCS FragMaps may be used for ligand optimization. It should be emphasized that the same FragMaps, in some cases supplemented with the SILCS-X halogen FragMaps, used for binding site identification along with pharmacophore development and database screening are also used for optimization efforts, showing the utility and extreme computational efficiency of the SILCS method. If an experimental structure of the ligand–protein complex is available, then the experimental structure may be used in the ligand optimization, with the experimental structure overlaid on the FragMaps and subjected to local SILCS-MC pose refinement. Visual analysis of the overlap of the various ligand moieties and the FragMaps, as well as the GFE contributions of the individual atoms on the ligand, may be used to identify which regions of the molecule are contributing to binding versus acting as scaffolding elements. Such analysis, especially visualizing the FragMaps adjacent to the ligand, can be used to generate ideas concerning the types of chemical modifications to make for improved affinity. In many cases, multiple FragMap types occupy the same region of 3D space, information that can be used to modify ring systems in the context of scaffold hopping. For example, if there is a phenyl ring occupying an apolar or aromatic FragMap and there is also an H-bond acceptor FragMap in that region, then one may consider modifying the ring into a heterocycle such as a pyridine. As stated above, this approach in general is of direct utility to medicinal chemists, as they can readily visualize modifications that will improve experimental activity while simultaneously accounting for synthetic accessibility.

Once possible potential modifications or sites on which functional groups may be substituted are identified, the SILCS-MC approach may then be used to obtain LGFE scores for the modified species, allowing them to be prioritized for synthesis and experimental assay. As stated above, the SILCS-MC pose refinement can be applied if the orientation of the lead compounds is known and only local relaxation of the compounds is desired, as with FEP calculations. However, full SILCS-MC docking may also be performed, allowing for reorientation of the ligand in the binding site, a phenomenon that often occurs as seen with inhibitors of the Mcl-1 protein [106]. Again, full SILCS-MC docking has been shown to be competitive with FEP methods but significantly more computationally efficient. Moreover, the BML approach may be used to optimize the FragMaps to improve their predictability as a medicinal chemistry campaign proceeds with additional BML optimizations as additional experimental data becomes available. This allows for continual improvements in the predictability of SILCS as the program proceeds, including the ability to do individual BML optimizations for specific chemical scaffolds. Numerous studies have used SILCS FragMaps for ligand optimization for diverse proteins such as Mcl-1, Bcl-xl [119], Bcl-6, Erk [132], mGluR5 [133, 134], Heme oxygenase [117], and, interestingly, the ribosome [135].

In the context of drug-like molecule ligand design and development, the SILCS method has been recently extended to RNA [111]. Given the polyanionic nature of RNA, this required special approaches for the SILCS simulations, leading to the need to perform separate SILCS simulations for the apolar and polar neutral solutes and for the charged solutes as methylammonium-dominated sampling of the RNA molecule. Using the modified protocol, it was shown that the GFE FragMaps recapitulated the functional groups of a number of RNA-binding ligands on multiple targets as identified in crystallographic studies. In addition, using the approach, new binding sites were identified in combination with SILCS-Hotspots. However, as discussed in the study, the significant impact of ligand binding on RNA conformation combined with the minimal amount of quantitative data on ligand–RNA interactions make RNA a challenging target for computational methods. Nevertheless, SILCS-RNA appears to offer information content to meet this challenge beyond that accessible to many of the other available methods.

Other areas in which the SILCS method has been shown to be useful are in the area of drug liability through the development of models to predict the hERG blockade [136, 137], where the approach can identify portions of the ligands that make the largest contribution to hERG binding. Motivated by the ability to apply SILCS to membrane-bound GPCRs, the method was applied to lipid bilayers and shown to be capable of mapping functional group free energy profiles across the membranes [102]. The approach was able to yield good agreement between bilayer/water and experimental octanol/water partition coefficients, predict resistances of the solutes in the bilayers, and calculate free energy profiles for drug-like molecules across the bilayers (PAMPA and POPC/cholesterol) in a computationally accessible manner [102]. The latter capability is currently being used to develop models for the prediction of membrane permeabilities of drug-like molecules in conjunction with deep neural nets [138]. Interestingly, the SILCS calculated free energy profiles of molecules across the bilayers represent absolute free energy profiles due to the use of GFE normalization based on the full volume of the simulation system, as the solutes and water are fully accessible to all regions of the simulation boxes.

Beyond small molecule-focused drug design, the SILCS technology has utility for studies of macromolecules themselves. The initial step in applying SILCS to proteins alone was its extension to calculate protein–protein interactions (PPI) [110]. SILCS-PPI uses a fast-Fourier transform (FFT) sampling approach in conjunction with the overlap of the SILCS FragMaps "receptor" protein with the distribution of functional groups on the "ligand" protein to score the orientations from which distribution of PPI orientations are obtained. The technology is competitive with available computational PPI methods, though the requirement to calculate the FragMaps makes it computationally demanding when only PPI analysis is needed. However, SILCS-PPI may be combined with SILCS-Hotspots to facilitate the formulation of biologics, including monoclonal antibodies (mAb). SILCS-Biologics [112, 113] combines the distribution of the probability of residues participating in PPI on the entire protein surface with the distribution of excipients, buffers, and monoions on the surface of the protein. This combination allows for excipients that may block PPI that contribute to aggregation or increased viscosity to be analyzed. In addition, information on excipients that may impact protein stability can be obtained. This combination of information may be used to facilitate the selection of excipients, especially in formulations that require a high protein concentration. In addition, analysis of the distribution of excipients, buffers, and monoions bound to the protein may be used to estimate the total effective charge and dipole of the protein [139], providing additional information of utility in biologics formulation.

The application of SILCS to proteins, including glycoproteins, has elucidated macromolecular interactions involved in immune function and downstream signaling. In one study, the interactions of the protein endoglycosidase S2 (EndoS2), a protein excreted by *Streptococcus pyogenes*, which deglycosylates the Fc of mAbs thereby limiting the host immune response, were investigated [140]. In the study, Fc glycans were docked to the carbohydrate binding module (CBM) and the glycoside hydrolase (GH) domains of EndoS2 using SILCS-MC, following which the remainder of the Fc was built from an ensemble of Fc-glycan conformations generated from extensive MD simulations. Following this docking and reconstruction procedure, MD simulations of selected docked complexes were performed from which details of the interaction of EndoS2 with the Fc were predicted. Notably, the study showed the importance of PPI between the Fc and EndoS2 rather than just interactions involving the glycan alone, an observation that was shown to be in agreement with subsequent experimental cryo-EM and crystallographic studies [141, 142]. In a second study, the role of clustering of the FcγRIIIa-FcεRIγ receptors upon multivalent binding of antibodies on phosphorylation events by the kinase LCK was investigated [143]. In the study, models of the transmembrane (TM) and intracellular (IC) regions of the FcγRIIIa-FcεRIγ complex in different spatial relationships mimicking different extents of clustering were investigated via MD simulations and SILCS-MC docking. Multiple long-time MD simulations of the complexes under the different extents of clustering were performed to generate large ensembles of conformations of the receptors. SILCS-MC docking of the Tyr-based activation motifs (ITAMs) of the IC regions of the receptor onto LCK was then performed, following which the full IC, TM, and membrane bilayer-LCK complexes were reconstructed. From the resulting ensemble of complexes, information on how LCK can perform multiple phosphorylations on individual ITAMS and on ITAMS in different FcγRIIIa-FcεRIγ complexes was obtained. Among these conformations, a number to which the kinase Syk could bind as required for downstream signaling were identified. This approach nicely showed how downstream signaling may be facilitated by "concentration" effects where ITAMs coming close to each other due to receptor clustering can occur. In combination, both studies highlight how large ensembles of conformations of flexible biomolecules such as glycans and disordered peptides can be generated and then docked to proteins using SILCS, from which ensembles of putative interactions responsible for biological events may be identified.

A key contributor to the success of SILCS technology is the use of the CHARMM General Force Field (CGenFF) [144–147] and the CGenFF program [145, 146]. CGenFF provides coverage for a wide range of chemical groups within biomolecules and drug-like molecules. The CGenFF program included with the SILCS software package (SilcsBio LLC) rapidly generates FF parameters for a ligand of interest according to analogy to known small molecule CGenFF parameters. These topologies and parameters are generated within fractions of a second, enabling the analysis of a multitude of ligands. In the context of SILCS simulations, CGenFF and the CGenFF program provide the empirical force field of the standard SILCS and SILCS-X solute molecules. In SILCS-X and for halogen-containing ligands, CGenFF represents the σ-hole as a positive point charge, a "lone pair," to improve the treatment of halogen bonding [148]. In conjunction with the CHARMM36 FF [149–157] and CHARMM TIP3P models [157, 158] describing the target macromolecule and surrounding water, CGenFF allows for highly accurate sampling of the solute-target interactions within the SILCS simulations. In the context of SILCS analyses performed after simulations, CGenFF is crucial for assigning FragMap types to solutes and ligands of interest based on an atom classification scheme for LGFE scoring. Additionally, in SILCS-MC, the detection of rotatable bonds is based on the topology of the ligand generated by the CGenFF program, and the intramolecular energy of the ligand is calculated using the CGenFF potential energy function.

5.3 SILCS Case Studies: Bovine Serum Albumin and Pembrolizumab

To illustrate the application of SILCS, here we provide two case studies of the SILCS method applied to bovine serum albumin (BSA) and Pembrolizumab, a humanized antibody used in cancer immunotherapy sold under the brand name Keytruda. In summary, given an initial structure of the target macromolecule, SILCS simulations are performed to sample the interaction pattern of solute molecules with the target macromolecule, in this case, BSA and pembrolizumab, and ultimately calculate pre-computed GFE FragMaps for use in a wide range of analyses. These analyses include SILCS-MC to sample spatial and conformational sampling of ligands, SILCS-Hotspots to identify allosteric sites, SILCS-PPI to map protein–protein interactions, and SILCS-Biologics to assess excipients for formulation development of protein-based drugs. Details of these SILCS-based analyses and their applications to BSA and pembrolizumab are presented in the following sections.

5.3.1 SILCS Simulations

SILCS simulations are performed to sample the interaction of solutes of different chemical classes and water with the target macromolecule and pre-compute SILCS FragMaps for subsequent SILCS analyses. For the SILCS simulations of BSA, the initial coordinates of BSA were extracted from its crystal structure (PDB: 4F5S [159]). For the SILCS simulations of pembrolizumab, the initial coordinates of pembrolizumab were extracted from the crystal structure of the full-length pembrolizumab mAb (PDB: 5DK3 [160]). For BSA, the entire protein was solvated in a water box. For pembrolizumab, due to its size, the Fc and Fab portions of the mAb were separated and then solvated and simulated independently. Additionally, only one Fab portion of pembrolizumab was simulated for computational expediency as both Fab portions of pembrolizumab are sequentially identical and structurally very similar. For each system, the dimensions of the water box were independently set such that the atoms of the target protein were separated from the box edge by at least 12 Å on all sides. Eight solutes representing different chemical classes (benzene, propane, dimethylether, methanol, formamide, imidazole, acetate, and methylammonium) were added into the system at a concentration of ~0.25 M, to probe the functional group preferences of the target proteins. For each protein, 10 independent simulation systems were built with different distributions of water and solute molecules as well as rotated protein sidechain configurations for the solvent-accessible residues.

Prior to the SILCS simulations, each simulation system was subjected to a 5000-step steepest descent energy minimization followed by a 250 ps MD equilibration. During the energy minimization and equilibration steps, all non-hydrogen atoms of the proteins were constrained through a harmonic positional restraint with a force constant of 2.4 kcal/mol-Å^2. Following the energy minimization and equilibration steps, each system was subjected to 25 GCMC cycles of 200 000 GCMC steps each to re-equilibrate the solute and water molecules around the target protein. Subsequently, a production run of 100 GCMC/MD cycles was performed. In each GCMC/MD cycle, 200 000 GCMC steps followed by 1 ns of MD simulation were executed. During the production run, Cα backbone atoms of the target protein were lightly constrained through a harmonic positional restraint with a force constant of 0.12 kcal/mol-Å^2. Additionally, to prevent the aggregation of solute molecules, a repulsive intermolecular wall was applied to select solute pairs. During the MD simulations, the Nosé–Hoover method was used to maintain the temperature at 298 K, and pressure was maintained at 1 bar using the Parrinello–Rahman barostat.

The CHARMM36m protein force field [157], CGenFF [144–147], and CHARMM TIP3P water model [157, 158] were used to describe protein, solutes, and water during the simulations, respectively. The GCMC portion of the runs was performed using SILCS software (SilcsBio LLC), and MD was conducted using the GROMACS [96] program. Upon completion of the 100 GCMC/MD cycles, 100 ns of simulation data was extracted per simulation system for a cumulative 1μs simulation time (100 ns * 10 simulation systems) per protein system (BSA, pembrolizumab Fab, and pembrolizumab Fc).

5.3.2 FragMap Construction

Using the SILCS simulation snapshots, probability distributions of the solutes and water molecules in and around the protein are calculated to produce FragMaps. The FragMaps are determined by binning selected atoms of the solute molecules into 1 Å × 1 Å × 1 Å cubic volume elements (voxels) of a grid spanning the entire system for each simulation snapshot, extracted every 10 ps. The voxel occupancies calculated in the presence of the target macromolecule are divided by the value in bulk as determined by the number of each solute in the system relative to the number of waters, assuming 55 M water, to obtain a normalized occupancy. The solute concentrations may also be set to an assumed concentration of 0.25 M or based on the total volume of the simulation system [102]. The normalized occupancies are then converted to GFE values by using the Boltzmann transformation. The GFE represents the free energy of functional groups, and GFE FragMaps can indicate both favorable and unfavorable interactions with the target macromolecule through negative and positive free energies, respectively. Additionally, an "exclusion map" is also calculated based on the regions not sampled by water or any solute molecules during the entire duration of the SILCS simulations. This exclusion map accounts for the conformational flexibility of the macromolecule. As such, the initial, rigid macromolecule experimental structure is not used for subsequent SILCS calculations aside from visualization purposes.

BSA, pembrolizumab Fab, and pembrolizumab Fc FragMaps were independently determined. The resulting FragMaps for BSA and pembrolizumab are shown in Figures 5.1a,b, respectively. For pembrolizumab, the FragMaps of the Fab and Fc domains of the mAb were determined independently as their SILCS simulations were also performed independently for computational efficiency. As only one Fab domain of pembrolizumab was simulated for computational expediency, the FragMaps resulting from the simulated Fab were reoriented to fit the second Fab domain. The independent sets of FragMaps for pembrolizumab Fc and Fab domains are overlaid on the full-length pembrolizumab structure in Figure 5.1b for clarity. The BSA FragMaps show abundant apolar and positively charge binding regions, shown by the green and cyan-colored FragMaps in Figure 5.1a at the protein surface. Additionally, negatively charged binding regions, shown by the orange-colored FragMap in Figure 5.1a, are also observed in pockets within the protein structure. The pembrolizumab FragMaps show the presence of apolar binding regions, as shown by the green-colored FragMap in Figure 5.1b, distributed across the Fc

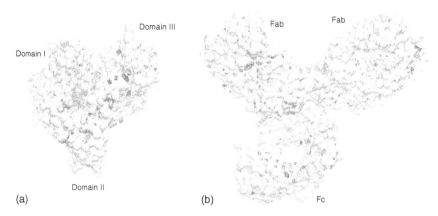

Figure 5.1 The SILCS FragMaps of (a) BSA and (b) pembrolizumab. BSA and pembrolizumab are shown in transparent surface representations. SILCS-FragMaps for generic apolar, generic H-bond donor, generic H-bond acceptor, negative, and positive groups are shown in green, blue, red, orange, and cyan mesh representations, respectively. Isocontour GFE FragMaps are shown at a contour level of −1.2 kcal/mol.

and Fab domains. Positively charged binding regions are primarily observed in the Fc domain of pembrolizumab due to the larger number of negatively charged residues in the Fc domain compared to the Fab domain. Additionally, a few pockets of hydrogen bond donor and acceptor binding regions in the Fab and Fc domains are also observed, as shown by the blue and red FragMaps in Figure 5.1b. Typically, hydrogen bond FragMaps occur at low contour levels, −0.6 kcal/mol vs. −1.2 kcal/mol shown in Figure 5.1, due to the balance of favorable solute–protein interactions and the desolvation penalty associated with such functional groups binding with the protein.

5.3.3 SILCS-MC

The SILCS FragMaps can be used to rapidly dock, score, and evaluate ligands for their binding to a target macromolecule through SILCS-MC. In the SILCS-MC method, selected atoms of the ligand are associated with a FragMap type, and a GFE score is assigned to each atom based on the value of the FragMap at that position. The atom FragMap type is translated from CGenFF atom types using an atom-classification scheme [106]. The coordinates of each atom are then used to determine its overlap with a FragMap voxel, with that atom being assigned the GFE value of the voxel. A ligand GFE (LGFE) is subsequently calculated based on a summation of the atomic GFE scores for classified atoms. The LGFEs serve as approximations to binding free energies but are not formal binding free energies due to additional factors, such as entropy loss of combining multiple smaller fragments into a larger ligand and the contribution of ligand and protein internal strain, among others, being omitted from the calculation.

The SILCS-MC docking procedure determines the most energetically favorable, or lowest LGFE, pose through a series of energy minimization, Markov chain MC, and simulated annealing steps. The initial pose of the ligand may be randomly generated for blind docking or taken from a predetermined set of coordinates for pose refinement. For SILCS-MC docking, the ligand of interest is placed randomly in a sphere centered at a user-defined coordinate of a user-defined size, typically 5 or 10 Å radius, at which five independent SILCS-MC runs are performed to sample ligand docked poses. Each of the SILCS-MC runs involves a minimum of 50 and a maximum of 250 cycles of Monte Carlo/Simulated Annealing (MC/SA) sampling of the molecule within the user-defined search space sphere. In each cycle, 10 000 steps of MC at room temperature are followed by 40 000 steps of SA, lowering the temperature are performed with the molecule reoriented at the beginning of each cycle. Subsequently, the ligand-docked poses are scored by LGFE and ligand efficiency (LE), with the LGFE divided by the number of heavy atoms.

For the case study, SILCS-MC was performed to dock divanillin to BSA. As divanillin has been experimentally determined to bind to binding site I of BSA [161] and Trp 212 quenching is commonly used to determine ligand binding to binding site I [162–165], the Cα coordinate of Trp 212 was used as the center of the 10 Å sphere search space. Figure 5.2 shows the most energetically favored, lowest LGFE,

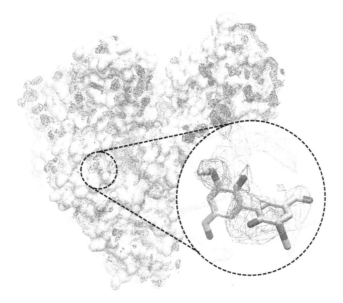

Figure 5.2 Most energetically favored conformation of divanillin binding to BSA in binding site I. Binding site I is encircled in black dotted lines with a zoomed-in view of divanillin's binding to BSA. BSA is shown in transparent surface representation and transparent gray tube representation within the zoomed-in view. SILCS-FragMaps for generic apolar, generic H-bond donor, generic H-bond acceptor, alcohol, negative, and positive groups are shown in green, blue, red, ochre, orange, and cyan mesh representation, respectively. Divanillin is shown in licorice representation. Isocontour GFE FragMaps are shown at a contour level of −0.5 kcal/mol for generic apolar, generic H-bond donor, generic H-bond acceptor, and alcohol groups and −1.2 kcal/mol for negative and positive groups. In the zoomed-in view, FragMaps for negative and positive groups are omitted, and the view is reoriented for clarity.

docked pose of divanillin binding to BSA. The LGFE of divanillin binding to BSA was −4.1 kcal/mol. In comparison, the LGFE of warfarin, a fluorescent marker of BSA binding site I, docked in the same manner using SILCS-MC was −2.3 kcal/mol. The predicted higher affinity, or more favorable LGFE, of divanillin compared to warfarin is in line with experiments, which showed that divanillin displaces warfarin within BSA binding site I [161].

5.3.4 SILCS-Hotspots

SILCS-Hotspots is an extension of SILCS-MC that identifies fragment-binding hotspots that are spatially distributed in and around the target molecule [107]. SILCS-Hotspots are performed by systematically partitioning the full 3D space of the target molecule into 14.14 Å × 14.14 Å × 14.14 Å subspaces in which fragments are independently docked using SILCS-MC. In each subspace, each fragment is randomly positioned in a sphere of 10 Å radius centered where the random variation of one rotatable bond of the fragment is generated through SILCS-MC. Subsequently, each fragment is subjected to 10 000 MC steps (at 300 K) followed by 40 000 MC annealing steps from 0 to 300 °K. This procedure is applied 1000 times for each fragment in each subspace. Subsequently, center-of-mass (COM)-based clustering, with a clustering radius of 3 Å, is performed for each fragment to identify orientations with the highest neighbor population. An additional round of clustering using a clustering radius of 4 Å is then performed on all poses selected in the first round of clustering across all fragments to identify hotspots, which may be populated by multiple fragments. The clustering radii of the first and second clustering rounds may be adjusted, with larger clustering radii typically yielding fewer, more spatially separated identified hotspots. The LGFE of each fragment in each Hotspot site (centers of predicted fragment binding sites) is averaged to determine the average LGFE of the hotspot and hotspots with LGFE scores greater than −2 kcal/mol are typically discarded. Other metrics, such as the number of fragments in a hotspot, may be used to rank order hotspots in addition to the average LGFE score.

In the case study, SILCS-Hotspots were applied to BSA to identify potential binding sites. For this study, 135 low-molecular-weight compounds from the Astex MiniFrags probing library [123] were used as the SILCS-Hotspots fragments. The resulting Hotspots identified are shown in Figure 5.3. As shown in Figure 5.3, the hotspots encompass the entire BSA protein, including interior pockets. The presence of multiple, energetically favorable adjacent hotspots throughout BSA indicates that BSA has a high propensity to bind several classes of ligands at different sites, in line with previous studies on serum albumin [169–171]. Aligning experimentally resolved structures of BSA bound to ligands (PDB IDs 4JK4 [166], 4OR0 [167], and 6QS9 [168]) with the structure used in the SILCS simulations and analyses shows that the experimentally determined binding sites of 3,5-diiodosalicylic acid (DIU), naproxen (NPS), and *R*-ketoprofen (JGE) are captured by the top (most favorable LGFE) 15 hotspots. Close-up views of the experimentally resolved binding of DIU, NPS, and JGE to BSA in relation to the hotspots and FragMaps are encircled in black dotted lines in Figure 5.3. These results confirm the ability of SILCS-Hotspots

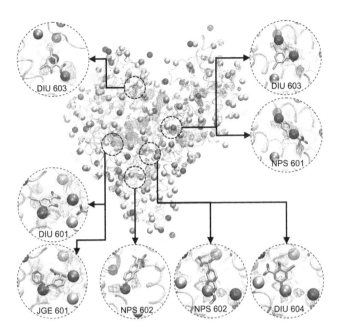

Figure 5.3 The SILCS-Hotspots and SILCS FragMaps of BSA, along with experimentally resolved binding sites of DIU, NPS, and JGE according to crystallography [166–168]. The crystallographic orientations of DIU, NPS, and JGE overlaid on BSA, the SILCS-Hotspots, and SILCS FragMaps, encircled in black dotted lines, are zoomed-in and reoriented for clarity. BSA is shown in gray, transparent tube representation; the SILCS-Hotspots are shown in VDW representation and are colored by their LGFE scores, with red indicating the most favorable and blue indicating the least favorable (−2 kcal/mol being the lowest LGFE shown); SILCS-FragMaps for generic apolar, generic H-bond donor, generic H-bond acceptor, alcohol, negative, and positive groups are shown in green, blue, red, ochre, orange, and cyan mesh representation, respectively.

to identify binding sites that correspond to known, experimentally resolved ligand binding sites. Note that in all cases, the ligand binding sites are occupied by multiple hotspots. This pattern is common for sites that are suitable for the binding of larger, drug-like molecules, including allosteric modulators [107].

5.3.5 SILCS-PPI

SILCS-PPI uses FragMaps, protein functional group probability grids, and FFTs [172] to perform protein–protein docking from which patterns of protein–protein interactions are identified. The protein functional group probability grid maps (PPGMaps) are extracted from the SILCS simulations and subsequently assigned to the corresponding FragMap types. The assignment is done such that the spatial overlap of the receptor protein FragMaps and the ligand-protein PPGMaps provides a rapid estimation of the protein receptor–protein–ligand interaction. SILCS-PPI performs protein–protein docking by maximizing the complementarity between the FragMaps of one protein and the PPGMaps through an FFT-based algorithm [110]. During the docking process, the receptor FragMaps and PPGMaps are fixed in space, and the ligand FragMaps and PPGMaps are translated and rotated systematically over all possible orientations. The docked poses are scored using protein grid free energies (PGFEs), which are calculated based on the overlap of FragMaps and PPGMaps. PPI preference (PPIP) maps are subsequently calculated using a two-step, COM and orientation-based clustering analysis of all docked poses. After the clustering, per-residue PPIP is computed as the number of contacts between the receptor and ligand–protein atoms, with any non-hydrogen atom within a 5 Å cutoff considered in contact, and summed over the top 2000 docked poses sorted by PGFE score. Each per-residue PPIP value is subsequently normalized by the maximum per-residue PPIP value, resulting in a PPIP score. The PPIP scores range from 0 to 1 with higher PPIP scores indicating that a residue is more likely to be involved in a PPI.

In the case study, regions of high PPIP in BSA and pembrolizumab were identified using SILCS-PPI. For the case study, BSA-BSA self-PPI and pembrolizumab Fab-Fc, Fc-Fc, and Fab-Fab were considered. As the pembrolizumab Fc and Fab domains were simulated independently, the full-length pembrolizumab structure (PDB: 5DK3 [160]) was overlaid on the receptor Fab or Fc, and any docked poses in which the ligand Fab/Fc resulted in steric clashes were discarded. In this way, poses that are sterically inaccessible in the full-length pembrolizumab were excluded. The predicted PPIP maps of BSA and pembrolizumab are shown in Figure 5.4. For

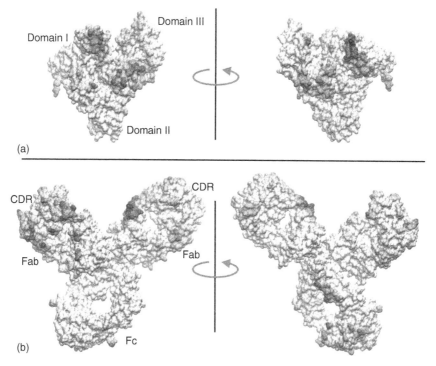

Figure 5.4 Predicted PPI preference maps of (a) BSA and (b) the full pembrolizumab (Fab and Fc). BSA and pembrolizumab are shown in surface representation with the highest PPI preference regions colored dark red and the lowest PPI preference regions colored white.

BSA, the strongest PPIP regions are in domains I and III (Figure 5.4a). The higher interaction preference of BSA in domains I and III is consistent with experimentally derived crystal structures of BSA dimers [159, 166–168] and experiments suggesting that BSA dimers are stabilized by residues Cys 34 and 513 [173], which are located at or near regions with predicted high PPIP. Interestingly, pembrolizumab does not show a particularly strong PPIP in its complementary determining region (CDR) over other domains of the mAb (Figure 5.4b). The regions with the strongest PPIP are distributed along the sides of the Fab and Fc domains of pembrolizumab and are not concentrated at the CDR. The relatively low PPIP of the CDR may explain experimental data showing the low propensity of pembrolizumab to aggregate even in refrigerated conditions, and the ability of pembrolizumab to retain its functional ability to bind with PD-1 after two weeks stored in saline solution at refrigerated conditions [174]. It is worth reiterating that the reported PPIP values are relative values and cannot be directly compared across different protein systems. Thus, comparisons of which proteins may be more prone to aggregation cannot be directly inferred from their self-PPIP values. Nevertheless, these PPIP maps may be used to inform the introduction of mutations to individual protein therapeutics to enhance their stability.

5.3.6 SILCS-Biologics

SILCS-Biologics combines SILCS-PPI and SILCS-Hotspots to guide excipient selection for therapeutic protein formulations. In SILCS-Biologics, SILCS-PPI is used to compute a protein–protein self-interaction map, and SILCS-Hotspots are used to identify binding sites of a set of excipients of interest. Subsequently, combining PPI self-interaction and excipient hotspot maps, SILCS-Biologics produces a range of data that can be processed through data science and machine learning approaches to predict various experimental properties. For example, the number of excipient binding sites overlapping with regions with high predicted PPIP has been shown to correlate with the experimentally determined viscosity for several excipients, including amino acids and sugars [112, 113]. Additionally, the number of binding sites with predicted high binding affinity may predict relative protein stability [112, 113].

In the case study, SILCS-Biologics was applied for pembrolizumab to examine how excipients interact with high PPIP regions of the mAb. For this case study, excipients histidine and sucrose were investigated as they are included in the commercial formulation of pembrolizumab (available at 25 mg/mL), Keytruda. Figure 5.5 shows the sites where the excipients are predicted to bind on the surface of pembrolizumab. According to the predicted binding sites, histidine and sucrose cooperatively bind to high PPIP regions of pembrolizumab (Figure 5.5). Histidine and sucrose exclusively bind to portions of pembrolizumab, particularly on the Fc domain of the mAb. The binding sites of histidine and sucrose covering high PPIP regions are hypothesized to prevent self-PPI, which would normally lead to aggregation and increased viscosity. Such atomistic-level understanding of how excipients interact with protein therapeutics in conjunction with how protein therapeutics may self-interact can

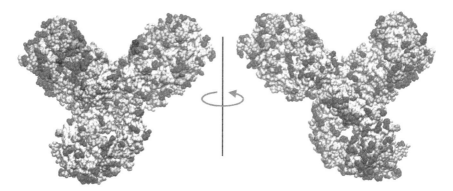

Figure 5.5 Predicted excipient binding sites and PPI preference map of the full pembrolizumab (Fab and Fc). Histidine and sucrose molecules are shown in blue and cyan VDW representations, respectively. Pembrolizumab is shown in surface representation with the highest PPI preference regions colored dark red and the lowest PPI preference regions colored white.

demystify the selection and optimization of excipient formulation to maximize protein stability and minimize aggregation and viscosity.

5.4 Conclusion

Co-solvent MD methods have become useful tools in CADD and have been successfully applied to a wide variety of macromolecular targets. These methods are advantageous to many other CADD methods as protein flexibility and competition with water are incorporated into the resulting predictions. SILCS, now commercialized in the form of SilcsBio LLC, represents one of the most extensively developed co-solvent MD methods, with the MacKerell laboratory making continual improvements and extensions of the SILCS technology. From one set of SILCS simulations, affinity patterns for diverse functional groups in the form of FragMaps are generated in and around the target macromolecule. These FragMaps are then used as the basis for a wide range of SILCS analyses, which include, among others, ligand docking through SILCS-MC, identification of allosteric binding sites through SILCS-Hotspots, protein–protein docking through SILCS-PPI, and protein therapeutic formulation through SILCS-Biologics. The included case studies for BSA and pembrolizumab show how a single pre-computed FragMaps of a target macromolecule can be used for a wide range of applications and analyses. Overall, the wide variety of analyses possible with SILCS sets it apart from other co-solvent methods and suggests that the technology may be expanded beyond its current uses.

Conflict of Interest

ADM Jr. is co-founder and Chief Scientific Officer of SilcsBio LLC.

Acknowledgments

The authors acknowledge financial support from NIH GM131710 and R44GM 130198 and computational resources provided by the Computer-Aided Drug Design (CADD) Center at the University of Maryland, Baltimore, as well as the Extreme Science and Engineering Discovery Environment (XSEDE).

References

1 Hansch, C., Maloney, P.P., Fujita, T., and Muir, R.M. (1962). Correlation of biological activity of phenoxyacetic acids with hammett substituent constants and partition coefficients. *Nature* 194 (4824): 178–180.
2 Hansch, C. and Fujita, T. (1964). P-Σ-Π analysis. A method for the correlation of biological activity and chemical structure. *J. Am. Chem. Soc.* 86 (8): 1616–1626.
3 Schultz, T.W., Lin, D.T., and Arnold, L.M. (1991). Qsars for monosubstituted anilines eliciting the polar narcosis mechanism of action. *Sci. Total Environ.* 109–110: 569–580.
4 Aptula, A.O., Netzeva, T.I., Valkova, I.V. et al. (2002). Multivariate discrimination between modes of toxic action of phenols. *Quant. Struct.-Activity Relat.* 21 (1): 12–22.
5 Ma, Q.-S., Yao, Y., Zheng, Y.-C. et al. (2019). Ligand-based design, synthesis and biological evaluation of xanthine derivatives as Lsd1/Kdm1a inhibitors. *Eur. J. Med. Chem.* 162: 555–567.
6 Mirabello, C. and Wallner, B. (2020). Interlig: improved ligand-based virtual screening using topologically independent structural alignments. *Bioinformatics* 36 (10): 3266–3267.
7 Jia, X., Ciallella, H.L., Russo, D.P. et al. (2021). Construction of a virtual opioid bioprofile: a data-driven Qsar modeling study to identify new analgesic opioids. *ACS Sustain. Chem. Eng.* 9 (10): 3909–3919.
8 Bajad, N.G., Swetha, R., Singh, R. et al. (2022). Combined structure and ligand-based design of dual Bace-1/Gsk-3β inhibitors for Alzheimer's disease. *Chem. Pap.* .
9 Perron, Q., Mirguet, O., Tajmouati, H. et al. (2022). Deep generative models for ligand-based de novo design applied to multi-parametric optimization. *J. Comput. Chem.* 43 (10): 692–703.
10 Koes, D.R. and Camacho, C.J. (2012). Zincpharmer: pharmacophore search of the zinc database. *Nucleic Acids Res.* 40 (Web Server issue): W409-14.
11 Ke, Y.-Y., Singh, V.K., Coumar, M.S. et al. (2015). Homology modeling of Dfg-in Fms-like tyrosine kinase 3 (Flt3) and structure-based virtual screening for inhibitor identification. *Sci. Rep.* 5 (1): 11702.
12 Parvaiz, N., Ahmad, F., Yu, W. et al. (2021). Discovery of β-lactamase Cmy-10 inhibitors for combination therapy against multi-drug resistant enterobacteriaceae. *PLoS ONE* 16 (1): e0244967.

13 Tabrez, S., Zughaibi, T.A., Hoque, M. et al. (2022). Targeting glutaminase by natural compounds: structure-based virtual screening and molecular dynamics simulation approach to suppress cancer progression. *Molecules* 27 (15): 5042.

14 Wang, L., Wu, Y., Deng, Y. et al. (2015). Accurate and reliable prediction of relative ligand binding potency in prospective drug discovery by way of a modern free-energy calculation protocol and force field. *J. Am. Chem. Soc.* 137 (7): 2695–2703.

15 Zhang, H., Jiang, W., Chatterjee, P., and Luo, Y. (2019). Ranking reversible covalent drugs: from free energy perturbation to fragment docking. *J. Chem. Inf. Model.* 59 (5): 2093–2102.

16 Cournia, Z., Allen, B.K., Beuming, T. et al. (2020). Rigorous free energy simulations in virtual screening. *J. Chem. Inf. Model.* 60 (9): 4153–4169.

17 Nikiforov, P.O., Blaszczyk, M., Surade, S. et al. (2017). Fragment-sized ethr inhibitors exhibit exceptionally strong ethionamide boosting effect in whole-cell mycobacterium tuberculosis assays. *ACS Chem. Biol.* 12 (5): 1390–1396.

18 Kessler, D., Gmachl, M., Mantoulidis, A. et al. (2019). Drugging an undruggable pocket on Kras. *Proc. Natl. Acad. Sci.* 116 (32): 15823–15829.

19 Abdul-Hammed, M., Adedotun, I.O., Falade, V.A. et al. (2021). Target-based drug discovery, admet profiling and bioactivity studies of antibiotics as potential inhibitors of Sars-Cov-2 main protease (Mpro). *Virusdisease* 32 (4): 642–656.

20 El Aissouq, A., Bouachrine, M., Ouammou, A., and Khalil, F. (2022). Homology modeling, virtual screening, molecular docking, molecular dynamic (Md) simulation, and admet approaches for identification of natural anti-parkinson agents targeting mao-B protein. *Neurosci. Lett.* 786: 136803.

21 Guedes, I.A., Pereira, F.S.S., and Dardenne, L.E. (2018). Empirical scoring functions for structure-based virtual screening: applications, critical aspects, and challenges. *Front. Pharmacol.* 9: 1089.

22 Maia, E.H.B., Assis, L.C., de Oliveira, T.A. et al. (2020). Structure-based virtual screening: from classical to artificial intelligence. *Front. Chem.* 8: 343.

23 Fischer, A., Smieško, M., Sellner, M., and Lill, M.A. (2021). Decision making in structure-based drug discovery: visual inspection of docking results. *J. Med. Chem.* 64 (5): 2489–2500.

24 Hussain, W., Rasool, N., and Khan, Y.D. (2021). Insights into machine learning-based approaches for virtual screening in drug discovery: existing strategies and streamlining through Fp-Cadd. *Curr. Drug Discov. Technol.* 18 (4): 463–472.

25 Sabe, V.T., Ntombela, T., Jhamba, L.A. et al. (2021). Current trends in computer aided drug design and a highlight of drugs discovered via computational techniques: a review. *Eur. J. Med. Chem.* 224: 113705.

26 Giordano, D., Biancaniello, C., Argenio, M.A., and Facchiano, A. (2022). Drug design by pharmacophore and virtual screening approach. *Pharmaceuticals (Basel)* 15 (5).

27 Lee, J.W., Maria-Solano, M.A., Vu, T.N.L. et al. (2022). Big data and artificial intelligence (Ai) methodologies for computer-aided drug design (Cadd). *Biochem. Soc. Trans.* 50 (1): 241–252.

References

28 Warwicker, J. and Watson, H.C. (1982). Calculation of the electric potential in the active site cleft due to α-helix dipoles. *J. Mol. Biol.* 157 (4): 671–679.

29 Klapper, I., Hagstrom, R., Fine, R. et al. (1986). Focusing of electric fields in the active site of Cu-Zn superoxide dismutase: effects of ionic strength and amino-acid modification. *Proteins: Struct. Funct. Bioinf.* 1 (1): 47–59.

30 Nicholls, A. and Honig, B. (1991). A rapid finite difference algorithm, utilizing successive over-relaxation to solve the poisson–boltzmann equation. *J. Comput. Chem.* 12 (4): 435–445.

31 Constanciel, R. and Contreras, R. (1984). Self consistent field theory of solvent effects representation by continuum models: introduction of desolvation contribution. *Theor. Chim. Acta* 65 (1): 1–11.

32 Still, W.C., Tempczyk, A., Hawley, R.C., and Hendrickson, T. (1990). Semianalytical treatment of solvation for molecular mechanics and dynamics. *J. Am. Chem. Soc.* 112 (16): 6127–6129.

33 Genheden, S. and Ryde, U. (2012). Comparison of End-Point Continuum-Solvation Methods for the Calculation of Protein-Ligand Binding Free Energies. *Proteins* 80 (5): 1326–1342.

34 Genheden, S. and Ryde, U. (2015). The Mm/Pbsa and Mm/Gbsa methods to estimate ligand-binding affinities. *Expert Opin. Drug Discovery* 10 (5): 449–461.

35 Wang, E., Sun, H., Wang, J. et al. (2019). End-point binding free energy calculation with Mm/Pbsa and Mm/Gbsa: strategies and applications in drug design. *Chem. Rev.* 119 (16): 9478–9508.

36 Orr, A.A., Yang, J., Sule, N. et al. (2020). Molecular mechanism for attractant signaling to dhma by *E. coli* Tsr. *Biophys. J.* 118 (2): 492–504.

37 Landau, L.D. (1938). *Statistical Physics*. Oxford: Clarendon.

38 Zwanzig, R.W. (1954). High-temperature equation of state by a perturbation method. I. Nonpolar gases. *J. Chem. Phys.* 22 (8): 1420–1426.

39 Hirono, S. and Kollman, P.A. (1990). Calculation of the relative binding free energy of 2'gmp and 2'amp to ribonuclease T1 using molecular dynamics/free energy perturbation approaches. *J. Mol. Biol.* 212 (1): 197–209.

40 Mutyala, R., Reddy, R.N., Sumakanth, M. et al. (2007). Calculation of relative binding affinities of fructose 1,6-bisphosphatase mutants with adenosine monophosphate using free energy perturbation method. *J. Comput. Chem.* 28 (5): 932–937.

41 Jiang, Z.-Y., Lu, M.-C., Xu, L.L. et al. (2014). Discovery of potent Keap1–Nrf2 protein–protein interaction inhibitor based on molecular binding determinants analysis. *J. Med. Chem.* 57 (6): 2736–2745.

42 Clark, A.J., Gindin, T., Zhang, B. et al. (2017). Free energy perturbation calculation of relative binding free energy between broadly neutralizing antibodies and the Gp120 glycoprotein of Hiv-1. *J. Mol. Biol.* 429 (7): 930–947.

43 Cournia, Z., Chipot, C., Roux, B. et al. (2021). Free energy methods in drug discovery—introduction. In: *Free Energy Methods in Drug Discovery: Current State and Future Directions*, vol. 1397, 1–38. American Chemical Society.

44 Mucs, D. and Bryce, R.A. (2013). The application of quantum mechanics in structure-based drug design. *Expert Opin. Drug Discovery* 8 (3): 263–276.

45 Cavasotto, C.N., Adler, N.S., and Aucar, M.G. (2018). Quantum chemical approaches in structure-based virtual screening and lead optimization. *Front. Chem.* 6.

46 Bryce, R.A. (2020). What next for quantum mechanics in structure-based drug discovery? In: *Quantum Mechanics in Drug Discovery* (ed. A. Heifetz), 339–353. New York, NY: Springer US.

47 Bissaro, M., Sturlese, M., and Moro, S. (2020). The rise of molecular simulations in fragment-based drug design (Fbdd): an overview. *Drug Discov. Today* 25 (9): 1693–1701.

48 Allen, K.N., Bellamacina, C.R., Ding, X. et al. (1996). An experimental approach to mapping the binding surfaces of crystalline proteins. *J. Phys. Chem.* 100 (7): 2605–2611.

49 Goodford, P.J. (1985). A computational procedure for determining energetically favorable binding sites on biologically important macromolecules. *J. Med. Chem.* 28 (7): 849–857.

50 Joseph-McCarthy, D., Hogle, J.M., and Karplus, M. (1997). Use of the multiple copy simultaneous search (Mcss) method to design a new class of picornavirus capsid binding drugs. *Proteins* 29 (1): 32–58.

51 Raman, E.P., Lakkaraju, S.K., Denny, R.A., and MacKerell, A.D. Jr., (2017). Estimation of relative free energies of binding using pre-computed ensembles based on the single-step free energy perturbation and the site-identification by ligand competitive saturation approaches. *J. Comput. Chem.* 38 (15): 1238–1251.

52 Seco, J., Luque, F.J., and Barril, X. (2009). Binding site detection and druggability index from first principles. *J. Med. Chem.* 52 (8): 2363–2371.

53 Alvarez-Garcia, D. and Barril, X. (2014). Molecular simulations with solvent competition quantify water displaceability and provide accurate interaction maps of protein binding sites. *J. Med. Chem.* 57 (20): 8530–8539.

54 Arcon, J.P., Defelipe, L.A., Modenutti, C.P. et al. (2017). Molecular dynamics in mixed solvents reveals protein–ligand interactions, improves docking, and allows accurate binding free energy predictions. *J. Chem. Inf. Model.* 57 (4): 846–863.

55 Arcon, J.P., Defelipe, L.A., Lopez, E.D. et al. (2019). Cosolvent-based protein pharmacophore for ligand enrichment in virtual screening. *J. Chem. Inf. Model.* 59 (8): 3572–3583.

56 Lexa, K.W. and Carlson, H.A. (2011). Full protein flexibility is essential for proper hot-spot mapping. *J. Am. Chem. Soc.* 133 (2): 200–202.

57 Lexa, K.W. and Carlson, H.A. (2013). Improving protocols for protein mapping through proper comparison to crystallography data. *J. Chem. Inf. Model.* 53 (2): 391–402.

58 Ghanakota, P. and Carlson, H.A. (2016). Driving structure-based drug discovery through cosolvent molecular dynamics: miniperspective. *J. Med. Chem.* 59 (23): 10383–10399.

59 Ung, P.M., Ghanakota, P., Graham, S.E. et al. (2016). Identifying binding hot spots on protein surfaces by mixed-solvent molecular dynamics: Hiv-1 protease as a test case. *Biopolymers* 105 (1): 21–34.

60 Graham, S.E., Leja, N., and Carlson, H.A. (2018). Mixmd probeview: robust binding site prediction from cosolvent simulations. *J. Chem. Inf. Model.* 58 (7): 1426–1433.

61 Ghanakota, P., DasGupta, D., and Carlson, H.A. (2019). Free energies and entropies of binding sites identified by mixmd cosolvent simulations. *J. Chem. Inf. Model.* 59 (5): 2035–2045.

62 Chan, W.K.B., DasGupta, D., Carlson, H.A., and Traynor, J.R. (2021). Mixed-solvent molecular dynamics simulation-based discovery of a putative allosteric site on regulator of G protein signaling 4. *J. Comput. Chem.* 42 (30): 2170–2180.

63 Smith, R.D. and Carlson, H.A. (2021). Identification of cryptic binding sites using mixmd with standard and accelerated molecular dynamics. *J. Chem. Inf. Model.* 61 (3): 1287–1299.

64 Prakash, P., Hancock, J.F., and Gorfe, A.A. (2015). Binding hotspots on K-Ras: consensus ligand binding sites and other reactive regions from probe-based molecular dynamics analysis. *Proteins: Struct. Funct. Bioinf.* 83 (5): 898–909.

65 Sayyed-Ahmad, A. and Gorfe, A.A. (2017). Mixed-probe simulation and probe-derived surface topography map analysis for ligand binding site identification. *J. Chem. Theory Comput.* 13 (4): 1851–1861.

66 Sayyed-Ahmad, A. (2018). Hotspot identification on protein surfaces using probe-based md simulations: successes and challenges. *Curr. Top. Med. Chem.* 18 (27): 2278–2283.

67 Yang, C.-Y. and Wang, S. (2010). Computational analysis of protein hotspots. *ACS Med. Chem. Lett.* 1 (3): 125–129.

68 Yang, C.-Y. and Wang, S. (2011). Hydrophobic binding hot spots of Bcl-Xl protein– protein interfaces by cosolvent molecular dynamics simulation. *ACS Med. Chem. Lett.* 2 (4): 280–284.

69 Yang, C.-Y. and Wang, S. (2012). Analysis of flexibility and hotspots in Bcl-Xl and Mcl-1 proteins for the design of selective small-molecule inhibitors. *ACS Med. Chem. Lett.* 3 (4): 308–312.

70 Yang, C.-Y. (2015). Identification of potential small molecule allosteric modulator sites on Il-1r1 ectodomain using accelerated conformational sampling method. *PLoS ONE* 10 (2): e0118671.

71 Privat, C., Granadino-Roldan, J.M., Bonet, J. et al. (2021). Fragment dissolved molecular dynamics: a systematic and efficient method to locate binding sites. *Phys. Chem. Chem. Phys.* 23 (4): 3123–3134.

72 Martinez-Rosell, G., Harvey, M.J., and De Fabritiis, G. (2018). Molecular-simulation-driven fragment screening for the discovery of new Cxcl12 inhibitors. *J. Chem. Inf. Model.* 58 (3): 683–691.

73 Martinez-Rosell, G., Lovera, S., Sands, Z.A., and De Fabritiis, G. (2020). Playmolecule crypticscout: predicting protein cryptic sites using mixed-solvent molecular simulations. *J. Chem. Inf. Model.* 60 (4): 2314–2324.

74 Kimura, S.R., Hu, H.P., Ruvinsky, A.M. et al. (2017). Deciphering cryptic binding sites on proteins by mixed-solvent molecular dynamics. *J. Chem. Inf. Model.* 57 (6): 1388–1401.

75 Zariquiey, F.S., de Souza, J.V., and Bronowska, A.K. (2019). Cosolvent analysis toolkit (Cat): a robust hotspot identification platform for cosolvent simulations of proteins to expand the druggable proteome. *Sci. Rep.* 9 (1): 1–14.

76 Kalenkiewicz, A., Grant, B.J., and Yang, C.Y. (2015). Enrichment of druggable conformations from apo protein structures using cosolvent-accelerated molecular dynamics. *Biology (Basel)* 4 (2): 344–366.

77 Bakan, A., Nevins, N., Lakdawala, A.S., and Bahar, I. (2012). Druggability assessment of allosteric proteins by dynamics simulations in the presence of probe molecules. *J. Chem. Theory Comput.* 8 (7): 2435–2447.

78 Lee, J.Y., Krieger, J.M., Li, H., and Bahar, I. (2020). Pharmmaker: pharmacophore modeling and hit identification based on druggability simulations. *Protein Sci.* 29 (1): 76–86.

79 Huang, D.Z. and Caflisch, A. (2011). The free energy landscape of small molecule unbinding. *PLoS Comput. Biol.* 7 (2).

80 Huang, D., Rossini, E., Steiner, S., and Caflisch, A. (2014). Structured water molecules in the binding site of bromodomains can be displaced by cosolvent. *ChemMedChem* 9 (3): 573–579.

81 Tan, Y.S., Śledź, P., Lang, S. et al. (2012). Using ligand-mapping simulations to design a ligand selectively targeting a cryptic surface pocket of polo-like kinase 1. *Angew. Chem.* 124 (40): 10225–10228.

82 Tan, Y.S., Spring, D.R., Abell, C., and Verma, C. (2014). The use of chlorobenzene as a probe molecule in molecular dynamics simulations. *J. Chem. Inf. Model.* 54 (7): 1821–1827.

83 Tan, Y.S., Spring, D.R., Abell, C., and Verma, C.S. (2015). The application of ligand-mapping molecular dynamics simulations to the rational design of peptidic modulators of protein–protein interactions. *J. Chem. Theory Comput.* 11 (7): 3199–3210.

84 Tan, Y.S. and Verma, C.S. (2020). Straightforward incorporation of multiple ligand types into molecular dynamics simulations for efficient binding site detection and characterization. *J. Chem. Theory Comput.* 16 (10): 6633–6644.

85 Yang, Y., Mahmoud, A.H., and Lill, M.A. (2018). Modeling of halogen–protein interactions in co-solvent molecular dynamics simulations. *J. Chem. Inf. Model.* 59 (1): 38–42.

86 Mahmoud, A.H., Yang, Y., and Lill, M.A. (2019). Improving atom-type diversity and sampling in cosolvent simulations using lambda-dynamics. *J. Chem. Theory Comput.* 15 (5): 3272–3287.

87 Yanagisawa, K., Moriwaki, Y., Terada, T., and Shimizu, K. (2021). Exprorer: rational cosolvent set construction method for cosolvent molecular dynamics using large-scale computation. *J. Chem. Inf. Model.* 61: 2744–2753.

88 Takemura, K., Sato, C., and Kitao, A. (2018). Coldock: concentrated ligand docking with all-atom molecular dynamics simulation. *J. Phys. Chem. B* 122 (29): 7191–7200.

89 Guvench, O. and MacKerell, A.D. Jr., (2009). Computational fragment-based binding site identification by ligand competitive saturation. *PLoS Comput. Biol.* 5 (7): e1000435.

90 Raman, E.P., Yu, W., Guvench, O., and MacKerell, A.D. (2011). Reproducing crystal binding modes of ligand functional groups using site-identification by ligand competitive saturation (Silcs) simulations. *J. Chem. Inf. Model.* 51 (4): 877–896.

91 Raman, E.P., Yu, W., Lakkaraju, S.K., and MacKerell, A.D. Jr., (2013). Inclusion of multiple fragment types in the site identification by ligand competitive saturation (Silcs) approach. *J. Chem. Inf. Model.* 53 (12): 3384–3398.

92 Goel, H., Hazel, A., Ustach, V.D. et al. (2021). Rapid and accurate estimation of protein-ligand relative binding affinities using site-identification by ligand competitive saturation. *Chem. Sci.* 12: 8844–8858.

93 Hamelberg, D., Mongan, J., and McCammon, J.A. (2004). Accelerated molecular dynamics: a promising and efficient simulation method for biomolecules. *J. Chem. Phys.* 120 (24): 11919–11929.

94 Case, D.A., Cheatham, T.E. 3rd, Darden, T. et al. (2005). The amber biomolecular simulation programs. *J. Comput. Chem.* 26 (16): 1668–1688.

95 Eastman, P., Swails, J., Chodera, J.D. et al. (2017). Openmm 7: rapid development of high performance algorithms for molecular dynamics. *PLoS Comput. Biol.* 13 (7): e1005659.

96 Van der Spoel, D., Lindahl, E., Hess, B. et al. (2005). Gromacs: fast, flexible, and free. *J. Comput. Chem.* 26 (16): 1701–1718.

97 Phillips, J.C., Braun, R., Wang, W. et al. (2005). Scalable molecular dynamics with Namd. *J. Comput. Chem.* 26 (16): 1781–1802.

98 Brooks, B.R., Bruccoleri, R.E., Olafson, B.D. et al. (1983). Charmm: a program for macromolecular energy, minimization, and dynamics calculations. *J. Comput. Chem.* 4 (2): 187–217.

99 Roe, D.R. and Cheatham, T.E. (2013). Ptraj and Cpptraj: software for processing and analysis of molecular dynamics trajectory data. *J. Chem. Theory Comput.* 9 (7): 3084–3095.

100 Foster, T.J., MacKerell, A.D. Jr., and Guvench, O. (2012). Balancing target flexibility and target denaturation in computational fragment-based inhibitor discovery. *J. Comput. Chem.* 33 (23): 1880–1891.

101 Goel, H., Hazel, A., Yu, W. et al. (2022). Application of site-identification by ligand competitive saturation in computer-aided drug design. *New J. Chem.* 46 (3): 919–932.

102 Lind, C., Pandey, P., Pastor, R.W., and MacKerell, A.D. Jr., (2021). Functional group distributions, partition coefficients, and resistance factors in lipid bilayers using site identification by ligand competitive saturation. *J. Chem. Theory Comput.* 17 (5): 3188–3202.

103 Humphrey, W., Dalke, A., and Schulten, K. (1996). Vmd: visual molecular dynamics. *J. Mol. Graph.* 14: 33–38.

104 DeLano, W.L. (2002). Pymol: an open-source molecular graphics tool. *CCP4 Newsletter Protein Crystallogr.* 40 (1): 82–92.

105 Lakkaraju, S.K., Raman, E.P., Yu, W., and MacKerell, A.D. Jr., (2014). Sampling of organic solutes in aqueous and heterogeneous environments using

oscillating μex grand canonical-like monte carlo-molecular dynamics simulations. *J. Chem. Theory Comput.* 10 (6): 2281–2290.

106 Ustach, V.D., Lakkaraju, S.K., Jo, S. et al. (2019). Optimization and evaluation of site-identification by ligand competitive saturation (Silcs) as a tool for target-based ligand optimization. *J. Chem. Inf. Model.* 59 (6): 3018–3035.

107 MacKerell, A.D. Jr., Jo, S., Lakkaraju, S.K. et al. (2020). Identification and characterization of fragment binding sites for allosteric ligand design using the site identification by ligand competitive saturation hotspots approach (silcs-hotspots). *Biochim. Biophys. Acta, Gen. Subj.* 1864 (4): 129519.

108 Yu, W., Lakkaraju, S.K., Raman, E.P., and MacKerell, A.D. Jr., (2014). Site-identification by ligand competitive saturation (Silcs) assisted pharmacophore modeling. *J. Comput. Aided Mol. Des.* 28 (5): 491–507.

109 Yu, W., Lakkaraju, S.K., Raman, E.P. et al. (2015). Pharmacophore modeling using site-identification by ligand competitive saturation (silcs) with multiple probe molecules. *J. Chem. Inf. Model.* 55 (2): 407–420.

110 Yu, W., Jo, S., Lakkaraju, S.K. et al. (2019). Exploring protein-protein interactions using the site-identification by ligand competitive saturation methodology. *Proteins: Struct. Funct. Bioinf.* 87 (4): 289–301.

111 Kognole, A.A., Hazel, A., and MacKerell, A.D. (2022). Silcs-Rna: toward a structure-based drug design approach for targeting rnas with small molecules. *J. Chem. Theory Comput.* 18 (9): 5672–5691.

112 Jo, S., Xu, A., Curtis, J.E. et al. (2020). Computational characterization of antibody-excipient interactions for rational excipient selection using the site identification by ligand competitive saturation (silcs)-biologics approach. *Mol. Pharm.* 17: 4323–4333.

113 Somani, S., Jo, S., Thirumangalathu, R. et al. (2021). Toward biotherapeutics formulation composition engineering using site-identification by ligand competitive saturation (silcs). *J. Pharm. Sci.* 110 (3): 1103–1110.

114 Yu, W., Weber, D.J., Shapiro, P., and MacKerell, A.D. (2020). Developing kinase inhibitors using computer-aided drug design approaches. In: *Next Generation Kinase Inhibitors: Moving Beyond the Atp Binding/Catalytic Sites* (ed. P. Shapiro), 81–108. Cham: Springer International Publishing.

115 Young, B.D., Yu, W., Rodríguez, D.J.V. et al. (2021). Specificity of molecular fragments binding to S100b versus S100a1 as identified by nmr and site identification by ligand competitive saturation (Silcs). *Molecules* 26 (2).

116 Jiang, W., Zhang, H., Yichun, L. et al. (2022). Binding free energies of piezo1 channel agonists at protein-membrane interface. *bioRxiv* 2022.06.27.497657.

117 Heinzl, G.A., Huang, W., Yu, W. et al. (2016). Iminoguanidines as allosteric inhibitors of the iron-regulated heme oxygenase (Hemo) of pseudomonas aeruginosa. *J. Med. Chem.* 59 (14): 6929–6942.

118 Cheng, H., Linhares, B.M., Yu, W. et al. (2018). Identification of thiourea-based inhibitors of the B-cell lymphoma 6 Btb domain via nmr-based fragment screening and computer-aided drug design. *J. Med. Chem.* 61: 7573–7588.

119 Lanning, M.E., Yu, W., Yap, J.L. et al. (2016). Structure-based design of N-substituted 1-hydroxy-4-sulfamoyl-2-naphthoates as selective inhibitors of the Mcl-1 oncoprotein. *Eur. J. Med. Chem.* 113: 273–292.

120 Lakkaraju, S.K., Yu, W., Raman, E.P. et al. (2015). Mapping functional group free energy patterns at protein occluded sites: nuclear receptors and G-protein coupled receptors. *J. Chem. Inf. Model.* 55: 700–708.

121 Zhao, M., Kognole, A.A., Jo, S. et al. (2023). GPU-specific algorithms for improved solute sampling in grand canonical Monte Carlo simulations. *J. Comput. Chem.* 44 (20): 1719. https://doi.org/10.1002/jcc.27121.

122 Taylor, R.D., MacCoss, M., and Lawson, A.D. (2014). Rings in drugs: miniperspective. *J. Med. Chem.* 57 (14): 5845–5859.

123 O'Reilly, M., Cleasby, A., Davies, T.G. et al. (2019). Crystallographic screening using ultra-low-molecular-weight ligands to guide drug design. *Drug Discov. Today* 24 (5): 1081–1086.

124 Gomes, A., da Silva, G.F., Lakkaraju, S.K. et al. (2021). Insights into glucose-6-phosphate allosteric activation of β-glucosidase A. *J. Chem. Inf. Model.* 61 (4): 1931–1941.

125 Shah, S.D., Lind, C., De Pascali, F. et al. (2022). In silico identification of a β2 adrenergic receptor allosteric site that selectively augments canonical β2ar-Gs signaling and function. *FASEB J.* 36 (S1).

126 Shah, S.D., Lind, C., De Pascalib, F. et al. (2022). In silico identification of a β2-adrenoceptor allosteric site that selectively augments cannical β2args signaling and function. *PNAS* .

127 Weston, S., Baracco, L., Keller, C. et al. (2020). The Ski complex is a broad-spectrum, host-directed antiviral drug target for coronaviruses, influenza, and filoviruses. *Proc. Natl. Acad. Sci.* 117 (48): 30687–30698.

128 Koes, D.R. and Camacho, C.J. (2011). Pharmer: efficient and exact pharmacophore search. *J. Chem. Inf. Model.* 51 (6): 1307–1314.

129 Oashi, T., Ringer, A.L., Raman, E.P., and MacKerell, J.A.D. (2011). Automated selection of compounds with physicochemical properties to maximize bioavailability and druglikeness. *J. Chem. Inf. Model.* 51: 148–158.

130 Macias, A.T., Mia, Y., Xia, G. et al. (2005). Lead validation and sar development via chemical similarity searching; application to compounds targeting the Py + 3 site of the Sh2 domain of P56lck. *J. Chem. Inf. Model.* 45: 1759–1766.

131 Solano-Gonzalez, E., Coburn, K.M., Yu, W. et al. (2021). Small molecules inhibitors of the heterogeneous ribonuclear protein A18 (Hnrnp A18): a regulator of protein translation and an immune checkpoint. *Nucleic Acids Res.* 49 (3): 1235–1246.

132 Samadani, R., Zhang, J., Brophy, A. et al. (2015). Small molecule inhibitors of Erk-mediated immediate early gene expression and proliferation of melanoma cells expressing mutated braf. *Biochem. J.* 467: 425–438.

133 He, X., Lakkaraju, S.K., Hanscom, M. et al. (2015). Acyl-2-aminobenzimidazoles: a novel class of neuroprotective agents targeting Mglur5. *Bioorg. Med. Chem.* 23 (9): 2211–2220.

134 Lakkaraju, S.K., Mbatia, H., Hanscom, M. et al. (2015). Cyclopropyl-containing positive allosteric modulators of metabotropic glutamate receptor subtype 5. *Bioorg. Med. Chem. Lett.* 25 (11): 2275–2279.

135 Glassford, I., Teijaro, C.N., Daher, S.S. et al. (2016). Ribosome-templated azide-alkyne cycloadditions: synthesis of potent macrolide antibiotics by in situ click chemistry. *J. Am. Chem. Soc.* 138 (9): 3136–3144.

136 Mousaei, M., Kudaibergenova, M., MacKerell, A.D. Jr., and Noskov, S. (2020). Assessing Herg1 blockade from bayesian machine-learning-optimized site identification by ligand competitive saturation simulations. *J. Chem. Inf. Model.* 60 (12): 6489–6501.

137 Goel, H., Yu, W., and MacKerell, A.D. Jr., (2022). Herg blockade prediction by combining site identification by ligand competitive saturation and physicochemical properties. *MDPI Chem.* 4: 630–645.

138 Pandey, P. and MacKerell, A. (2023). Combining SILCS and artificial intelligence for high-throughput prediction of drug molecule passive permeability. *J. Chem. Inf. Model.* 63 (18): 5903–5915. https://doi.org/10.1021/acs.jcim.3c00514.

139 Orr, A.A., Tao, A., Guvench, O., and MacKerell, A.D. Jr., (2023). Site identification by ligand competitive saturation-biologics approach for structure-based protein charge prediction. *Mol. Pharmaceutics.* 20 (5): 2600–2611. https://doi.org/10.1021/acs.molpharmaceut.3c00064.

140 Aytenfisu, A.H., Deredge, D., Klontz, E.H. et al. (2021). Insights into substrate recognition and specificity for Igg by endoglycosidase S2. *PLoS Comput. Biol.* .

141 Sudol, A.S.L., Butler, J., Ivory, D.P., and Crispin, M. (2022). Extensive substrate recognition by the streptococcal antibody-degrading enzymes ides and endos. *BioRxiv* https://doi.org/10.1101/2022.06.19.496714.

142 Trastoy, B., Du, J., Cifuente, J. et al. (2022). Mechanism of antibody-specific deglycosylation and immune evasion by streptococcal Igg-specific endoglycosidases. *Res. Square* https://doi.org/10.21203/rs.3.rs-1774503/v1.

143 Chong, G. and MacKerell, A.D. Jr., (2022). Spatial requirements for itam signaling in an intracellular natural killer cell model membrane. *Biochim. Biophys. Acta Gen. Subj.* 1866 (11): 130221.

144 Vanommeslaeghe, K., Hatcher, E., Acharya, C. et al. (2010). Charmm general force field: a force field for drug-like molecules compatible with the charmm all-atom additive biological force fields. *J. Comput. Chem.* 31 (4): 671–690.

145 Vanommeslaeghe, K. and MacKerell, A.D. Jr., (2012). Automation of the charmm general force field (Cgenff) I: bond perception and atom typing. *J. Chem. Inf. Model.* 52 (12): 3144–3154.

146 Vanommeslaeghe, K., Raman, E.P., and MacKerell, A.D. Jr., (2012). Automation of the charmm general force field (Cgenff) Ii: assignment of bonded parameters and partial atomic charges. *J. Chem. Inf. Model.* 52 (12): 3155–3168.

147 Yu, W., He, X., Vanommeslaeghe, K., and MacKerell, A.D. Jr., (2012). Extension of the charmm general force field to sulfonyl-containing compounds and its utility in biomolecular simulations. *J. Comput. Chem.* 33 (31): 2451–2468.

148 Soteras Gutierrez, I., Lin, F.Y., Vanommeslaeghe, K. et al. (2016). Parametrization of halogen bonds in the charmm general force field: improved treatment of ligand-protein interactions. *Bioorg. Med. Chem.* 24 (20): 4812–4825.

149 Guvench, O., Hatcher, E., Venable, R.M. et al. (2009). Charmm additive all-atom force field for glycosidic linkages between hexopyranoses. *J. Chem. Theory Comput.* 5 (9): 2353–2370.

150 Klauda, J.B., Venable, R.M., Freites, J.A. et al. (2010). Update of the charmm all-atom additive force field for lipids: validation on six lipid types. *J. Phys. Chem. B* 114 (23): 7830–7843.

151 Raman, E.P., Guvench, O., and MacKerell, A.D. (2010). Charmm additive all-atom force field for glycosidic linkages in carbohydrates involving furanoses. *J. Phys. Chem. B* 114 (40): 12981–12994.

152 Denning, E.J., Priyakumar, U.D., Nilsson, L., and Mackerell, A.D. Jr., (2011). Impact of 2'-hydroxyl sampling on the conformational properties of Rna: update of the charmm all-atom additive force field for RNA. *J. Comput. Chem.* 32 (9): 1929–1943.

153 Guvench, O., Mallajosyula, S.S., Raman, E.P. et al. (2011). Charmm additive all-atom force field for carbohydrate derivatives and its utility in polysaccharide and carbohydrate–protein modeling. *J. Chem. Theory Comput.* 7 (10): 3162–3180.

154 Best, R.B., Zhu, X., Shim, J. et al. (2012). Optimization of the additive charmm all-atom protein force field targeting improved sampling of the backbone Φ, Ψ and side-chain X1 and X2 dihedral angles. *J. Chem. Theory Comput.* 8 (9): 3257–3273.

155 Hart, K., Foloppe, N., Baker, C.M. et al. (2012). Optimization of the charmm additive force field for DNA: improved treatment of the Bi/Bii conformational equilibrium. *J. Chem. Theory Comput.* 8 (1): 348–362.

156 Mallajosyula, S.S., Guvench, O., Hatcher, E., and MacKerell, A.D. (2012). Charmm additive all-atom force field for phosphate and sulfate linked to carbohydrates. *J. Chem. Theory Comput.* 8 (2): 759–776.

157 Huang, J., Rauscher, S., Nawrocki, G. et al. (2017). Charmm36m: an improved force field for folded and intrinsically disordered proteins. *Nat. Methods* 14 (1): 71–73.

158 Durell, S.R., Brooks, B.R., and Ben-Naim, A. (1994). Solvent-induced forces between two hydrophilic groups. *J. Phys. Chem.* 98 (8): 2198–2202.

159 Bujacz, A. (2012). Structures of bovine, equine and leporine serum albumin. *Acta Crystallogr. D Biol. Crystallogr.* 68 (Pt 10): 1278–1289.

160 Scapin, G., Yang, X., Prosise, W.W. et al. (2015). Structure of full-length human anti-Pd1 therapeutic Igg4 antibody pembrolizumab. *Nat. Struct. Mol. Biol.* 22 (12): 953–958.

161 Venturini, D., de Souza, A.R., Caracelli, I. et al. (2017). Induction of axial chirality in divanillin by interaction with bovine serum albumin. *PLoS ONE* 12 (6): e0178597.

162 Papadopoulou, A., Green, R.J., and Frazier, R.A. (2005). Interaction of flavonoids with bovine serum albumin: a fluorescence quenching study. *J. Agric. Food Chem.* 53 (1): 158–163.

163 Zhao, H., Ge, M., Zhang, Z. et al. (2006). Spectroscopic studies on the interaction between riboflavin and albumins. *Spectrochim. Acta A Mol. Biomol. Spectrosc.* 65 (3): 811–817.

164 Cheng, Z. and Zhang, Y. (2008). Spectroscopic investigation on the interaction of salidroside with bovine serum albumin. *J. Mol. Struct.* 889 (1): 20–27.

165 Meti, M.D., Nandibewoor, S.T., Joshi, S.D. et al. (2015). Multi-spectroscopic investigation of the binding interaction of fosfomycin with bovine serum albumin. *J. Pharm. Anal.* 5 (4): 249–255.

166 Sekula, B., Zielinski, K., and Bujacz, A. (2013). Crystallographic studies of the complexes of bovine and equine serum albumin with 3,5-diiodosalicylic acid. *Int. J. Biol. Macromol.* 60: 316–324.

167 Bujacz, A., Zielinski, K., and Sekula, B. (2014). Structural studies of bovine, equine, and leporine serum albumin complexes with naproxen. *Proteins* 82 (9): 2199–2208.

168 Castagna, R., Donini, S., Colnago, P. et al. (2019). Biohybrid electrospun membrane for the filtration of ketoprofen drug from water. *ACS Omega* 4 (8): 13270–13278.

169 Karush, F. (1950). Heterogeneity of the binding sites of bovine serum albumin1. *J. Am. Chem. Soc.* 72 (6): 2705–2713.

170 Fasano, M., Curry, S., Terreno, E. et al. (2005). The extraordinary ligand binding properties of human serum albumin. *IUBMB Life* 57 (12): 787–796.

171 Velez Rueda, A.J., Benítez, G.I., Sommese, L.M. et al. (2022). Structural and evolutionary analysis unveil functional adaptations in the promiscuous behavior of serum albumins. *Biochimie* 197: 113–120.

172 Katchalski-Katzir, E., Shariv, I., Eisenstein, M. et al. (1992). Molecular surface recognition: determination of geometric fit between proteins and their ligands by correlation techniques. *Proc. Natl. Acad. Sci. U. S. A.* 89 (6): 2195–2199.

173 Ameseder, F., Biehl, R., Holderer, O. et al. (2019). Localised contacts lead to nanosecond hinge motions in dimeric bovine serum albumin. *Phys. Chem. Chem. Phys.* 21 (34): 18477–18485.

174 Sundaramurthi, P., Chadwick, S., and Narasimhan, C. (2020). Physicochemical stability of pembrolizumab admixture solution in normal saline intravenous infusion bag. *J. Oncol. Pharm. Pract.* 26 (3): 641–646.

Part II

Quantum Mechanics Application for Drug Discovery

6

QM/MM for Structure-Based Drug Design: Techniques and Applications

Marc W. van der Kamp and Jaida Begum

University of Bristol, School of Biochemistry, Cantock's Close, Bristol BS8 1TS, United Kingdom

6.1 Introduction

Hybrid quantum mechanics/molecular mechanics (QM/MM) modeling is a multiscale modeling concept whereby a molecular system is divided into two different regions. One (typically small) "active" region is described using a quantum mechanical method, so that electronic changes can be captured, including polarization, charge transfer, and bond rearrangement. Another region, typically large and surrounding the smaller QM region, is treated using an MM force field. This allows a computationally efficient but detailed description of, for example, biomolecular macromolecules (e.g. enzymes/receptors and the surrounding solvent) with their small molecule ligands (e.g. substrates, inhibitors). QM/MM approaches, therefore, have many potential applications in computational biomolecular science, including those relevant to drug design, such as the calculation of spectroscopic properties [1] and pK_a values [2, 3], scoring and pose refinement in small molecule docking (see Section 6.2.2), and prediction of ligand binding affinities [4]. In particular, however, QM/MM has emerged as a reliable computational method to investigate chemical reactions in enzymes [5, 6]. Indeed, in the seminal work from 1976 that introduced the QM/MM concept [7], Warshel and Levitt applied it to perform energy calculations on an enzyme reaction mechanism (hen egg white lysozyme). Several years later, the QM/MM approach was then used for geometry optimization [8], and soon after also for QM/MM molecular dynamics simulation [9], and applied to study enzyme reactions [10].

Since those early days, QM/MM calculations have become increasingly popular, especially for applications on enzymes and indeed simulations of enzyme reactions [5, 11], with over 100 research articles that detail the use of QM/MM methods for enzymes appearing each year since 2007. Such studies have addressed fundamental general questions regarding enzyme catalysis, questions regarding the precise nature of enzymatic reaction mechanisms, and revealed differences in the efficiency of catalysis between enzyme variants (e.g. related to enzyme design), among other insights. Naturally, when such studies are performed on enzymes that are drug

Computational Drug Discovery: Methods and Applications, First Edition.
Edited by Vasanthanathan Poongavanam and Vijayan Ramaswamy.
© 2024 WILEY-VCH GmbH. Published 2024 by WILEY-VCH GmbH.

targets, the QM/MM studies can assist structure-based drug design by gaining detailed mechanistic knowledge (e.g. key catalytic interactions in active sites), especially for the transient chemical transition states, which can inspire the design of tight-binding ligands. In addition, QM/MM studies can also provide insights into drug activation or breakdown, enzyme-mediated adverse reactions, and the effectivity of so-called warheads for covalent inhibitors. Over the past decades, several reviews have highlighted the use of QM/MM in relation to drug design and development [12–16], including the use of QM/MM studies that aided the synthesis of new covalent inhibitors [17].

In this chapter, we first introduce the QM/MM approach, alongside QM/MM modeling methods that can be used for modeling protein–ligand interactions as well as reactions. Then, we highlight examples of where QM/MM studies have provided insights into existing covalent drugs as well as covalent inhibitors with potential as drugs in several important therapeutic areas, including cancer treatments based on tyrosine kinase inhibition, emerging resistance of bacteria against treatment with β-lactam antibiotics, and potential treatments with covalent inhibitors of viral infections such as SARS-CoV-2.

6.2 QM/MM Approaches

6.2.1 Combined Quantum Mechanical/Molecular Mechanical Energy Calculations

The principle of QM/MM approaches is to treat a small part of the system, the QM region (e.g. a drug and key surrounding residues in the binding site), with a quantum mechanical (QM) method (describing the electronic structure of molecules), and the rest of the system with a molecular mechanical (MM) method (where interactions between atoms are described using a simple potential energy function, a "force field," with no direct inclusion of electrons). The QM treatment then allows modeling of the electronic rearrangements, including polarization, charge transfer, and importantly, the breaking and making of chemical bonds. Simultaneous inclusion of the environment with a more computationally efficient MM treatment allows consideration of the influence of this environment on the QM region, e.g. including polarization and effects on the energetics of reactions. It is also possible to model the latter without directly representing the electronic structure, such as with the empirical valence bond (EVB) approach [18, 19] (which combines MM representations of different states along a reaction) and is sometimes also described as a QM/MM approach, but will not be discussed here. There are several reviews where QM/MM methods as applied to enzymes have been described in detail (e.g. Ref. [6]). Here, we give a brief overview of the main approaches and considerations.

The most commonly employed QM/MM approach uses an additive scheme, where the total energy of the system is calculated as:

$$E = E_{QM}(\mathbf{R}_{QM}, \mathbf{R}_{MM}) + E_{QM/MM}(\mathbf{R}_{QM}, \mathbf{R}_{MM}) + E_{MM}(\mathbf{R}_{MM}) \qquad (6.1)$$

where E_{QM} is the energy of the QM subsystem $\boldsymbol{R}_{QM}$ polarized by the MM environment, E_{MM} is the MM force field energy for $\boldsymbol{R}_{MM}$ and $E_{QM/MM}$ is the interaction energy between the QM and MM parts. The first two terms in Eq. (6.1) depend on both the QM and MM parts. In the first term, the wave function (or electronic density in the case of density functional theory [DFT]) is obtained in the presence of MM point charges. The second term includes the electrostatic force of the QM electron density that is acting on the MM atoms. Van der Waals interactions between the QM and MM regions are typically modeled through the use of a (standard) Lennard-Jones function, with QM atoms being assigned MM parameters for this purpose. When using polarizable force fields, it is further possible to allow polarization of the MM environment by the QM region [20], although this is not yet commonly employed. The additive approach was implemented in the CHARMM simulation package already in the late '80s [21], and in Amber in the late '00s [22], including "internal" support for several popular semi-empirical QM methods, as well as interfaces to existing QM programs. Additive QM/MM calculations can also be performed with Schrödinger's QSite program [23] (using Jaguar for QM calculations) [24]. More recently, implementations of additive QM/MM have also been developed for nanoscale molecular dynamics (NAMD) [25] and Gromacs [26]. In addition, many QM programs include options to perform QM/MM calculations with the additive approach, such as NWChem [27], ORCA [28], and CP2K [29], and independent interface programs that can combine various QM and MM software packages also exist, such as ChemShell [30, 31], PUPIL [32], and QMCube (QM^3) [33].

An alternative to the additive QM/MM approach is a subtractive scheme, where the total energy of the system is described using:

$$E = E_{QM}(\boldsymbol{R}_{QM}) + E_{MM}(\boldsymbol{R}_{QM}, \boldsymbol{R}_{MM}) - E_{MM}(\boldsymbol{R}_{QM}) \tag{6.2}$$

where $E_{MM}(\boldsymbol{R}_{QM}, \boldsymbol{R}_{MM})$ indicates the energy of the total system at the MM level. Note that here, the QM region is also described by an MM force field, and thus the choice of suitable MM parameters (including for different states in a reaction) should be an important consideration. This scheme is employed by the ONIOM method [34] implemented in Gaussian, for example. Details and applications of this subtractive approach, including the important difference between so-called "mechanical embedding" and "electrostatic embedding," are discussed by Ryde [35]. Notably, "electrostatic embedding," where the QM system is polarized by the MM partial charges, is important to include the influence of the biomolecular environment (and therefore for calculations of reasonable accuracy, unless very large QM regions are used). As indicated above, implementations of the additive QM/MM scheme essentially always include such "electrostatic embedding."

With an expression for the energy of a QM/MM system (as in Eq. 6.1 or Eq. 6.2), it is now possible to calculate gradients and forces for all atoms, and thus perform energy minimization and molecular dynamics simulation (using the Born-Oppenheimer approximation, i.e. movement of the QM atoms is propagated classically). Thus, many "standard" simulation techniques now become available, with the benefit that covalent bond making and breaking, alongside other electronic effects, can be included in the QM region.

6.2.2 QM/MM Methods for the Evaluation of Non-Covalent Inhibitor Binding

The potential benefits of more accurate inclusion of ligand descriptions and, in particular, polarization effects in small molecule-target binding have led to the development of QM/MM approaches to help predict binding conformation, evaluate binding energies, and even correct calculations of drug binding kinetics based on MM molecular dynamics (MD) simulation [36]. For small molecule docking, QM/MM methods can aid in the scoring of previously generated poses, refinement of such poses, as well as pose generation itself. For recovering native poses from co-crystal structures, Cho et al. demonstrated that (re)calculating ligand charges based on QM (B3LYP/6-31G*) in the MM protein environment can lead to significantly higher accuracy in predicted binding modes [37]. More recently, a similar approach to include QM/MM calculations in the docking protocol, with both the ligand and residues in the binding site treated QM, was shown to improve results in tests on GPCR co-crystal structures [38]. Ligand charges derived from QM calculations in the (MM) protein environment were also shown to improve both docking geometry and scoring for protein–ligand complexes that involve halogen bonding [39]. Further, a version of the attracting cavities docking approach with on-the-fly QM/MM calculations (using the efficient SCC-DFTB QM method) showed clear benefits for metalloenzymes: significant improvements were obtained for zinc metalloproteins and heme proteins (the latter with covalent binding between ligand and heme iron) [40]. Docking pose refinement (after standard generation) using short MM MD simulations and QM/MM minimization was also shown to improve pose accuracy and ranking, including for the heme protein cytochrome c peroxidase [41].

For a detailed evaluation of relative binding energies in small molecule series (e.g. in lead optimization), methods involving conformational sampling of the protein–ligand system are often used. In alchemical free energy perturbation (FEP) simulations, the application of QM/MM is not straightforward due to the nonphysical intermediate states involved with, e.g. insertion or deletion of atoms. However, applying an MM to QM/MM free energy correction to the end states (sometimes referred to as a "reference potential" or "book-ending" approach) is perfectly feasible, and methods continue to be actively developed and tested [42–46]. Early tests based on a structurally diverse set of fructose 1,6-bisphosphatase inhibitors indicate that such QM/MM-based free-energy perturbation approaches have potential for structure-based drug design [47].

The popular MM-PB(GB)SA end-point method is commonly applied based on molecular dynamics sampling of protein–ligand complexes. It provides an intermediate in terms of accuracy and computational resource between docking-based scoring and alchemical FEP [48]. After the (MM-based) sampling, binding energy calculations can make use of QM/MM, resulting in a QM/MM-PB(GB)SA approach, which can improve the resulting binding affinity ranking. For example, for ranking of inhibitors of polo-like kinase 1, it was found that when ligands were ranked using QM/MM-GBSA using semi-empirical QM methods instead of MM-GBSA, the ranking of congeneric compounds was significantly improved [49]. An alternative to QM/MM-GBSA or -PBSA, with a similar philosophy of incorporating polarization, etc., is the application of QM calculations in conjunction with a continuum solvent model following MM MD sampling. Examples include the so-called MM/QM-COSMO method [50], which was applied to obtain insights into phosphopeptide binding to the BRCA1 C-terminal domain [51]. A similar approach, named SQM/COSMO [52], applies semi-empirical QM calculations and was recently shown to be sufficiently computationally efficient to be used in virtual screening, improving on other scoring functions [53].

The above-mentioned QM polarized ligand docking approach originating from Cho et al. [37] has been implemented in Schrödinger's Glide software as QM-polarized ligand docking (QPLD) [54], which has helped popularize this approach. The combination of QPLD with QM/MM-PBSA scoring (without further MD-based conformational sampling) has been shown to be advantageous, for example, in correctly ranking and obtaining insight into halogen-substituted ligands targeting phosphorylase kinases [55]. QM/MM in rescoring can be advantageous also when standard (MM-based) docking (and refinement) is performed. For example, for the identification of Myt1 kinase inhibitors, QM/MM-GBSA rescoring was shown to improve upon docking scores or MM-GBSA rescoring in the classification of active and inactive inhibitors [56]. The predictive power of such rescoring was indicated by testing five compounds predicted as active, of which three showed significant inhibition of Myt1. Further examples demonstrate the direct use of (QM/MM) pose refinement and QM/MM-PBSA scoring in guiding the synthesis of inhibitors of human O-GlcNAc hydrolase [57] (a target for Alzheimer's disease) and glycogen phosphorylase [58] (a target for type 2 diabetes).

As described in this section, the ability of QM/MM methods to capture polarization (and charge transfer) effects in protein–ligand interactions is often beneficial for the evaluation of protein–ligand binding. It is possible that the continuing active development of polarizable force fields [59], including automation of small molecule parameterization for such force fields [60, 61], will yield similar fundamental advantages as QM/MM or QM-based binding energy evaluation approaches [62], i.e. a more accurate description of polarization, at a significantly lower computational cost. Nevertheless, QM(/MM) approaches may still be required when electronically complex phenomena, such as sigma holes, cation-π interactions, or metal coordination, are important [63], as these remain challenging also for polarizable force field approaches.

6.2.3 QM/MM Reaction Modeling

Modeling of reactions in biomolecular systems can yield important structural insights (e.g. transition state structures, which can provide inspiration for effective enzyme inhibitors) and energetic insights (e.g. related to differences in reactivity within a series of covalent drug leads). For QM/MM modeling of reactions in biomolecular systems, typically either a QM/MM potential energy profile or minimum energy pathway (MEP) is (first) obtained, or QM/MM molecular dynamics simulations are performed, biased along the reaction path, to obtain free energy profiles. Many techniques are available for both options. In this section, we will briefly describe some of the more commonly applied techniques for biomolecular systems and related considerations.

As for small molecule systems, a potential energy profile (or MEP) can be obtained by minimizing the energy of the system at several points along a reaction coordinate (or "collective variable"). This reaction coordinate could be based directly on (a combination of) geometric features describing bond making and breaking, or methods that optimize an MEP based on providing initial (reactant) and final (product) states can be used. The latter includes Nudged Elastic Band optimization (with the Climbing Image variant being able to provide a "true" transition state, the saddle point on the potential energy surface), which has been adapted specifically for QM/MM [64, 65], and the similar Replica Path method [66]. (A brief overview of several reaction path methods and their application in biomolecules is included in Ref. [21].) Due to the large configuration space available in biomolecular systems, one should be aware of changes in conformations along the optimized reaction pathway, which may not be directly related to the reaction. For example, when relying on iterative minimizations along a geometric reaction coordinate, sudden "jumps" in the QM/MM potential energy can be caused by a small change (e.g. rotation of a water molecule or side chain) further away from the active site where bond changing takes place. To avoid this from happening, part of the MM region could be constrained, and minimizations backwards and forwards along the reaction may resolve such jumps. Notably, due to the many possible conformations available to biomolecular (e.g. drug-target) systems, a single optimized reaction path (or a single set of stationary points optimized along this path) may not be representative, or indeed not allow for a confident prediction of differences between related systems (e.g. different covalent drugs, enzyme variants). Indeed, many different starting conformations may need to be considered (and reaction energies exponentially averaged) to obtain converged reaction barrier energies [67]. The direct output from optimized reaction paths, alongside structures, will be the QM/MM potential energies. Hybrid DFT functionals are suitable for high-accuracy structure optimization, which can then allow energies to be calculated with "gold standard" wavefunction methods, such as CCSD(T) or variants thereof. Performing such single-point energy calculations is also popular for correcting energy profiles obtained with more approximate methods (e.g. semi-empirical QM corrected by hybrid DFT). To estimate *free* energy profiles from minimum energy paths (beyond estimating entropic contributions through frequency calculations of optimized stationary points) one can use QM/MM-FEP approaches, which incorporate conformational sampling of the MM environment. These range from approximate to more sophisticated, e.g. from keeping the QM region completely fixed, to re-introducing some sampling in the QM region [68, 69], to re-optimizing the reaction path [70].

With QM/MM molecular dynamics simulations, conformations can be directly sampled along the reaction path by using enhanced sampling techniques (without the need of calculating a potential energy profile). Due to the need to compute forces every femtosecond, these simulations are significantly more computationally demanding than optimizing potential energy profiles and thus may require the use of semi-empirical QM methods or more limited sampling (e.g. with DFT functionals). A commonly used enhanced sampling approach is umbrella sampling, where (mild) restraints are applied to particular reaction coordinate values (or "windows") along the reaction. The sampling can then be subsequently unbiased using statistical methods to obtain the potential of mean force (or free energy profile). Popular methods for this include the well-known Weighted Histogram Analysis Method [71], Umbrella Integration [72], and the more recently developed Dynamic Histogram Analysis Method [73, 74]. The position along the reaction coordinate can be determined by sampling along one or two reaction coordinates (generating 1D or 2D free energy profiles along these) or by approaches that "dynamically" optimize the minimum free energy path, such as finite-temperature string methods [75–78]. The advantage of the latter approach is that, in principle, a minimum free energy path can be determined in the space of a large number of interdependent collective variables (distances, angles, etc.). A path collective variable combining several relevant degrees of freedom for the reaction process can then be used, which allows multi-event mechanisms with (partially) concerted pathways to be captured accurately. Other enhanced sampling approaches commonly used for QM/MM reaction simulations are steered MD [79] and metadynamics [80]. The latter is also popular in combination with Car-Parinello MD (CP-MD) in a QM/MM scheme for enzyme reaction simulations. CP-MD provides an alternative to the common "Born-Oppenheimer MD" approach, where the QM energy and forces come from a converged SCF calculation in each step. Instead, in CP-MD, wave functions are treated as fictitious dynamic variables and follow the motion of the nuclei "on the fly." In the latter, a plane-wave basis is (almost always) used together with a full DFT functional (e.g. PBE, BLYP), which may limit the accuracy. QM/MM modeling of enzyme reactions with CP-MD was established in the 2000s, and an efficient implementation (with GROMACS for the MM part) is now available via MiMiC [81].

Once free energy profiles of enzyme reactions are obtained, the resulting reaction barrier can, in principle, be directly compared to an (apparent) activation free energy obtained from an experimental apparent first-order reaction rate constant (k_{cat}, or similarly, k_{off}/k_{on} rates if these are dominated by the chemical reaction studied), using transition state theory (TST) [82]. Most approaches for QM/MM reaction modeling will have some shortcomings to arrive at highly accurate reaction barriers, such as limited conformational sampling or limited accuracy of the QM method employed. One strategy is to model the QM region using cheap, approximate methods (e.g. semi-empirical QM) during conformational sampling, and then correct the resulting barrier through the use of high-level QM(/MM) calculations. For such an approach, one should ensure that the more approximate methods sample the same mechanism [83, 84]. In favorable cases, QM/MM simulations can thereby calculate enzyme reaction barriers in quantitative agreement with experimental kinetics [85]. It is important to note, however, that qualitative agreement for the reaction barrier is often sufficient to distinguish between alternative mechanisms, determine differences in reactivity for drug series, or assess the effect of mutations; differences between barriers for comparable systems are often still captured accurately (see, e.g. Section 6.3.2).

6.3 Applications of QM/MM for Covalent Drug Design and Evaluation

Historically, the pharmaceutical industry was reluctant to focus its efforts on covalent drug design [86]. Given that toxicity (together with efficacy) is the leading cause of drug attrition in clinical development [87], covalent inhibitors were disfavored due to safety concerns: covalent binding may occur on off-target proteins. However, attitudes toward covalent drugs have changed in the past 3 decades, and the FDA has approved over 50 covalent drugs to date [88, 89]. The majority of these are treatments for cancer as well as bacterial and viral infections [89]. In this section, we discuss examples of the use of QM/MM simulations for obtaining a detailed understanding of key targets and their covalent inhibitors, which can be used in the development of covalent drugs. These calculations can provide information on the non-covalent interactions as well as the formation and breakdown of covalent bonds between the drug and its target. Many covalent inhibitors have two distinct motifs: one for providing specific recognition of the target through non-covalent interactions and one that provides reactivity to form the covalent link, commonly referred to as the "warhead." Related to this, the binding process can be described using a two-step model, where first an enzyme-inhibitor non-covalent complex is established, before the covalent link is formed (Scheme 6.1). The chemical step leading to covalent inhibition can be described either as reversible or irreversible, depending on the magnitude of rate k_{-2}, i.e., the free energy barrier for bond breaking. To aid the design of covalent inhibitors, covalent docking is already a recognized tool [89], useful, for example, as a high-throughput screening method and to suggest binding modes (see Chapter 25). For more detailed insights into the energetics and the process of covalent drug binding, atomistic simulations can be used. Whereas the energetics of non-covalent complex formation can be captured by MM (or QM/MM) approaches, differences in the rate constants for covalent complex formation will typically require QM/MM reaction simulations.

$$E + I \underset{k_{-1}}{\overset{k_1}{\rightleftharpoons}} E \cdot I \underset{k_{-2}}{\overset{k_2}{\rightleftharpoons}} E - I$$

Scheme 6.1 General two-step model for covalent inhibition.
E = enzyme, I = covalent inhibitor, E · I = non-covalent complex, E−I = covalent complex. In addition to the non-covalent steps (i.e. k_1 and k_{-1}), the rate constant k_2 is relevant for irreversible covalent inhibition, whereas the balance between k_2 and k_{-2} determines reversible covalent inhibition.

6.3.1 Covalent Tyrosine Kinase Inhibitors for Cancer Treatment

Receptor tyrosine kinases (RTKs) are key regulators of cellular processes and can thereby govern cell growth and metastasis in cancer [90]. It is therefore no surprise that inhibitors of RTKs have been actively developed over the past decades. Several covalent inhibitors, binding to a cysteine in the RTK active site, have been approved as drugs and are in clinical use for the epidermal growth factor receptor (EGFR, e.g. afatinib, neratinib, osimertinib) and Bruton tyrosine kinase (BTK, e.g. acalabrutinib, ibrutinib). In this section, we discuss how QM/MM modeling has provided insights relevant for covalent drug design and understanding emerging resistance for these two RTKs, after introducing some of the biological context.

The binding of cognate ligands at the extracellular site of the transmembrane protein EGFR can trigger signaling networks leading to cellular proliferation, differentiation, and survival [91]. Mutations in EGFR, which can cause it to be in a prolonged state of activation [92], are associated with various types of cancer, including non-small-cell lung cancer (NSCLC), which is responsible for 85% of all lung cancers [93]. Interfering with EGFR signaling using small-molecule inhibitors is therefore an appealing strategy for oncogenic treatment. First-generation ATP-competitive inhibitors for EGFR, such as gefitinib and erlotinib, were developed for NSCLC. While initial treatments were positively received, secondary drug resistance mutations, such as T790M, significantly reduced the clinical efficacy of these reversible non-covalent drugs [94, 95]. This led to the development of second-generation inhibitors such as afatinib and dacomitinib to inhibit EGFR T790M [96]. These compounds contain a distinctive electrophilic acrylamide warhead, which reacts with the nucleophilic Cys797 in the ATP-binding site to form a covalent adduct [97, 98]. Although these drugs can arrest EGFR T790M activity, selectivity issues arise as reactions also occur with Cys797 in wild type (WT) EGFR, leading to unwanted side effects.

Soon after detailed kinetic investigations of several covalent EGFR inhibitors were reported [97], Capoferri et al. [99] investigated the reaction mechanism for a prototypical irreversible covalent inhibitor with an acrylamide warhead against WT EGFR. In their approach, QM/MM (SCC-DFTB/ff99SB) MD umbrella sampling with a path collective variable was used to simulate the mechanism of covalent binding to the targeted Cys797. The simulations revealed that Asp800 likely acts as a general acid/base catalyst. Once it deprotonates Cys797, a concerted mechanism occurs in which the Cα of the acrylamide inhibitor is protonated by Asp800, with concomitant formation of the saturated β-substituted product. The covalently bound product was calculated to be ~12 kcal/mol more stable than the non-covalent complex, emphasizing that the binding is spontaneous and exergonic, consistent with experimental findings [97, 100, 101]. The semi-empirical SCC-DFTB method, known to typically underestimate reaction barriers [102–104], predicted a reaction barrier of 14.6 kcal/mol, lower than the barrier of ~20 kcal/mol derived from the experimental rate [97]. The authors further concluded that desolvation of the Cys797 thiolate is key, suggesting that intrinsic acrylamide reactivity should only have a minor impact on the potency of inhibitors. Overall, this study represents the likely mechanism and energetics involved in second-generation EGFR inhibitors, which are strongly exergonic upon binding (i.e. k_2 is greater than k_{-2}, see Scheme 6.1) [105]. The issue with such reactive, irreversible inhibitors is that specificity for the cancer-related EGFR variant is often low, leading to toxicity. For this reason, third-generation EGFR inhibitors use a more weakly reactive warhead, leading to a reversible process (k_{-2} is not negligible). Indeed, such third-generation inhibitors as osimertinib, lazertinib, and rociletinib do essentially not target WT EGFR [106, 107]. Unfortunately, however, further EGFR mutations can confer resistance to these third-generation inhibitors. One such mutation is L718Q [108], but the precise molecular mechanism by which glutamine at position 718 reduces the efficacy of osimertinib, for example, was not known. This motivated the work of Callegari et al. [105], where a combination of QM/MM reaction simulations, binding affinity calculations (using the WaterSwap method [109]), and MM MD simulations were used to investigate possible effects on the Cys797 pK_a, the reaction mechanism, binding affinity, and non-covalent conformational dynamics of osimertinib. Throughout, comparison between EGFR T790M and EGFR T790M/L718Q was performed to identify which properties were significantly affected by the L718Q mutation (Figure 6.1). Initially, both approximate pK_a calculations on MM MD

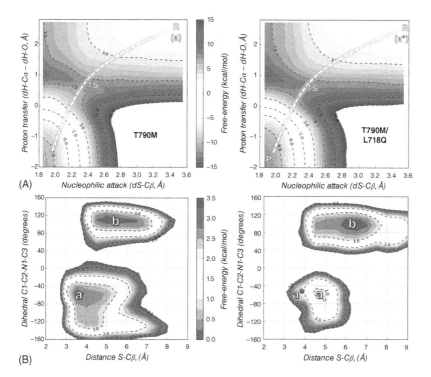

Figure 6.1 Assessment of osimertinib interactions with EGFR T790M (*left*) and EGFR T790M/L718Q (*right*). (A) Free energy surfaces based on 2D QM/MM (SCC-DFTB/ff99SB) umbrella sampling simulations indicate similar reaction barriers between the non-covalent reactant (R) and the covalently bound (Cys797 alkylated) state (P). The proton transfer reaction coordinate is defined as $d(H_{Asp800}-C\alpha_{acrylamide}) - d(H_{Asp800}-O_{Asp800})$ and nucleophilic attack as $d(S_{Cys797}-C\beta_{acrylamide})$. (B) 2D free energy surfaces in the space of $d(S_{Cys797}-C\beta_{acrylamide})$ and the acrylamide $C_1-C_2-N_1-C_3$ dihedral angle. Free energies are calculated from the frequency distribution of conformations observed in four independent 300 ns MM MD simulations for each complex. For EGFR T790M/L718Q, *a** represents the region of reactive conformations of osimertinib (which approximately corresponds to basin *a* for EGFR T790M). Source: Reproduced with permission of Callegari et al. [105]/ Royal Society of Chemistry / Public Domain CC BY 3.0.

snapshots and QM/MM (SCC-DFTB/ff99SB) umbrella sampling simulations of the Cys797–S$^-$/Asp800–COOH to Cys797–SH/Asp800–COO$^-$ reaction were performed to ascertain that the resistance to osimertinib is not due to an increase in the Cys797 pK_a. Equally, the Cys-alkylation reaction energetics are not altered significantly, as indicated by the energy profiles obtained through 2D QM/MM umbrella sampling (Figure 6.1A and B). The binding affinity of osimertinib prior to the reaction was also not affected. However, MM MD simulations (4x 300 ns) revealed that there is a significant difference in the preferred binding conformation of osimertinib in the EGFR active site. In EGFR T790M, configurations in which the nucleophile (Cys797 S) and electrophile (acrylamide Cβ) are within reacting distance occur for >20% of the simulation time. This reduces to only 0.7% for EGFR T790M/L718Q, representing a free energy difference of ~3 kcal/mol (Figure 6.1A and B). This is primarily due to the stabilization of an alternative conformational "basin" (b), through hydrogen bonding of the Gln718 side-chain -NH$_2$ moiety with the acrylamide carbonyl. The combination of methods thus allows a detailed explanation of how the L718Q mutation leads to osimertinib resistance. This highlights how detailed multiscale modeling can help rationalize the effects of a mutation in the binding site on interactions with a covalent drug.

Not only can QM/MM simulations (combined with other MM modeling techniques) be used to obtain detailed insight into the origins of resistance to existing drugs in EGFR variants, but they can also contribute to the design of new covalent drugs for EGFR. For example, Castelli et al. used QM/MM simulations in their detailed study comparing heteroarylthioacetamide derivates as EGFR inhibitors [110]. The main aim was to select alternative (weaker) warheads than acrylamide, which may help design selective inhibitors. To compare warhead effectivity, QM/MM simulations were performed to compare a chloroacetamide and two different 2-(imidazol-2-ylthio)acetamide derivatives as warheads. QM/MM (PDDG-PM3/ff99SB) steered MD was used to obtain approximate free energy profiles for the Cys797 alkylation reaction, with Cys797 and Asp800 (alongside the inhibitor) in the QM region. Initial simulations indicated that protonation of the imidazole ring, which is in close proximity to Asp800, is required to form a stable Cys797 alkylation product. The subsequent free energy profiles (obtained using the Jarzynski equality) suggested that only one of the two 2-(imidazol-2-ylthio)acetamide derivatives, without N-methylation on the imidazole, led to an energetically favorable reaction (similar to that for the chloroacetamide warhead). The simulations were thus used to prioritize the synthesis and testing of an EGFR-targeting compound with this warhead. This compound showed promising results in several different assays (with negligible reactivity with cysteine in solution), indicating that the imidazol-2-ylthioacetamide warhead could be used for more selective targeting of cysteines in kinases.

Bruton's tyrosine kinase (BTK) is another RTK that plays an essential role in multiple signaling pathways [111]. In the B-cell receptor signal transduction pathway, for which BTK's function is best understood [112], it is responsible for normal B-cell survival and proliferation. Downregulation of BTK activity causes immunodeficiency diseases [113], whereas overexpression of BTK leads to autoimmune disorders [114] and various blood cancers [115]. Targeting BTK using small molecule inhibitors is therefore a therapeutic strategy for treating pathologies involved in the upregulation of BTK [116]. The first BTK inhibitor developed for the treatment of multiple B-cell cancers was ibrutinib [117]. Ibrutinib irreversibly inactivates BTK through its acrylamide warhead, which is responsible for binding covalently to Cys481 in the active site. Unlike EGFR, in which an aspartate base activates the nucleophilic cysteine residue [99], BTK does not have a suitable residue to act as a base in close proximity to the targeted cysteine. Hence, the mechanism of covalent binding is different, but not known, which prompted Voice et al. [118] to investigate this using QM/MM simulations. The semi-empirical DFTB3 method was used for the QM region, after benchmarking indicated that this method can predict reaction pathways for thiol addition mechanisms that are structurally in close agreement with higher-level methods such as ωB97-XD and MP2. QM/MM MD umbrella sampling at the DFTB3/ff14SB level was then used to investigate several possible covalent bond formation mechanisms of ibrutinib with BTK. This indicated that the most plausible mechanism, in good agreement with experimental kinetics [119], proceeds through three key steps: (i) activation of the nucleophilic Cys481 by the carbonyl oxygen atom of the Michael acceptor, (ii) enol-intermediate formation as a result of the nucleophilic attack of the cysteine thiolate onto the electrophilic warhead, and (iii) solvent-assisted tautomerization from the enol to the covalently bound keto product. The final step was indicated to be rate-limiting in BTK, with a water molecule present for solvent-assisted tautomerization. The reaction free energy of ~ -37 kcal/mol is consistent with the irreversible inhibition of BTK by ibrutinib. The authors further note that changing the heterocyclic core and/or warhead should not improve the covalent reactivity (i.e. k_{inact} rates). Voice et al. [118] thus suggest that an alternative strategy for tuning reactivity and increasing specificity could be to design inhibitors that modulate the Cys481 pK_a and/or its conformational behavior. This emphasizes the importance of considering the influence of the protein environment, as this may govern the reactivity of inhibitors. Indeed, Voice et al. previously showed that although ligand-only metrics, such as predicted proton affinity and reaction energies, work well for predicting the reactivity of small reactive fragments (e.g. acrylamide warheads), they fail for larger drug-like compounds [120]. This was attributed to the lack of such metrics to incorporate binding conformations, solvation, and intermolecular interactions.

Awoonor-Williams and Rowley [121] characterized both the non-covalent and covalent components of the covalent binding of a t-butyl inhibitor with a cyanoacrylamide warhead to BTK, using a combination of methods. First, they used constant-pH MD to predict the pK_a of Cys481, which defines the cost of thiolate formation (avoiding having to model a specific deprotonation mechanism). Second, absolute binding free energies for the non-covalent interactions were calculated using alchemical free-energy perturbation at the MM level, indicating that Van der Waals dispersion plays an important role. Conformational energies were also factored in, as the ligand can adopt multiple states in the bulk solution compared to when it is bound to the enzyme. Third, covalent bond formation was simulated using QM/MM MD umbrella sampling (ωB97X-D3BJ/def2-TZVP for the

QM region, CHARMM36 for MM), indicating an activation barrier of 3.4 kcal/mol. Finally, subtractive ONIOM QM/MM calculations were used to estimate the relative stability of the covalent adduct compared to the non-covalent complex with Cys481 as a thiol (indicating a ~31 kcal/mol difference, consistent with irreversible inhibition). Similar to the work by Callegari et al. on EGFR [105], this is another example where combining several different MM and QM/MM calculations for the relevant processes involved in covalent-drug binding shows promise for their use in the evaluation and development of inhibitors.

6.3.2 Evaluation of Antibiotic Resistance Conferred by β-Lactamases

Antimicrobial resistance (AMR) in bacteria is one of the leading health threats of the twenty-first century, affecting healthcare systems and economies. It has been estimated that in 2019, ~1.3 million people died due to bacterial AMR (with ~5 million deaths associated with it) [122], a number that was previously projected to soar to 10 million per year by 2050 if current trends continue [122, 123]. Several of the leading pathogenic infections for these AMR-related deaths are caused by Gram-negative bacteria, which are usually treated with β-lactam antibiotics. These antibiotics exert their antibacterial activity by covalently inhibiting penicillin-binding proteins, which are crucial enzymes involved in bacterial cell wall maintenance. The main cause of resistance toward β-lactam drugs in Gram-negative bacteria, which are the most frequently prescribed antibiotics worldwide, is the presence of β-lactamases (BLs). BLs can be split into two main categories: Class A, C, and D are serine BLs (SBLs) and Class B are metallo-BLs. Collectively, they can efficiently catalyze the breakdown of all four classes of β-lactams: penicillins, cephalosporins, carbapenems, and monobactams. To combat β-lactamase-mediated antibiotic resistance, so-called combination therapy is now available for treating complicated infections. This involves co-administering a β-lactamase inhibitor along with a β-lactam antibiotic, thus restoring the antibacterial action. Examples of clinically approved β-lactamase inhibitors include classical inhibitors (with a β-lactam core) sulbactam, tazobactam, and clavulanic acid, and nonclassical inhibitors (without a β-lactam core) avibactam, relebactam, and vaborbactam.

QM/MM simulations have played an important role in establishing the detailed mechanisms of β-lactam antibiotic breakdown in the different β-lactamase classes. The mechanism for all SBLs, similar to serine hydrolases, involves an acylation step to form a covalently bound acyl-enzyme intermediate (where the β-lactam ring is broken), followed by deacylation to release the inactive antibiotic. The details of these steps differ between the classes, however. For the Class A enzymes, QM/MM simulations helped confirm that Glu166 (absent in Class C or D enzymes) is acting as the base in deacylation [124]. This was based on potential energy profiles obtained for the reaction between the archetypal TEM-1 β-lactamase and (benzyl)penicillin, using AM1/CHARMM22 energy minimizations corrected by B3LYP/6-31G+(d). In acylation, there are two feasible options in Class A SBLs for the base that abstracts the proton from Ser70 (which performs the nucleophilic attack to form the covalent acyl-enzyme intermediate): Glu166 and Lys73 (in its neutral state). Initially, findings based on potential energy surfaces from different groups and techniques led to different conclusions: either Glu166 acts as the base (AM1/CHARMM27 optimizations using the additive QM/MM scheme corrected by single point B3LYP/6-31G(d) calculations) [125, 126], or Lys73 is preferred as the base (ONIOM calculations with structure optimization at the HF/3-21G-OPLS-AA level and energy calculations at the MP2/6–31 + G(d)-OPLSA-AA level) [127], with a slightly higher barrier indicated for Glu166. Subsequent additive QM/MM calculations at higher levels of theory (structure optimization with B3LYP/6–31+G(d)-CHARMM27 and energies with SCS-MP2/aug-cc-pVTZ-CHARMM27) [128] indicated that Glu166 was the more likely base to remove the proton from Ser70 (via a bridging water molecule), with Lys73 stabilizing the transition state for this step. However, for the acylation reaction between KPC-2 and the β-lactamase inhibitor avibactam, several different recent QM/MM studies report that, in this case, Lys73 is the likely base [129–131]. It is therefore possible that the nature of the base in Class A SBL acylation differs depending on the enzyme/drug combination.

The mechanisms involved in Class B metallo-β-lactamases are still debated, complicated by the involvement of the (typically) two zinc ions in the active site and uncertainties regarding the detailed structures of on-pathway intermediates, but QM/MM studies have contributed to insights here also, further highlighting the mutual benefits between protein crystallography and QM/MM simulation (see, e.g. discussion in Ref. [5]). In Class C SBLs (less extensively studied than Class A and D SBLs), QM/MM studies on the AmpC enzyme from *Escherichia coli* with cephalotin (metadynamics using CPMD with PBE and plane-wave basis sets, with GROMOS as MM force field) indicate that Lys67 (equivalent to Lys73 in Class A SBLs) acts as the base in acylation instead of Tyr150, which was proposed to be involved in protonation of the β-lactam ring N atom [132]. The same authors also studied deacylation for this system and concluded that Tyr150 performs a similar role as Glu166 in Class A SBLs in this step (as established previously using a different QM/MM approach on the deacylation of penicillin by P99) [133]: it abstracts a proton from the deacylating water (DW) during its nucleophilic attack [134]. Comparison between their two studies led to the prediction that the acylation reaction is the rate-determining step in cephalothin hydrolysis for *E. coli* AmpC. Notably, Class D SBLs feature an unusual carboxylated lysine (Lys73), which was proposed to act as the base in both acylation and deacylation. QM/MM studies based on umbrella sampling simulation at the PM3-PDDG/ff12SB level [135], alongside the inability of Lys73 mutants to support deacylation, helped confirm this.

Establishing the mechanistic details of the reactions involved in β-lactamases is important to gain further understanding of why certain antibiotics are less effective than others and how β-lactamase-mediated resistance against antibiotics can arise. This knowledge can then aid in the further (re)design of antibiotics and β-lactamase inhibitors to help combat such resistance. Once mechanisms are known, QM/MM reaction simulation protocols that correctly capture the difference in β-lactamase efficiency between enzyme-antibiotic combinations can provide such insights in fine atomic detail. For the breakdown of the important carbapenem antibiotics by Class A SBLs, deacylation is expected to be rate-limiting [136].

Chudyk et al. therefore focused on this step and showed how for eight different Class A SBLs, QM/MM MD simulations can correctly distinguish between those that can efficiently break down meropenem and those that cannot [137]. These simulations were based on SCC-DFTB/ff12SB umbrella sampling simulations along two reaction coordinates, one describing the nucleophilic attack of the DW onto the acyl oxygen and another the proton transfer between Glu166 and the DW, thus arriving at the tetrahedral intermediate. Although the semi-empirical SCC-DFTB method (required for the extensive conformational sampling performed in this work) again underestimates the absolute barriers for this reaction, the difference in barriers between carbapenemases (KPC-2, SFC-1, NMC-A, and SME-1) and carbapenem-inhibited SBLs (TEM-1, SHV-1, BlaC, and CTX-M-16) was very clear, consistent with kinetic data. Subsequently, it was shown that sampling can be significantly reduced, by focusing umbrella sampling just on an approximate minimum free energy path and reducing simulation length. This led to a computationally efficient assay to assess the carbapenem hydrolysis efficiency of Class A SBL enzymes [138]. To establish the origins of the difference in carbapenem hydrolysis in these eight enzymes, Chudyk et al. later conducted a detailed analysis and performed further "computational experiments" by simulating the reaction (using the same QM/MM MD protocols) with specific restraints or mutations [139]. This indicated that efficiency for carbapenem hydrolysis by Class A SBLs is influenced by a range of factors, including optimal stabilization by the oxyanion hole, the presence or absence of the active site Cys69-Cys238 disulfide bridge, the orientation of the 6α-hydroxyethyl group of the carbapenem scaffold (including its interaction with Asn132), and the interaction of Asn170 with Glu166, the base in deacylation. Disrupting any of these factors away from their optimal values can lead to loss of carbapenemase activity, whereas the introduction of single individual factors (e.g. by conformational constraints or mutation) is not sufficient to introduce carbapenemase activity in carbapenem-inhibited Class A SBLs. The identified interactions could be exploited in the development of new β-lactam antibiotics able to evade resistance conferred by Class A carbapenemases. The transferability of the QM/MM MD umbrella sampling simulation protocols was further highlighted by their application to reaction simulations of KPC-2 and TEM-1 acyl enzymes formed by interaction with the classical covalent SBL inhibitor clavulanic acid [140]. This revealed that, of several possible adducts, the decarboxylated *trans*-enamine species is responsible for inhibition.

An alternative approach to using QM/MM MD reaction simulations and analysis to obtain insights into the breakdown efficiency is to obtain many different potential energy profiles and analyze their structural features to determine which acyl-enzyme conformations/interactions lead to efficient breakdown. The analysis of such features can be done using machine learning (ML) approaches, as was recently demonstrated for the deacylation of imipenem by the Class A GES-5 SBL [141], based on 500 semi-automatically generated pathways per reaction, obtained using DFTB3/CHARMM36 optimization and B3LYP-D3/6–31+G**-CHARMM36 energy calculations. Interpretation of the edge-conditioned graph convolutional neural network trained to predict the QM/MM barriers based on the initial acyl enzyme conformation, helped provide insights into the difference in deacylation efficiency between the Δ^2 and Δ^1 tautomers of the pyrroline ring. As established in the '80s [142, 143] (and assumed in other simulation studies), the Δ^1 tautomer is significantly more stable; hence, deacylation will occur with the Δ^2 tautomer. Among the features that determine the reactivity of Δ^2 tautomer acyl enzyme conformations, the importance of the 6α-hydroxyethyl orientation for efficient deacylation was again highlighted.

Aside from the Class A carbapenemases described above, another common route for carbapenem resistance is hydrolysis by Class D SBLs. Within this diverse class, OXA-48-like enzymes have emerged as a particular concern for clinical microbiologists [144]. These plasmid-borne SBLs, which are often difficult to detect, typically hydrolyze carbapenems fairly efficiently, while retaining activity against other antibiotic substrates (such as oxacillin) [145]. Differences in the efficiency of carbapenem hydrolysis do exist, however, with OXA-48 (and related variants) showing higher efficiency for hydrolyzing imipenem than meropenem and other 1β-methyl carbapenems (such as ertapenem and doripenem) [145]. This difference was investigated by Hirvonen et al. using MM MD simulations and QM/MM MD umbrella sampling [146]. MM MD conformational sampling (with ff14SB) of the OXA-48 acyl enzymes with imipenem and meropenem revealed that the 6α-hydroxyethyl moiety can adopt three distinct orientations. Subsequently, free energy profiles for the deacylation step (formation of the tetrahedral intermediate), known to be rate-limiting, were calculated using DFTB3/ff14SB umbrella sampling with each of the three distinct 6α-hydroxyethyl orientations, as well as different hydration of the carboxylate group attached to Lys73 (carboxy-Lys73) for each (Figure 6.2a). As expected (and indicated by the benchmarking calculations performed), reaction barriers are underestimated. However, the difference in reaction barriers between imipenem and meropenem ($\Delta\Delta^{\ddagger}G$) agrees very well with experimental kinetics, which indicate this should be between 1.4 and 3.5 kcal/mol. For the active site conformations with the lowest barriers, the difference is 2.2 kcal/mol (Figure 6.2a), rising to 2.8 kcal/mol when the lower likelihood of sampling the 6α-hydroxyethyl group in orientation I for imipenem (based on 5x 200 ns MM MD simulations) is considered. The barriers with different active site conformations demonstrate that increasing hydration around the carboxy-Lys73 base impairs deacylation: an additional water molecule next to the DW increases the barrier by at least 2 kcal/mol (Figure 6.2a). Further inspection of reaction simulations with the 6α-hydroxyethyl group in orientation I provides an explanation for the difference in deacylation efficiency between imipenem (lacking a 1β-methyl group) and meropenem. In imipenem deacylation, the DW donates hydrogen bonds to both carboxy-Lys73 and the 6α-hydroxyethyl moiety (Figure 6.2b). In meropenem, the latter interaction is reversed; the DW now accepts a hydrogen bond from the 6α-hydroxyethyl moiety instead. This difference, leading to a higher barrier as well as a change in the location of the transition state (Figure 6.2b), is likely due to a modification of the hydrogen bonding network around the 6α-hydroxyethyl, ultimately caused by the presence of the 1β-methyl group. Hirvonen et al. also used similar QM/MM (DFTB3/ff14SB) reaction simulations to compare the efficiency of

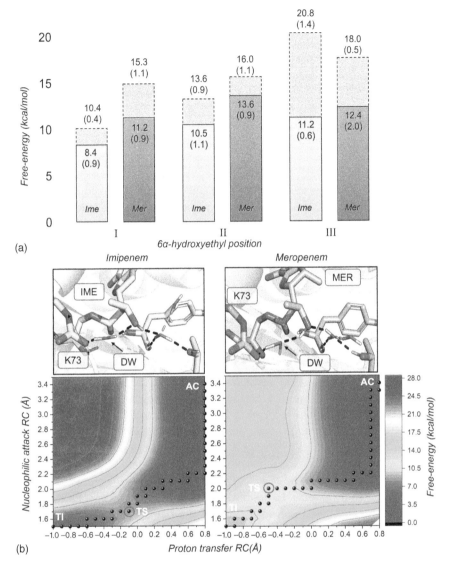

Figure 6.2 Dissection of the deacylation of imipenem and meropenem by the Class D β-lactamase OXA-48 using QM/MM (DFTB3/ff14SB) reaction simulations. (a) Free energy barriers obtained from 2D umbrella sampling for the three different 6α-hydroxyethyl orientations observed in MM MD. Each bar includes the barrier obtained with a single water molecule hydrogen bonded to the carboxy-Lys73 (lowest barrier, outlined as *solid black line*) or with two water molecules (highest barrier, outlined as *dashed lines*). Each barrier is derived from three individual umbrella sampling runs, with standard deviations in parenthesis. Ime = imipenem, Mer = meropenem. (b) Free energy surfaces and transition state locations for 6α-hydroxyethyl orientation I (lowest energy barriers in panel A), with alternative active site hydrogen bond configurations. DW = deacylating water, AC = acyl enzyme, TS = transition state, TI = tetrahedral intermediate. *Left*: Free energy surface for imipenem deacylation. The DW is donating a hydrogen bond to the carbapenem 6α-hydroxyl group. *Right*: Free energy surface for meropenem deacylation. The carbapenem 6α-hydroxyl donates a hydrogen bond to the DW. Source: Reproduced with permission of Hirvonen et al. [146]/ American Chemical Society / Public Domain CC BY 3.0.

the important cephalosporin antibiotic ceftazidime [147]. Whereas OXA-48 does not hydrolyze ceftazidime (and therefore will not confer significant resistance to bacterial infections treated with it), the variant OXA-163 does [145]. After distinguishing between reactive and nonreactive acyl-enzyme complex conformations of ceftazidime with DFTB3/ff14SB umbrella sampling, the barriers for deacylation obtained again indicated relative differences ($\Delta\Delta^{\ddagger}G$) consistent with experimental kinetics for OXA-48, OXA-163, and OXA-181 (another OXA-48-like enzyme not capable of ceftazidime hydrolysis). Further investigations of these simulations then indicated that subtle changes in the active site interactions lead to differences in hydration of carboxy-Lys73, which cause the difference in deacylation efficiency (rather than, e.g. differences in stabilization by the oxyanion hole). Overall, these works on dissecting the efficiency of β-lactam antibiotic breakdown by SBLs indicate how detailed QM/MM reaction simulations (combined with MM MD simulation) that consider alternative active site configurations provide detailed insights into the molecular origins of antibiotic resistance.

6.3.3 Covalent SARS-CoV-2 Inhibitors: Mechanism and Insights for Design

Severe acute respiratory syndrome coronavirus 2 (SARS-CoV-2) was first identified in December 2019 [148, 149] and then rapidly spread worldwide, with the World Health Organization declaring a global pandemic in March 2020. The virus had claimed the lives of more than 6.5 million people worldwide as of October 2022 [150]. Despite the success of the rapid development of COVID-19 vaccines, vaccination does not eradicate severe illness due to the emergence of COVID-19 variants [151, 152]. It is thus of high importance to develop effective treatment strategies that can prevent disease progression. One such strategy is to develop effective inhibitors of the main protease, M[pro] (also known as 3-chymotrypsin-like protease, 3CL[pro]), a key enzyme involved in the viral replication and transcription processes [153].

The mechanism of the cysteine protease M[pro] is expected to feature a two-step acylation–deacylation process [154], similar to the serine β-lactamases, for example. Several groups investigated the details of this mechanism in SARS-CoV-2 M[pro] soon after its structure was determined. Ramos-Guzman et al. [155] simulated the reaction with a representative peptide substrate at the B3LYP-D3/6-31G*-ff14SB QM-MM level. After characterizing the Michaelis complex between M[pro] and the peptide, the adaptive string method [77] was used to perform umbrella sampling for both the acylation and deacylation reaction steps, using the relevant distances between the substrate and the Cys145/His41 catalytic dyad as collective variables. This indicated a free energy barrier of 14.6 kcal/mol for acylation and a concerted but asynchronous mechanism, where the catalytic dyad (Cys145$^-$/His41H$^+$) forms an ion pair, followed by the nucleophilic addition of the cysteine residue and protonation of the amide nitrogen at the P1′ position of the substrate by His41. This mechanism, featuring an initial Cys145$^-$/His41H$^+$ ion pair formation, was also supported for a similar peptide substrate, using transition state optimization followed by intrinsic reaction coordinate optimizations with the subtractive ONIOM method [156], with geometries optimized at the B3LYP/6-31G(d,p)-ff14SB level and energy calculations up to DLPNO-CCSD(T)/cc-pVTZ (with estimation of free energies using zero-point energy, thermal, and entropic corrections). However, the first QM/MM reaction simulations of SARS-CoV-2 M^{pro}, performed by Swiderek and Moliner [157], indicated that in acylation, proton transfer between Cys145 and His41 may be concerted with nucleophilic attack of Cys145 to the carbonyl atom of the substrate. This was based on QM/MM umbrella sampling along two reaction coordinates with AM1/ff03, corrected to M06-2X/6-31+G(d, p)-MM by comparison of potential energy surfaces. The three studies agree that for M^{pro}, the Cys145$^-$/His41H$^+$ ion pair is higher in energy than the corresponding neutral state, but whether or not this ion pair is a stable intermediate preceding nucleophilic attack by the Cys145 thiolate may depend on the substrate: in Refs. [155, 156], a serine is in position P1' (which is expected to stabilize the ion pair through its side-chain hydroxyl), whereas in Ref. [157], the chromophore 7-amino-4-carbamoylmethylcoumarin is present there.

Several groups used QM/MM reaction simulations early in the pandemic to provide insights into M^{pro} inhibitor design [158–161], often based on the structure of SARS-CoV-2 in complex with a known covalent coronavirus M^{pro} inhibitor, initially deposited already in January 2020 [162]. This peptidyl inhibitor, named N3, features a Michael acceptor warhead (see Figure 6.3). Ramos-Guzmán et al. used MM and QM/MM (B3LYP-D3/6-31G*-ff14SB) MD simulations to investigate the reaction between M^{pro} and N3 [158]. Based on these simulations and their previous simulations with a peptide substrate [155], they suggested possible modifications to improve the non-covalent affinity (by restoring interactions that the P2' group of the peptide substrate forms with M^{pro}), as well as lower the barrier to covalent bond formation (by focusing on facilitating Cys145$^-$/His41H$^+$ ion pair formation). Awoonor-Williams and Abu-Saleh [159] compared the same N3 inhibitor with an alternative α-ketoamide inhibitor, using (MM) alchemical binding free energy calculations (for the non-covalent complex) and ONIOM QM/MM optimizations for covalent complex formation (using M06-2X/def2-TZVP for the QM region). This indicated that for N3, the covalent adduct formation is not very efficient, consistent with Ref. [158]. (Notably, this study modeled the Cys145/His41 catalytic dyad as neutral rather than ionic; the protonation states of these two residues can be affected by the pH or presence of a substrate/inhibitor [163, 164].) Ramos-Guzmán et al. further used their MM and QM/MM MD simulation protocols, combined with alchemical binding free energy calculations, to study the interaction between M^{pro} and PF-00835231 [165]. This is a hydroxymethylketone derivative drug candidate (formed upon phosphate hydrolysis of lufotrelvir) that inhibits M^{pro} with low nM affinity and suitable pharmacokinetics [166]. MM (ff14SB) MD simulations indicated that the P1' hydroxymethyl group mimics the serine residue interactions present in natural substrates, but also forms additional interactions with Gly143 and Asn142. Moreover, the B3LYP-D3/6–31+G*-ff14SB free energy profile obtained using the adaptive string method indicated that this group also participates in the rate-limiting step of the reaction (with an activation barrier consistent with experiment [167]): nucleophilic attack of the Cys145 thiolate on the carbonyl carbon of the inhibitor is concerted with proton transfer to the carbonyl oxygen from His41 via the P1' hydroxyl. Based on the analysis of the simulations, the authors suggested that the addition of a chloromethyl moiety to the P1' hydroxymethyl group should lower the reaction barrier for the formation of the covalent complex. This was then confirmed by simulation: thermodynamic integration (at the MM level) indicated that the modification does not affect the non-covalent binding free energy, whereas the reaction barrier calculated by their QM/MM MD simulations was significantly reduced, primarily due to the expected stabilization of the ionic dyad pair. This work thus indicates how QM/MM simulations (combined with MM MD) can suggest possible routes to further improve M^{pro} covalent inhibitors.

The use of combining QM/MM simulations directly with the design of covalent SARS-CoV-2 M^{pro} inhibitors was first demonstrated by Arafet et al. [160], who employed their QM/MM approach based on semi-empirical AM1/ff03 umbrella sampling corrected to the M06-2X/6–31+G(d,p) level (as used previously for investigating a peptide substrate [157], see above). First, the covalent reaction between M^{pro} and the N3 inhibitor was characterized, starting from the covalent complex, indicating an exergonic reaction (with the covalent complex ~18 kcal/mol more stable than the initial non-covalent complex). Then, based on these and previous simulations alongside medicinal chemistry experience, two inhibitors were designed: **B1**, where the warhead of **N3** was retained, and **B2**, where both the recognition portion (with glutamine at P1, consistent with the M^{pro} substrate specificity) and the warhead (now a nitroalkene) were changed (Figure 6.3). The subsequently calculated QM/MM free energy profiles, in which the catalytic dyad residues and the **P1'**, **P1**, and **P2** fragments of the ligands were modeled in the

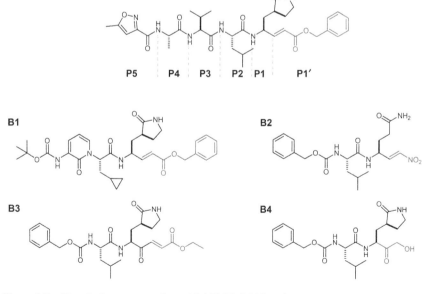

Figure 6.3 Chemical structures of peptidyl SARS-CoV-2 main protease (M^{pro}) inhibitor **N3** (with the different fragments indicated by *gray dashed lines*) and proposed derivatives (**B1, B2, B3, B4**) [160, 161]. Warheads are highlighted in *orange*.

QM region, indicated the same mechanism as for N3. The barrier toward covalent inhibition was essentially the same between N3 and B1 and somewhat lower for B2. Further, the B1 compound resulted in the most stable Cys145$^-$/His41H$^+$ ion pair. The main difference between the two, however, was that B1 inhibition was predicted to be clearly irreversible (covalent complex ~28 kcal/mol more stable than the non-covalent complex) and B2 reversible (~11 kcal/mol difference). Overall, the QM/MM calculations indicate that interactions between the recognition portion and M^{pro} affect the energetics of the formation of the covalent complex, as these determine the orientation of the inhibitor in the active site.

Marti et al. [161] then further built on this work to design and evaluate two further peptidyl inhibitor compounds, B3 and B4 (Figure 6.3). The recognition portion was selected based on the M^{pro} interactions from their previous QM/MM simulations [160, 161], using the P1 moiety of B1 and the P2 and P3 moieties of B2. The warheads differ: B3 has an ethyl oxo-enoate warhead and B4 has a hydroxymethyl ketone warhead (the same as the PF-00835231 inhibitor). Starting from models of the non-covalent complex, free energy profiles for each inhibitor were obtained using umbrella sampling at the M06-2X/6–31+G(d,p)-ff03 level along a path collective variable combining key distances. After confirming the mechanism, the reaction was divided into two steps: formation of the Cys145$^-$/His41H$^+$ ion pair and the subsequent covalent complex formation by nucleophilic attack of Cys145 on the inhibitor and proton transfer from His41 to the (former) warhead. The resulting free energy profile indicates that the barrier and reaction energy for ion pair formation are essentially identical, with the ion pair ~8 kcal/mol higher in energy than the initial neutral catalytic dyad (Figure 6.4a). The rate-limiting covalent bond formation step (TS2 in Figure 6.4a) is influenced by the type of warhead present. For B3, nucleophilic attack by Cys145 and proton transfer from His41 to Cα were predicted to take place concertedly (Figure 6.4b). For B4, the reactive thiolate already approaches the carbonyl carbon in the Cys145$^-$/His41H$^+$ ion pair state, and then completes the nucleophilic attack with concomitant protonation of the carbonyl oxygen by His41 through the B4 warhead hydroxyl moiety (Figure 6.4c). Although the energetics of the rate-limiting step are similar, B3 is indicated as a more promising lead for M^{pro} inhibitor design, as it is somewhat more reactive than B4 (activation barrier of 13.5 vs. 15.2 kcal/mol, Figure 6.4a) and leads to a ~2 kcal/mol more stable covalent complex. The mechanism for B4 covalent complex formation (alongside a relatively low barrier), however, indicates that modulating the pK_a of the warhead hydroxyl group can potentially lead to increased potency.

As well as the detailed QM/MM studies highlighted above, others have also developed QM/MM-based approaches to help evaluate and design M^{pro} inhibitors. Mondal and Warshel [168] first studied a reversible α-ketoamide inhibitor using the EVB approach for reaction simulation (with parameters to a reference reaction calculated at the B3LYP/6–31+G** level) together with a protein dipole Langevin dipole method they previously developed (PDLD/S-LRA/β) for the non-covalent binding energy. They noted that, in addition to the electrophilicity of the warhead, the last step of the mechanism (protonation of the covalent complex) can be tuned to control the level of exothermicity, resulting in either reversible or

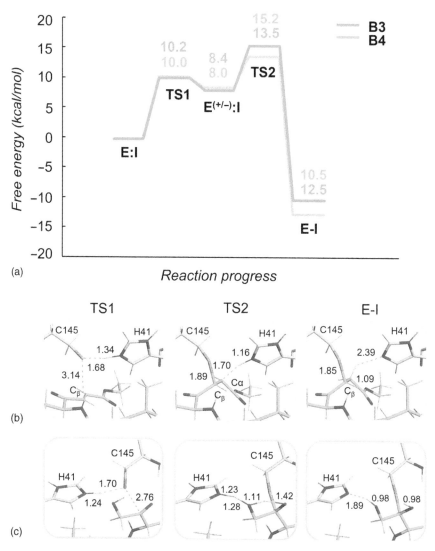

Figure 6.4 QM/MM (M06-2X/6–31+G(d,p)-ff03) reaction simulations for the designed B3 and B4 SARS-CoV-2 M^{pro} inhibitors. (a) Free energy profiles of B3 (*blue line*) and B4 (*green line*) inhibitors. (b) Optimized structures of key states in the inhibition process of M^{pro} by B3 (*blue*) and (c) B4 (*green*). Carbon atoms of the inhibitor are shown in *green* while those of the catalytic residues Cys145 and His41 are in *cyan*. Important hydrogen bonding interactions are indicated as *green dashed lines*, and key distances are given in Å. Adapted from Ref. [161].

irreversible covalent inhibition. The same group then developed an approach using a thermodynamic cycle with PDLD/S-LRA/β calculations for covalent inhibitor binding free energy calculations, avoiding more time-consuming QM/MM or EVB reaction simulations [169]. Calculations for covalent inhibitors against M^{pro} and the 20S proteasome showed excellent agreement with experimental results, indicating that the method is effective for inhibitors with different warheads (such as aldehyde and α-ketoamide) and can be applied to both reversible and irreversible inhibitors. Chan and co-workers [170] investigated a series of M^{pro} natural substrates and covalent inhibitors generated by the COVID Moonshot project [171] using a range of biomolecular simulation methods. QM/MM umbrella sampling simulations following the proton transfer between Cys145 and His41 proved to be useful in determining that the preferred state of the catalytic dyad was neutral. Notably, both extensive MM MD simulations [172] and QM calculations [173] have also highlighted the dependence of inhibitor binding on M^{pro} His protonation and tautomer preferences.

The simulation studies discussed in this section indicate that both the catalytic and inhibition mechanisms of M^{pro} depend on several factors that can influence the formation and stability of the covalent complex: pK_a of active site residues, solvent accessibility, induced fit effects, and the nature of the substrate/inhibitor [170]. However, for both substrates and inhibitors, the rate-limiting step is indicated to be covalent bond formation. Different studies can reach different conclusions for the detailed mechanistic pathway, even for the same inhibitor (e.g. N3); this can be partly due to the QM level, sampling method, and QM region used [158, 160, 161]. For instance, neglecting key residues involved in the stabilization of the oxyanion can lead to higher activation energies [156]. Nevertheless, QM/MM studies have shown to be able to aid the design of potent and selective inhibitors as lead compounds against SARS-CoV-2 M^{pro}.

6.4 Conclusions and Outlook

The practical application of QM/MM simulations of biomolecular systems is becoming increasingly routine, aided by the availability of accessible QM/MM software, as well as increases in computational power. The ability of QM methods to provide accurate descriptions of small molecules and their interactions in target binding sites is attractive, and methodological developments make their use increasingly feasible in the context of drug design [174], with QM/MM offering a tractable approach to include the wider biomolecular environment. Further, the determination of reaction pathways of pharmaceutically relevant enzymes with QM/MM methods can provide valuable insights: atomistic understanding of the interactions of transition states and reaction intermediates can suggest new scaffolds or modifications of existing compounds, leading to the design of both noncovalent and covalent inhibitors as starting points for drug discovery campaigns. For covalent inhibitors, QM/MM methods can be directly used to obtain insights into their reactivity, mechanism, and stability, providing priorities for synthesis and suggestions for further development. Many such applications now exist, as highlighted in this chapter.

A key trend, observed in studies across all target families discussed here, is to combine QM/MM reaction simulations with other simulation and modeling techniques. This allows for computationally tractable extensive (MM) conformational sampling of reactive complexes and stable intermediates, as well as the determination of relative or absolute non-covalent binding free energies, prior to modeling chemical reactions with QM/MM. For example, for kinases and cysteine proteases, the pK_a of the key active site cysteine can first be determined using constant pH MD simulations or indeed machine learning algorithms [175]. Then, MM MD simulations, perhaps combined with alchemical free energy methods (thermodynamic integration or free-energy perturbation), can provide quantitative information on the non-covalent step of the binding process. For the covalent step, QM/MM reaction simulations would be used. The resulting complete characterization can highlight key aspects and interactions to focus on in further rounds of design, tuning potency, and selectivity. In the near future, we expect that such combinations of methods, including QM/MM reaction simulations, will be applied more routinely to aid the design of covalent drugs, likely alongside computational methods capable of high-throughput screening, such as covalent docking [176] (see Chapter 25).

It should be noted that QM/MM reaction simulations to accurately estimate differences in energy barriers are not yet straightforward to apply. However, the first demonstrations of the use of QM/MM as a practical computational assay for covalent drugs have now been reported [138, 177]. For efficiency, semi-empirical QM methods are typically used to make screening of tens of compounds (or target variants) tractable. Such methods ideally still require benchmarking and/or extensive testing against experimental data for each new chemical application to ensure reliability. A potential future solution for this could be the use of machine learning molecular potentials, which promise the accuracy of high-level QM at speeds faster than semi-empirical QM [178], so that accurate assays become feasible for larger amounts of compounds and/or more conformational sampling. For such machine learning potentials to be used in simulations in a biomolecular context, however, further methodological developments are still required, such as adequate solutions for electrostatic embedding in an ML/MM setting. Such developments, combined with improvements in automation, can make QM/MM reaction simulation a practical tool in the coming years, extending the existing experimental/computational toolbox [89] for covalent drug design.

References

1 Morzan, U.N., de Armino, D.J.A., Foglia, N.O. et al. (2018). Spectroscopy in complex environments from QM-MM simulations. *Chem Rev* 118 (7): 4071–4113.

2 Uddin, N., Choi, T.H., and Choi, C.H. (2013). Direct absolute pK_a predictions and proton transfer mechanisms of small molecules in aqueous solution by QM/MM-MD. *J Phys Chem B* 117 (20): 6269–6275.

3 Nelson, J.G., Peng, Y.X., Silverstein, D.W., and Swanson, J.M.J. (2014). Multi-scale reactive molecular dynamics for absolute pK(a) predictions and amino acid deprotonation. *J Chem Theory Comput* 10 (7): 2729–2737.

4 Steinmann, C., Olsson, M.A., and Ryde, U. (2018). Relative ligand-binding free energies calculated from multiple short QM/MM MD simulations. *J Chem Theory Comput* 14 (6): 3228–3237.

5 van der Kamp, M.W. and Mulholland, A.J. (2013). Combined quantum mechanics/molecular mechanics (QM/MM) methods in computational enzymology. *Biochemistry* 52 (16): 2708–2728.

6 Senn, H.M. and Thiel, W. (2009). QM/MM methods for biomolecular systems. *Angew Chem Int Ed Engl* 48 (7): 1198–1229.

7 Warshel, A. and Levitt, M. (1976). Theoretical studies of enzymic reactions: dielectric, electrostatic and steric stabilization of the carbonium ion in the reaction of lysozyme. *J Mol Biol* 103 (2): 227–249.

8 Singh, U.C. and Kollman, P.A. (1986). A combined ab initio quantum mechanical and molecular mechanical method for carrying out simulations on complex molecular systems: applications to the $CH_3Cl + Cl-$ exchange reaction and gas phase protonation of polyethers. *J Comput Chem* 7 (6): 718–730.

9 Field, M.J., Bash, P.A., and Karplus, M. (1990). A combined quantum mechanical and molecular mechanical potential for molecular dynamics simulations. *J Comput Chem* 11 (6): 700–733.

10 Bash, P.A., Field, M.J., Davenport, R.C. et al. (1991). Computer simulation and analysis of the reaction pathway of triosephosphate isomerase. *Biochemistry* 30 (24): 5826–5832.

11 Ranaghan, K.E. and Mulholland, A.J. (2017). Chapter 11 QM/MM methods for simulating enzyme reactions. In: *Simulating enzyme reactivity: computational methods in enzyme catalysis*, 375–403. The Royal Society of Chemistry.

12 Mulholland, A.J. (2005). Modelling enzyme reaction mechanisms, specificity and catalysis. *Drug Discov Today* 10 (20): 1393–1402.

13 Menikarachchi, L.C. and Gascon, J.A. (2010). QM/MM approaches in medicinal chemistry research. *Curr Top Med Chem* 10 (1): 46–54.

14 Lodola, A. and De Vivo, M. (2012). The increasing role of QM/MM in drug discovery. *Adv Protein Chem Struct Biol* 87: 337–362.

15 Barbault, F. and Maurel, F. (2015). Simulation with quantum mechanics/molecular mechanics for drug discovery. *Expert Opin Drug Discov* 10 (10): 1047–1057.

16 Kulkarni, P.U., Shah, H., and Vyas, V.K. (2022). Hybrid quantum mechanics/molecular mechanics (QM/MM) simulation: a tool for structure-based drug design and discovery. *Mini Rev Med Chem* 22 (8): 1096–1107.

17 Lodola, A., Callegari, D., Scalvini, L. et al. (2020). Design and SAR analysis of covalent inhibitors driven by hybrid QM/MM simulations. *Methods Mol Biol* 2114: 307–337.

18 Warshel, A. (1991). *Computer modeling of chemical reactions in enzymes and solutions*. New York: J. Wiley & Sons, Inc. ISBN: 0-47-1533955.

19 Kamerlin, S.C.L. and Warshel, A. (2010). The EVB as a quantitative tool for formulating simulations and analyzing biological and chemical reactions. *Faraday Discuss* 145: 71–106.

20 Loco, D., Lagardere, L., Caprasecca, S. et al. (2017). Hybrid QM/MM molecular dynamics with AMOEBA polarizable embedding. *J Chem Theory Comput* 13 (9): 4025–4033.

21 Brooks, B.R., Brooks, C.L. 3rd, Mackerell, A.D. Jr. et al. (2009). CHARMM: the biomolecular simulation program. *J Comput Chem* 30 (10): 1545–1614.

22 Thibault, J.C., Cheatham, T.E. 3rd, and Facelli, J.C. (2014). iBIOMES lite: summarizing biomolecular simulation data in limited settings. *J Chem Inf Model* 54 (6): 1810–1819.

23 Schrödinger Release 2021-4: QSite, Schrödinger, LLC, New York, NY (2021).

24 Murphy, R.B., Philipp, D.M., and Friesner, R.A. (2000). A mixed quantum mechanics/molecular mechanics (QM/MM) method for large-scale modeling of chemistry in protein environments. *J Comput Chem* 21 (16): 1442–1457.

25 Melo, M.C.R., Bernardi, R.C., Rudack, T. et al. (2018). NAMD goes quantum: an integrative suite for hybrid simulations. *Nat Methods* 15 (5): 351–354.

26 Kubar, T., Welke, K., and Groenhof, G. (2015). New QM/MM implementation of the DFTB3 method in the gromacs package. *J Comput Chem* 36 (26): 1978–1989.

27 Valiev, M., Yang, J., Adams, J.A. et al. (2007). Phosphorylation reaction in cAPK protein kinase-free energy quantum mechanical/molecular mechanics simulations. *J Phys Chem B* 111 (47): 13455–13464.

28 Neese, F., Wennmohs, F., Becker, U., and Riplinger, C. (2020). The ORCA quantum chemistry program package. *J Chem Phys* 152 (22): 224108.

29 Kuhne, T.D., Iannuzzi, M., Del Ben, M. et al. (2020). CP2K: an electronic structure and molecular dynamics software package – quickstep: efficient and accurate electronic structure calculations. *J Chem Phys* 152 (19): 194103.

30 Sherwood, P., Vries, A.H., Guest, M.F. et al. (2003). QUASI: a general purpose implementation of the QM/MM approach and its application to problems in catalysis. *J Mol Struct THEOCHEM* 632 (1–3): 1–28.

31 Lu, Y., Farrow, M.R., Fayon, P. et al. (2019). Open-source, Python-based redevelopment of the ChemShell multiscale QM/MM environment. *J Chem Theory Comput* 15 (2): 1317–1328.

32 Torras, J., Roberts, B.P., Seabra, G.M., and Trickey, S.B. (2015). PUPIL: a software integration system for multi-scale QM/MM-MD simulations and its application to biomolecular systems. *Adv Protein Chem Struct Biol* 100: 1–31.

33 Marti, S. (2021). QMCube (QM[3]): an all-purpose suite for multiscale QM/MM calculations. *J Comput Chem* 42 (6): 447–457.

34 Vreven, T., Byun, K.S., Komaromi, I. et al. (2006). Combining quantum mechanics methods with molecular mechanics methods in ONIOM. *J Chem Theory Comput* 2 (3): 815–826.

35 Ryde, U. (2016). QM/MM calculations on proteins. *Methods Enzymol* 577: 119–158.

36 Haldar, S., Comitani, F., Saladino, G. et al. (2018). A multiscale simulation approach to modeling drug-protein binding kinetics. *J Chem Theory Comput* 14 (11): 6093–6101.

37 Cho, A.E., Guallar, V., Berne, B.J., and Friesner, R. (2005). Importance of accurate charges in molecular docking: quantum mechanical/molecular mechanical (QM/MM) approach. *J Comput Chem* 26 (9): 915–931.

38 Kim, M. and Cho, A.E. (2016). Incorporating QM and solvation into docking for applications to GPCR targets. *Phys Chem Chem Phys* 18 (40): 28281–28289.

39 Kurczab, R. (2017). The evaluation of QM/MM-driven molecular docking combined with MM/GBSA calculations as a halogen-bond scoring strategy. *Acta Crystallogr B Struct Sci Cryst Eng Mater* 73 (Pt 2): 188–194.

40 Chaskar, P., Zoete, V., and Rohrig, U.F. (2017). On-the-fly QM/MM docking with attracting cavities. *J Chem Inf Model* 57 (1): 73–84.

41 Burger, S.K., Thompson, D.C., and Ayers, P.W. (2011). Quantum mechanics/molecular mechanics strategies for docking pose refinement: distinguishing between binders and decoys in cytochrome C peroxidase. *J Chem Inf Model* 51 (1): 93–101.

42 Lee, T.S., Allen, B.K., Giese, T.J. et al. (2020). Alchemical binding free energy calculations in AMBER20: advances and best practices for drug discovery. *J Chem Inf Model* 60 (11): 5595–5623.

43 Hudson, P.S., Boresch, S., Rogers, D.M., and Woodcock, H.L. (2018). Accelerating QM/MM free energy computations via intramolecular force matching. *J Chem Theory Comput* 14 (12): 6327–6335.

44 Kearns, F.L., Warrensford, L., Boresch, S., and Woodcock, H.L. (2019). The good, the bad, and the ugly: "HiPen", a new dataset for validating (S)QM/MM free energy simulations. *Molecules* 24 (4): 681.

45 Olsson, M.A. and Ryde, U. (2017). Comparison of QM/MM methods to obtain ligand-binding free energies. *J Chem Theory Comput* 13 (5): 2245–2253.

46 Giese, T.J. and York, D.M. (2019). Development of a robust indirect approach for MM --> QM free energy calculations that combines force-matched reference potential and Bennett's acceptance ratio methods. *J Chem Theory Comput* 15 (10): 5543–5562.

47 Rathore, R.S., Sumakanth, M., Reddy, M.S. et al. (2013). Advances in binding free energies calculations: QM/MM-based free energy perturbation method for drug design. *Curr Pharm Des* 19 (26): 4674–4686.

48 Genheden, S. and Ryde, U. (2015). The MM/PBSA and MM/GBSA methods to estimate ligand-binding affinities. *Expert Opin Drug Discov* 10 (5): 449–461.

49 Pu, C., Yan, G., Shi, J., and Li, R. (2017). Assessing the performance of docking scoring function, FEP, MM-GBSA, and QM/MM-GBSA approaches on a series of PLK1 inhibitors. *MedChemComm* 8 (7): 1452–1458.

50 Anisimov, V.M. and Cavasotto, C.N. (2011). Quantum mechanical binding free energy calculation for phosphopeptide inhibitors of the Lck SH2 domain. *J Comput Chem* 32 (10): 2254–2263.

51 Anisimov, V.M., Ziemys, A., Kizhake, S. et al. (2011). Computational and experimental studies of the interaction between phospho-peptides and the C-terminal domain of BRCA1. *J Comput Aided Mol Des* 25 (11): 1071–1084.

52 Pecina, A., Meier, R., Fanfrlik, J. et al. (2016). The SQM/COSMO filter: reliable native pose identification based on the quantum-mechanical description of protein-ligand interactions and implicit COSMO solvation. *Chem Commun (Camb)* 52 (16): 3312–3315.

53 Pecina, A., Eyrilmez, S.M., Kopruluoglu, C. et al. (2020). SQM/COSMO scoring function: reliable quantum-mechanical tool for sampling and ranking in structure-based drug design. *ChemPlusChem* 85 (11): 2362–2371.

54 Glide, S. (2021). *Schrödinger release 2021-4: QM-polarized ligand docking protocol*. New York, NY: LLC.

55 Begum, J., Skamnaki, V.T., Moffatt, C. et al. (2015). An evaluation of indirubin analogues as phosphorylase kinase inhibitors. *J Mol Graph Model* 61: 231–242.

56 Wichapong, K., Rohe, A., Platzer, C. et al. (2014). Application of docking and QM/MM-GBSA rescoring to screen for novel Myt1 kinase inhibitors. *J Chem Inf Model* 54 (3): 881–893.

57 Kiss, M., Szabo, E., Bocska, B. et al. (2021). Nanomolar inhibition of human OGA by 2-acetamido-2-deoxy-d-glucono-1,5-lactone semicarbazone derivatives. *Eur J Med Chem* 223: 113649.

58 Chetter, B.A., Kyriakis, E., Barr, D. et al. (2020). Synthetic flavonoid derivatives targeting the glycogen phosphorylase inhibitor site: QM/MM-PBSA motivated synthesis of substituted 5,7-dihydroxyflavones, crystallography, in vitro kinetics and ex-vivo cellular experiments reveal novel potent inhibitors. *Bioorg Chem* 102: 104003.

59 Jing, Z., Liu, C., Cheng, S.Y. et al. (2019). Polarizable force fields for biomolecular simulations: recent advances and applications. *Annu Rev Biophys* 48: 371–394.

60 Walker, B., Liu, C., Wait, E., and Ren, P. (2022). Automation of AMOEBA polarizable force field for small molecules: Poltype 2. *J Comput Chem* 43 (23): 1530–1542.

61 Rupakheti, C.R., MacKerell, A.D. Jr., and Roux, B. (2021). Global optimization of the Lennard-Jones parameters for the drude polarizable force field. *J Chem Theory Comput* 17 (11): 7085–7095.

62 Amezcua, M., El Khoury, L., and Mobley, D.L. (2021). SAMPL7 host-guest challenge overview: assessing the reliability of polarizable and non-polarizable methods for binding free energy calculations. *J Comput Aided Mol Des* 35 (1): 1–35.

63 Crespo, A., Rodriguez-Granillo, A., and Lim, V.T. (2017). Quantum-mechanics methodologies in drug discovery: applications of docking and scoring in lead optimization. *Curr Top Med Chem* 17 (23): 2663–2680.

64 Liu, H., Lu, Z., Cisneros, G.A., and Yang, W. (2004). Parallel iterative reaction path optimization in ab initio quantum mechanical/molecular mechanical modeling of enzyme reactions. *J Chem Phys* 121 (2): 697–706.

65 Cisneros, G.A., Liu, H., Lu, Z., and Yang, W. (2005). Reaction path determination for quantum mechanical/molecular mechanical modeling of enzyme reactions by combining first order and second order "chain-of-replicas" methods. *J Chem Phys* 122 (11): 114502.

66 Woodcock, H.L., Hodošček, M., Sherwood, P. et al. (2003). Exploring the quantum mechanical/molecular mechanical replica path method: a pathway optimization of the chorismate to prephenate Claisen rearrangement catalyzed by chorismate mutase. *Theoret Chem Acc* 109 (3): 140–148.

67 Ryde, U. (2017). How many conformations need to be sampled to obtain converged QM/MM energies? The curse of exponential averaging. *J Chem Theory Comput* 13 (11): 5745–5752.

68 Guimaraes, C.R., Udier-Blagovic, M., Tubert-Brohman, I., and Jorgensen, W.L. (2005). Effects of Arg90 neutralization on the enzyme-catalyzed rearrangement of Chorismate to prephenate. *J Chem Theory Comput* 1 (4): 617–625.

69 Rosta, E., Klahn, M., and Warshel, A. (2006). Towards accurate ab initio QM/MM calculations of free-energy profiles of enzymatic reactions. *J Phys Chem B* 110 (6): 2934–2941.

70 Hu, H., Lu, Z.Y., and Yang, W.T. (2007). QM/MM minimum free-energy path: methodology and application to triosephosphate isomerase. *J Chem Theory Comput* 3 (2): 390–406.

71 Kumar, S., Bouzida, D., Swendsen, R.H. et al. (1992). The weighted histogram analysis method for free-energy calculations on biomolecules. 1. The method. *J Comput Chem* 13 (8): 1011–1021.

72 Kästner, J. (2012). Umbrella integration with higher-order correction terms. *J Chem Phys* 136 (23): 234102.

73 Rosta, E. and Hummer, G. (2015). Free energies from dynamic weighted histogram analysis using unbiased Markov state model. *J Chem Theory Comput* 11 (1): 276–285.

74 Stelzl, L.S., Kells, A., Rosta, E., and Hummer, G. (2017). Dynamic histogram analysis to determine free energies and rates from biased simulations. *J Chem Theory Comput* 13 (12): 6328–6342.

75 Vanden-Eijnden, E. and Venturoli, M. (2009). Revisiting the finite temperature string method for the calculation of reaction tubes and free energies. *J Chem Phys* 130 (19): 194103.

76 Rosta, E., Nowotny, M., Yang, W., and Hummer, G. (2011). Catalytic mechanism of RNA backbone cleavage by ribonuclease H from quantum mechanics/molecular mechanics simulations. *J Am Chem Soc* 133 (23): 8934–8941.

77 Zinovjev, K. and Tunon, I. (2017). Adaptive finite temperature string method in collective variables. *J Phys Chem A* 121 (51): 9764–9772.

78 Zinovjev, K., Ruiz-Pernia, J.J., and Tunon, I. (2013). Toward an automatic determination of enzymatic reaction mechanisms and their activation free energies. *J Chem Theory Comput* 9 (8): 3740–3749.

79 Park, S., Khalili-Araghi, F., Tajkhorshid, E., and Schulten, K. (2003). Free energy calculation from steered molecular dynamics simulations using Jarzynski's equality. *J Chem Phys* 119 (6): 3559–3566.

80 Laio, A. and Parrinello, M. (2002). Escaping free-energy minima. *Proc Natl Acad Sci U S A* 99 (20): 12562–12566.

81 Bolnykh, V., Olsen, J.M.H., Meloni, S. et al. (2020). MiMiC: multiscale modeling in computational chemistry. *Front Mol Biosci* 7: 45.

82 Garcia-Viloca, M., Gao, J., Karplus, M., and Truhlar, D.G. (2004). How enzymes work: analysis by modern rate theory and computer simulations. *Science* 303 (5655): 186–195.

83 Serapian, S.A. and van der Kamp, M.W. (2019). Unpicking the cause of stereoselectivity in actinorhodin ketoreductase variants with atomistic simulations. *ACS Catal* 9 (3): 2381–2394.

84 Mlynsky, V., Banas, P., Sponer, J. et al. (2014). Comparison of ab initio, DFT, and semiempirical QM/MM approaches for description of catalytic mechanism of hairpin ribozyme. *J Chem Theory Comput* 10 (4): 1608–1622.

85 Claeyssens, F., Harvey, J.N., Manby, F.R. et al. (2006). High-accuracy computation of reaction barriers in enzymes. *Angew Chem Int Ed* 45 (41): 6856–6859.

86 Bauer, R.A. (2015). Covalent inhibitors in drug discovery: from accidental discoveries to avoided liabilities and designed therapies. *Drug Discov Today* 20 (9): 1061–1073.

87 Smith, G.F. (2011). Designing drugs to avoid toxicity. *Prog Med Chem* 50: 1–47.

88 Sutanto, F., Konstantinidou, M., and Domling, A. (2020). Covalent inhibitors: a rational approach to drug discovery. *RSC Med Chem* 11 (8): 876–884.

89 Boike, L., Henning, N.J., and Nomura, D.K. (2022). Advances in covalent drug discovery. *Nat Rev Drug Discov* 1-18.

90 Pottier, C., Fresnais, M., Gilon, M. et al. (2020). Tyrosine kinase inhibitors in cancer: breakthrough and challenges of targeted therapy. *Cancer* 12 (3): 731.

91 Seshacharyulu, P., Ponnusamy, M.P., Haridas, D. et al. (2012). Targeting the EGFR signaling pathway in cancer therapy. *Expert Opin Ther Targets* 16: 15–31.

92 Bethune, G.C., Bethune, D.C., Ridgway, N.D., and Xu, Z. (2010). Epidermal growth factor receptor (EGFR) in lung cancer: an overview and update. *J Thorac Dis* 2 (1): 48–51.

93 Molina, J.R., Yang, P., Cassivi, S.D. et al. (2008). Non-small cell lung cancer: epidemiology, risk factors, treatment, and survivorship. *Mayo Clin Proc* 352 (8): 584–594.

94 Kobayashi, S.S., Boggon, T.J., Dayaram, T. et al. (2005). EGFR mutation and resistance of non-small-cell lung cancer to gefitinib. *N Engl J Med* 352 (8): 786–792.

95 Morgillo, F., Della Corte, C.M., Fasano, M., and Ciardiello, F. (2016). Mechanisms of resistance to EGFR-targeted drugs: lung cancer. *ESMO Open* 1 (3): e000060.

96 Yu, H.A. and Riely, G. (2013). Second-generation epidermal growth factor receptor tyrosine kinase inhibitors in lung cancers. *J Natl Compr Canc Netw* 11 (2): 161–169.

97 Schwartz, P.A., Kuzmič, P., Solowiej, J.E. et al. (2013). Covalent EGFR inhibitor analysis reveals importance of reversible interactions to potency and mechanisms of drug resistance. *Proc Natl Acad Sci* 111 (1): 173–178.

98 Hossam, M., Lasheen, D.S., and Abouzid, K.A.M. (2016). Covalent EGFR inhibitors: binding mechanisms, synthetic approaches, and clinical profiles. *Arch Pharm* 349 (8): 573–593.

99 Capoferri, L., Lodola, A., Rivara, S., and Mor, M. (2015). Quantum mechanics/molecular mechanics modeling of covalent addition between EGFR-cysteine 797 and N-(4-anilinoquinazolin-6-yl) acrylamide. *J Chem Inf Model* 55 (3): 589–599.

100 Blair, J.A., Rauh, D., Kung, C. et al. (2007). Structure-guided development of affinity probes for tyrosine kinases using chemical genetics. *Nat Chem Biol* 3 (4): 229–238.

101 Carmi, C., Galvani, E., Vacondio, F. et al. (2012). Irreversible inhibition of epidermal growth factor receptor activity by 3-aminopropanamides. *J Med Chem* 55 (5): 2251–2264.

102 Lence, E., van der Kamp, M.W., González-Bello, C., and Mulholland, A.J. (2018). QM/MM simulations identify the determinants of catalytic activity differences between type II dehydroquinase enzymes. *Org Biomol Chem* 16 (24): 4443–4455.

103 Yao, J., Guo, H.-B., Chaiprasongsuk, M. et al. (2015). Substrate-assisted catalysis in the reaction catalyzed by salicylic acid binding protein 2 (SABP2), a potential mechanism of substrate discrimination for some promiscuous enzymes. *Biochemistry* 54 (34): 5366–5375.

104 Demapan, D., Kussmann, J., Ochsenfeld, C., and Cui, Q. (2022). Factors that determine the variation of equilibrium and kinetic properties of QM/MM enzyme simulations: QM region, conformation, and boundary condition. *J Chem Theory Comput* 18 (4): 2530–2542.

105 Callegari, D., Ranaghan, K.E., Woods, C.J. et al. (2018). L718Q mutant EGFR escapes covalent inhibition by stabilizing a non-reactive conformation of the lung cancer drug osimertinib. *Chem Sci* 9 (10): 2740–2749.

106 Gao, X., Le, X., and Costa, D.B. (2016). The safety and efficacy of osimertinib for the treatment of EGFR T790M mutation positive non-small-cell lung cancer. *Expert Rev Anticancer Ther* 16 (4): 383–390.

107 He, J., Huang, Z., Han, L. et al. (2021). Mechanisms and management of 3rd-generation EGFR-TKI resistance in advanced non-small cell lung cancer (review). *Int J Oncol* 59 (5).

108 Bersanelli, M., Minari, R., Bordi, P. et al. (2016). L718Q mutation as new mechanism of acquired resistance to AZD9291 in EGFR-mutated NSCLC. *J Thorac Oncol* 11 (10): e121–e123.

109 Woods, C.J., Malaisree, M., Hannongbua, S., and Mulholland, A.J. (2011). A water-swap reaction coordinate for the calculation of absolute protein-ligand binding free energies. *J Chem Phys* 134 (5): 054114.

110 Castelli, R., Bozza, N., Cavazzoni, A. et al. (2019). Balancing reactivity and antitumor activity: heteroarylthioacetamide derivatives as potent and time-dependent inhibitors of EGFR. *Eur J Med Chem* 162: 507–524.

111 Weber, A.N.R., Bittner, Z.A., Liu, X. et al. (2017). Bruton's tyrosine kinase: an emerging key player in innate immunity. *Front Immunol* 8.

112 Wang, Q., Pechersky, Y., Sagawa, S. et al. (2019). Structural mechanism for Bruton's tyrosine kinase activation at the cell membrane. *Proc Natl Acad Sci U S A* 116 (19): 9390–9399.

113 López-Herrera, G., Vargas-Hernández, A., Gonzalez-Serrano, M.E. et al. (2014). Bruton's tyrosine kinase—an integral protein of B cell development that also has an essential role in the innate immune system. *J Leukoc Biol* 95 (2): 243–250.

114 Crofford, L.J., Nyhoff, L.E., Sheehan, J.H., and Kendall, P.L. (2016). The role of Bruton's tyrosine kinase in autoimmunity and implications for therapy. *Expert Rev Clin Immunol* 12 (7): 763–773.

115 Kil, L.P., de Bruijn, M.J.W., van Nimwegen, M. et al. (2012). Btk levels set the threshold for B-cell activation and negative selection of autoreactive B cells in mice. *Blood* 119 (16): 3744–3756.

116 Wen, T., Wang, J., Shi, Y. et al. (2021). Inhibitors targeting Bruton's tyrosine kinase in cancers: drug development advances. *Leukemia* 35 (2): 312–332.

117 Gayko, U., Fung, M.-C., Clow, F. et al. (2015). Development of the Bruton's tyrosine kinase inhibitor ibrutinib for B cell malignancies. *Ann N Y Acad Sci* 1358: 82–94.

118 Voice, A., Tresadern, G., Twidale, R.M. et al. (2021). Mechanism of covalent binding of ibrutinib to Bruton's tyrosine kinase revealed by QM/MM calculations. *Chem Sci* 12 (15): 5511–5516.

119 Kaptein, A., de Bruin, G., Emmelot-van Hoek, M. et al. (2019). Potency and selectivity of BTK inhibitors in clinical development for B-cell malignancies. *Clin Lymphoma Myeloma Leuk* 132: 1871.

120 Voice, A., Tresadern, G., Hv, V., and Mulholland, A.J. (2019). Limitations of ligand-only approaches for predicting the reactivity of covalent inhibitors. *J Chem Inf Model* 59 (10): 4220–4227.

121 Awoonor-Williams, E. and Rowley, C.N. (2021). Modeling the binding and conformational energetics of a targeted covalent inhibitor to Bruton's tyrosine kinase. *J Chem Inf Model* 61 (10): 5234–5242.

122 Murray, C.J.L., Ikuta, K.S., Sharara, F. et al. (2022). Global burden of bacterial antimicrobial resistance in 2019: a systematic analysis. *Lancet (London, England)* 399 (10325): 629–655.

123 O'Neill, J. (2016). Tackling Drug-Resistant Infections Globally: Final Report and Recommendations. Government of the United Kingdom.

124 Hermann, J.C., Ridder, L., Höltje, H.-D., and Mulholland, A.J. (2006). Molecular mechanisms of antibiotic resistance: QM/MM modelling of deacylation in a class a beta-lactamase. *Org Biomol Chem* 4 (2): 206–210.

125 Hermann, J.C., Ridder, L., Mulholland, A.J., and Holtje, H.D. (2003). Identification of Glu166 as the general base in the acylation reaction of class A beta-lactamases through QM/MM modeling. *J Am Chem Soc* 125 (32): 9590–9591.

126 Hermann, J.C., Hensen, C., Ridder, L. et al. (2005). Mechanisms of antibiotic resistance: QM/MM modeling of the acylation reaction of a class A beta-lactamase with benzylpenicillin. *J Am Chem Soc* 127 (12): 4454–4465.

127 Meroueh, S.O., Fisher, J.F., Schlegel, H.B., and Mobashery, S. (2005). Ab initio QM/MM study of class A beta-lactamase acylation: dual participation of Glu166 and Lys73 in a concerted base promotion of Ser70. *J Am Chem Soc* 127 (44): 15397–15407.

128 Hermann, J.C., Pradon, J., Harvey, J.N., and Mulholland, A.J. (2009). High level QM/MM modeling of the formation of the tetrahedral intermediate in the

acylation of wild type and K73A mutant TEM-1 class A beta-lactamase. *J Phys Chem A* 113 (43): 11984–11994.

129 Choi, H., Paton, R.S., Park, H., and Schofield, C.J. (2016). Investigations on recyclisation and hydrolysis in avibactam mediated serine β-lactamase inhibition. *Org Biomol Chem* 14 (17): 4116–4128.

130 Das, C.K. and Nair, N.N. (2020). Elucidating the molecular basis of avibactam-mediated inhibition of class A beta-lactamases. *Chemistry* 26 (43): 9639–9651.

131 Lizana, I., Uribe, E.A., and Delgado, E.J. (2021). A theoretical approach for the acylation/deacylation mechanisms of avibactam in the reversible inhibition of KPC-2. *J Comput Aided Mol Des* 35 (9): 943–952.

132 Tripathi, R.C. and Nair, N.N. (2013). Mechanism of acyl-enzyme complex formation from the Henry-Michaelis complex of class C β-lactamases with β-lactam antibiotics. *J Am Chem Soc* 135 (39): 14679–14690.

133 Gherman, B.F., Goldberg, S.D., Cornish, V.W., and Friesner, R.A. (2004). Mixed quantum mechanical/molecular mechanical (QM/MM) study of the deacylation reaction in a penicillin binding protein (PBP) versus in a class C beta-lactamase. *J Am Chem Soc* 126 (24): 7652–7664.

134 Tripathi, R.C. and Nair, N.N. (2016). Deacylation mechanism and kinetics of acyl-enzyme complex of class C β-lactamase and cephalothin. *J Phys Chem B* 120 (10): 2681–2690.

135 Sgrignani, J., Grazioso, G., and De Amici, M. (2016). Insight into the mechanism of hydrolysis of meropenem by OXA-23 serine-β-lactamase gained by quantum mechanics/molecular mechanics calculations. *Biochemistry* 55 (36): 5191–5200.

136 Swarén, P., Maveyraud, L., Raquet, X. et al. (1998). X-ray analysis of the NMC-A β-lactamase at 1.64-Å resolution, a class A carbapenemase with broad substrate specificity. *J Biol Chem* 273 (41): 26714–26721.

137 Chudyk, E.I., Limb, M.A.L., Jones, C.E.S. et al. (2014). QM/MM simulations as an assay for carbapenemase activity in class A β-lactamases. *Chem Commun* 50 (94): 14736–14739.

138 Hirvonen, V.H.A., Hammond, K., Chudyk, E.I. et al. (2019). An efficient computational assay for β-lactam antibiotic breakdown by class A β-lactamases. *J Chem Inf Model* 59 (8): 3365–3369.

139 Chudyk, E.I., Beer, M., Limb, M.A.L. et al. (2022). QM/MM simulations reveal the determinants of carbapenemase activity in class A β-lactamases. *ACS Infect Dis* 8 (8): 1521–1532.

140 Fritz, R.A., Alzate-Morales, J.H., Spencer, J. et al. (2018). Multiscale simulations of clavulanate inhibition identify the reactive complex in class A β-lactamases and predict the efficiency of inhibition. *Biochemistry* 57 (26): 3560–3563.

141 Song, Z. and Tao, P.-C. (2022). Graph-learning guided mechanistic insights into imipenem hydrolysis in GES carbapenemases. *Electron Struct* 4 (3).

142 Charnas, R.L. and Knowles, J.R. (1981). Inhibition of the RTEM beta-lactamase from Escherichia coli. Interaction of enzyme with derivatives of olivanic acid. *Biochemistry* 20 (10): 2732–2737.

143 Easton, C.J. and Knowles, J.R. (1982). Inhibition of the RTEM beta-lactamase from Escherichia coli. Interaction of the enzyme with derivatives of olivanic acid. *Biochemistry* 21 (12): 2857–2862.

144 Poirel, L., Potron, A., and Nordmann, P. (2012). OXA-48-like carbapenemases: the phantom menace. *J Antimicrob Chemother* 67 (7): 1597–1606.

145 Hirvonen, V.H.A., Spencer, J., and van der Kamp, M.W. (2021). Antimicrobial resistance conferred by OXA-48 β-lactamases: towards a detailed mechanistic understanding. *Antimicrob Agents Chemother* 65 (6): e00184–e00121.

146 Hirvonen, V.H.A., Weizmann, T.M., Mulholland, A.J. et al. (2022). Multiscale simulations identify origins of differential carbapenem hydrolysis by the OXA-48 β-lactamase. *ACS Catal* 12 (8): 4534–4544.

147 Hirvonen, V.H.A., Mulholland, A.J., Spencer, J., and van der Kamp, M.W. (2020). Small changes in hydration determine cephalosporinase activity of OXA-48 β-lactamases. *ACS Catal* 10 (11): 6188–6196.

148 Huang, C., Wang, Y.-m., Li, X.-w. et al. (2020). Clinical features of patients infected with 2019 novel coronavirus in Wuhan, China. *Lancet (London, England)* 395 (10223): 497–506.

149 Li, Q., Guan, X.-h., Wu, P. et al. (2020). Early transmission dynamics in Wuhan, China, of novel coronavirus-infected pneumonia. *N Engl J Med* 382: 1199–1207.

150 WHO (2022). *COVID-19 dashboard*. Geneva: World Health Organization [updated 2022 Oct; cited 2022 Oct 20]. Available from: https://covid19.who.int.

151 Cevik, M., Grubaugh, N.D., Iwasaki, A., and Openshaw, P. (2021). COVID-19 vaccines: keeping pace with SARS-CoV-2 variants. *Cell* 184 (20): 5077–5081.

152 Mahase, E. (2021). Covid-19: what new variants are emerging and how are they being investigated? *BMJ* 372: n158.

153 Ullrich, S. and Nitsche, C. (2020). The SARS-CoV-2 main protease as drug target. *Bioorg Med Chem Lett* 30 (17): 127377.

154 Solowiej, J., Thomson, J.A., Ryan, K. et al. (2008). Steady-state and pre-steady-state kinetic evaluation of severe acute respiratory syndrome coronavirus (SARS-CoV) 3CLpro cysteine protease: development of an ion-pair model for catalysis. *Biochemistry* 47 (8): 2617–2630.

155 Ramos-Guzman, C.A., Ruiz-Pernia, J.J., and Tunon, I. (2020). Unraveling the SARS-CoV-2 main protease mechanism using multiscale methods. *ACS Catal* 10: 12544–12554.

156 Fernandes, H.S., Sousa, S.F., and Cerqueira, N. (2022). New insights into the catalytic mechanism of the SARS-CoV-2 main protease: an ONIOM QM/MM approach. *Mol Divers* 26 (3): 1373–1381.

157 Swiderek, K. and Moliner, V. (2020). Revealing the molecular mechanisms of proteolysis of SARS-CoV-2 M(pro) by QM/MM computational methods. *Chem Sci* 11 (39): 10626–10630.

158 Ramos-Guzman, C.A., Ruiz-Pernia, J.J., and Tunon, I. (2021). A microscopic description of SARS-CoV-2 main protease inhibition with Michael acceptors. Strategies for improving inhibitor design. *Chem Sci* 12 (10): 3489–3496.

159 Awoonor-Williams, E. and Abu-Saleh, A.A.A. (2021). Covalent and non-covalent binding free energy calculations for peptidomimetic inhibitors of SARS-CoV-2 main protease. *Phys Chem Chem Phys* 23 (11): 6746–6757.

160 Arafet, K., Serrano-Aparicio, N., Lodola, A. et al. (2020). Mechanism of inhibition of SARS-CoV-2 M(pro) by N3 peptidyl Michael acceptor explained by QM/MM simulations and design of new derivatives with tunable chemical reactivity. *Chem Sci* 12 (4): 1433–1444.

161 Marti, S., Arafet, K., Lodola, A. et al. (2022). Impact of warhead modulations on the covalent inhibition of SARS-CoV-2 M(pro) explored by QM/MM simulations. *ACS Catal* 12 (1): 698–708.

162 Jin, Z., Du, X., Xu, Y. et al. (2020). Structure of M(pro) from SARS-CoV-2 and discovery of its inhibitors. *Nature* 582 (7811): 289–293.

163 Zanetti-Polzi, L., Smith, M.D., Chipot, C. et al. (2021). Tuning proton transfer thermodynamics in SARS-CoV-2 main protease: implications for catalysis and inhibitor design. *J Phys Chem Lett* 12 (17): 4195–4202.

164 Kneller, D.W., Phillips, G., Weiss, K.L. et al. (2020). Unusual zwitterionic catalytic site of SARS-CoV-2 main protease revealed by neutron crystallography. *J Biol Chem* 295 (50): 17365–17373.

165 Ramos-Guzman, C.A., Ruiz-Pernia, J.J., and Tunon, I. (2021). Inhibition mechanism of SARS-CoV-2 main protease with ketone-based inhibitors unveiled by multiscale simulations: insights for improved designs. *Angew Chem Int Ed Engl* 60 (49): 25933–25941.

166 Hoffman, R.L., Kania, R.S., Brothers, M.A. et al. (2020). Discovery of ketone-based covalent inhibitors of coronavirus 3CL proteases for the potential therapeutic treatment of COVID-19. *J Med Chem* 63 (21): 12725–12747.

167 Ma, C., Sacco, M.D., Hurst, B. et al. (2020). Boceprevir, GC-376, and calpain inhibitors II, XII inhibit SARS-CoV-2 viral replication by targeting the viral main protease. *Cell Res* 30 (8): 678–692.

168 Mondal, D. and Warshel, A. (2020). Exploring the mechanism of covalent inhibition: simulating the binding free energy of alpha-ketoamide inhibitors of the main protease of SARS-CoV-2. *Biochemistry* 59 (48): 4601–4608.

169 Zhou, J., Saha, A., Huang, Z., and Warshel, A. (2022). Fast and effective prediction of the absolute binding free energies of covalent inhibitors of SARS-CoV-2 main protease and 20S proteasome. *J Am Chem Soc* 144 (17): 7568–7572.

170 Chan, H.T.H., Moesser, M.A., Walters, R.K. et al. (2021). Discovery of SARS-CoV-2 M[pro] peptide inhibitors from modelling substrate and ligand binding. *Chem Sci* 12 (41): 13686–13703.

171 Achdout, H., Aimon, A., Bar-David, E. et al. (2022). Open science discovery of oral non-covalent SARS-CoV-2 main protease inhibitor therapeutics. *bioRxiv*.

172 Pavlova, A., Lynch, D.L., Daidone, I. et al. (2021). Inhibitor binding influences the protonation states of histidines in SARS-CoV-2 main protease. *Chem Sci* 12 (4): 1513–1527.

173 Poater, A. (2020). Michael acceptors tuned by the pivotal aromaticity of histidine to block COVID-19 activity. *J Phys Chem Lett* 11 (15): 6262–6265.

174 Bryce, R.A. (2020). What next for quantum mechanics in structure-based drug discovery? 2114: 339–353.

175 Gokcan, H. and Isayev, O. (2022). Prediction of protein pK_a with representation learning. *Chem Sci* 13 (8): 2462–2474.

176 Schirmeister, T., Kesselring, J., Jung, S. et al. (2016). Quantum chemical-based protocol for the rational design of covalent inhibitors. *J Am Chem Soc* 138 (27): 8332–8335.

177 Galvani, F., Scalvini, L., Rivara, S. et al. (2022). Mechanistic modeling of monoglyceride lipase covalent modification elucidates the role of leaving group expulsion and discriminates inhibitors with high and low potency. *J Chem Inf Model* 62 (11): 2771–2787.

178 Smith, J.S., Nebgen, B.T., Zubatyuk, R. et al. (2019). Approaching coupled cluster accuracy with a general-purpose neural network potential through transfer learning. *Nat Commun* 10 (1): 2903.

7

Recent Advances in Practical Quantum Mechanics and Mixed-QM/MM-Driven X-Ray Crystallography and Cryogenic Electron Microscopy (Cryo-EM) and Their Impact on Structure-Based Drug Discovery

Oleg Borbulevych and Lance M. Westerhoff

QuantumBio Inc, 2790 West College Ave, State College, PA 16801, United States

7.1 Introduction

X-ray crystallography is a core experimental procedure used to determine the three-dimensional (3D) atomic structure of biomolecular systems and inform structure-based drug discovery (SBDD) and fragment-based drug discovery (FBDD) efforts in most pharmaceutical companies and laboratories. SBDD utilizes protein (or DNA or RNA) and protein–ligand and protein–protein 3D structures to provide insights in lead discovery and optimization campaigns. In these campaigns, bonded and nonbonded interactions are explored and optimized to find compounds that best fit (chemically) the protein binding pocket. Therefore, obtaining an accurate representation of these structures is critical to the successful execution of SBDD projects. Successful FBDD screening also depends on accurate structure, but with the added complexity that it is typically carried out by soaking protein crystals with a cocktail of up to 10 small molecule compounds prior to the data collection, structure solution, and ligand placement utilized in SBDD [1]. Given the size of fragment compounds, this cocktail presents a challenge in that the density can accommodate any number of fragments in any number of orientations. Therefore, determining the correct orientation(s) of a particular fragment within an electron density blob, as such density is often weak or partial in the fragment area, becomes more difficult. Furthermore, in both SBDD and FBDD, ligands containing "flippable" functional groups, such as an amide group, are particularly susceptible to placement uncertainties since light elements (e.g. N and O) are often not readily distinguishable in macromolecular crystallography. Given the quality of the crystal structures is essential for the success of high-throughput screening, docking, and scoring (e.g. rank ordering) of potential drug candidates, this uncertainty can impact the entire drug discovery process.

Due to recent advances in data collection, processing, structure solution, and refinement automation, X-ray crystallography has become a routine method.

Computational Drug Discovery: Methods and Applications, First Edition.
Edited by Vasanthanathan Poongavanam and Vijayan Ramaswamy.
© 2024 WILEY-VCH GmbH. Published 2024 by WILEY-VCH GmbH.

Despite that, the majority of crystal structures are still determined at modest or low resolutions, which generally leads to significant uncertainties in atomic coordinates and other structural errors [2, 3]. It has been argued that those structural errors adversely impact ligand binding affinity predictions [2], which are critical to SBDD/FBDD applications. A significant drawback of traditional macromolecular refinement stems from the fact that conventional stereochemical restraints – which are used almost exclusively for the refinement process – are rudimentary in nature and do not account for nonbonded interactions such as electrostatics, polarization, hydrogen bonds, dispersion, and charge transfer [4–6]. Moreover, conventional refinement methods rely entirely on a detailed, ex situ description of the molecular geometry for each ligand or cofactor in the model as captured in a Crystallographic Information File (CIF). Unfortunately, the creation of *accurate* CIFs is a nontrivial task, and this process often leads to bound ligand structures with less than desirable quality [5] due to an incomplete a priori understanding of in situ bound bond lengths and angles and a lack of intermolecular interactions in conventional refinement functionals [7].

One way to improve X-ray models is to utilize quantum mechanics (QM) during the crystallographic refinement; however, traditionally, the size of virtually all biological systems prohibited a straightforward application of the QM methods. Nevertheless, in 2002, with the aid of the program COMQUM-X [8], the first mixed-quantum mechanics/molecular mechanics (QM/MM) X-ray refinement was conducted using a small QM portion of the system (around 25 heavy atoms). Since then, several examples of the QM-refined structures against X-ray data have been reported [9–17], emphasizing the ligand geometry improvement and protonation state determination [18]. In 2014, QuantumBio Inc. – building on the previous work of the Merz laboratory [13, 15–17] – introduced a new, much more automated QM refinement technique that works by replacing the conventional stereochemical restraints of the ligand(s), cofactor(s), active sites(s), residue(s), or even the entire protein–ligand complex with accurate quantum-based energy functionals in "real-time" during the refinement [19, 20] as computed by the linear-scaling QM semiempirical quantum mechanics (SE-QM) method [20–22]. Prior to this work, it was demonstrated that such QM linear scaling calculations can capture the critical interactions between a target and its ligand(s), such as hydrogen bonds, electrostatics, polarization, charge transfer, and metal coordination [23–27], and because this QM refinement protocol explicitly skips any information provided by CIF, the method gives rise to better, more accurate in situ ligand and active site geometries. This earlier work gave rise to an even more performant, QM/MM methodology based on the ONIOM formalism [28] as implemented in DivCon to treat most any macromolecular structure using a single functional [29]. It is this primary work that has led to routine, high-throughput QM/MM X-ray refinement (and more recently cryo-EM refinement), which has a direct impact on the models used in SBDD, and this impact will be discussed in detail in this publication.

7.2 Feasibility of Routine and Fast QM-Driven X-Ray Refinement

One of the first practical approaches to incorporating QM/MM functional into X-ray refinement was the program COMQUM-X [8], implemented to integrate a QM/MM algorithm with the crystallographic software crystallography and NMR system (CNS) [30]. With that method, a ligand and approximately 25 atoms around it were treated at the ab initio Becke-Perdew86/6-31G* or B3LYP/6-31G* QM level of theory, and the rest of the residues were computed with molecular mechanics (MM) with the AMBER force field. During these refinements, the bulk of the structure was fixed to reduce computational costs. Merz's group then made a significant advance in the field by implementing the divide-and-conquer (D&C), linear scaling, and SE-QM methods previously described [21, 22, 31–33]. D&C SE-QM utilizes an approximate solution of the Schrödinger equation that can be written using the Hartee–Fock–Roothaan formalism as

$$\mathbf{FC} = \mathbf{CE} \tag{7.1}$$

where $\mathbf{F}$ is the Fock matrix, $\mathbf{C}$ is the matrix of molecular orbital (MO) coefficients, and $\mathbf{E}$ is the eigenvalue energy matrix. D&C SE-QM divides the protein–ligand complex into subsystems generally corresponding to the amino acid residues in the protein, and Eq. (7.1) is solved for each subsystem. Therefore, matrix diagonalization – the most expensive part of the QM calculation – is performed on each subsystem (along with a buffer region) instead of the entire complex, leading to significant savings in CPU time and memory use. Obtained MOs and density matrixes for the subsystems are then combined to yield a solution for the whole system. As a result, the calculation's memory and CPU time requirements scale ~linearly or $\sim O(n)$, where n is the number of atoms in the system. This is contrasted to traditional QM methods in which the memory requirements and CPU time exhibit $O(n^3)$ scaling. Thus, when this linear scaling formalism is joined with the already fast semiempirical level of theory used in DivCon, D&C makes routine QM calculations – including all-atom model optimization/refinement – possible on very large biological systems. In this early work, this method was applied to X-ray crystallography via integration with the CNS [30] package [13], in which all atoms in the structure were treated using the AM1 Hamiltonian [34]. But even with the linear scaling, applying the all-atom SE-QM in the X-ray refinement regime is more computationally expensive than conventional refinement methods. Furthermore, SE-QM when applied to protein systems can cause systematic deviations from the standard geometry in the backbone [13, 19]. Finally, the lack of d-orbital support in AM1 led to limits in compatibility with metal-containing complexes. Therefore, the next step in the evolution of QM refinement was to combine linear scaling SE-QM using the more modern PM6 Hamiltonian [35, 36] for the ligand(s), cofactor(s), metals, active site(s), and chosen residue regions, with MM using the AMBER ff14sb [37] force field as implemented in DivCon for the remainder of the macromolecular system [29]. Finally, instead of CNS, which has been largely superseded in SBDD organizations, the DivCon module or plugin was integrated first with PHENIX [4] and then with BUSTER [38] to deploy the method on more modern platforms [19, 29].

To increase the accessibility of QM and QM/MM refinement for the community and support high-throughput crystallographic refinement, QuantumBio [39] went beyond this core development and implemented a user-friendly, **fully automated** molecular perception and preparation protocol that supports almost any protein/DNA/RNA/ligand structure. This development addresses a long-standing need to perform QM, MM, and QM/MM calculations quickly and easily with much fewer convergence problems or setup issues. Using the following protocol, models are determined and refined, which are not only chemically correct but chemically complete as well (with likely protonation states, residue rotamer states, and so on):

- Fast structure protonation, including optimization of the hydrogen network and flip states and pH effects implemented based on [40].
- Automated molecular perception [41] and formal charge determination of the entire system based on graph theory algorithms, including any unknown species, e.g. ligands, cofactors, metal coordination, nonstandard amino acids, and truncated residues.
- Automatic assignment of MM types for the entire system, including ligands, etc., based on molecular perception, and hence corresponding MM parameters for any AMBER forcefield chosen.
- Automatic residue-based selection of any number of QM regions extended by a given radius from any center, such as ligands, etc.
- Automatic link-atom (proton) addition for any internally "broken" bonds at the QM:MM interface.

Finally, to address traditional convergence problems in macromolecular QM calculations, this new DivCon uses several modern QM convergence optimization algorithms combined with Extended Hückle theory [42].

7.3 Metrics to Measure Improvement

To gage the performance of the QM-driven refinement in comparison to conventional approaches, several metrics reviewed below in detail have been used [15, 19, 29, 43–45].

7.3.1 Ligand Strain Energy

Ligand strain – which shows how much strain the ligand must accept to bind with the target protein – is an industry-standard method to define the quality of refined ligand structural models [15, 19, 45–47]. We calculate [15, 48] the local ligand strain energy or E_{Strain} as the difference between the energy of the isolated ligand

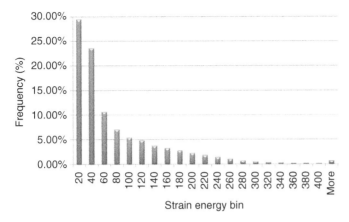

Figure 7.1 The distribution of strain energy values for 134 345 ligand poses calculated at the PM6 level.

conformation (optimized) and the protein-bound ligand conformation according to Eq. (7.2),

$$E_{Strain} = E_{SinglePoint} - E_{Optimized} \qquad (7.2)$$

where $E_{SinglePoint}$ is the single-point energy computed for the ligand X-ray geometry, and $E_{Optimized}$ is the energy of the optimized ligand that corresponds to the local minimum. In 2012, we explored the strain energy distribution of over 134 345 ligand poses deposited in PDB [49]. As shown in (Figure 7.1), about 55% of all ligand poses belong in a 0–40 kcal/mol bin, ~25% of poses have strain energy above 100 kcal/mol, and the balance falls into three bins between 40 and 100 kcal/mol.

7.3.2 ZDD of Difference Density

In crystallography, the difference density ($\Delta\rho$) reveals the disagreement between the experimental data and the refined model. The positive (+) and negative (−) difference density peaks indicate not only missing or incorrectly placed ligands, fragments, and water molecules, but they also show more subtle details such as incorrectness in geometry or conformations of a molecule or positions of individual atoms. To quantitively characterize the amount of the difference density around any residue in the crystal structure, a new quality indicator – the real-space Z score of difference density (ZDD) – was proposed by Tickle [43]. The benefit of this indicator is that it measures the *accuracy* of the model. In contrast, the often-used real space correlation coefficient (RSCC) correlates with *both* the accuracy *and* precision of the model, making RSCC less useful for measuring model accuracy (vs. experimental density).

A detailed derivation of ZDD can be found elsewhere [43, 50], but briefly, the Z score of difference density values at the point r is expressed as follows,

$$Z(\Delta\rho(r)) = \frac{\Delta\rho(r)}{\sigma(\Delta\rho(r))} \qquad (7.3)$$

where $\sigma(\Delta\rho(\mathbf{r}))$ is the standard deviation of the difference density, corresponds to the random error of the model, and is pure **precision**. In contrast, the Z score of the difference density measures the residual, nonrandom error and is pure **accuracy**. Positive and negative grid density points are analyzed separately, leading to two values, ZDD− and ZDD+). The maximum absolute number of those two values gives a final ZDD score (7.4).

$$ZDD = \max(\text{abs}(ZDD-), ZDD+) \tag{7.4}$$

7.3.3 Overall Crystallographic Structure Quality Metrics: MolProbity Score and Clashscore

MolProbity is a macromolecular model validation tool, which uses multiple quality criteria [44] to calculate a MolProbity score (MPScore). This logarithm-based score combines three key component metrics: Clashscore, Ramachandran plot statistics, and rotamer outliers [51]. The lower the MPScore, the better the quality of the model. Out of these three metrics, the Clashscore is a useful metric in itself and is the number of clashes per 1000 atoms. Clashes are determined by the construction of a nonbonded atom contact surface around each atom using the rolling probe algorithm [52]. A clash occurs when the nonbonded surface around one atom overlaps with the surface of another atom by more than 0.4 Å [3]. Overall, a crystal structure with problematic geometry results in many clashes and a high Clashscore value [44].

7.4 QM Region Refinement

Our first approach to the integration of SE-QM with the PHENIX [4] suite (PHENIX/DivCon) is called *Region-QM* refinement [19]. In this algorithm, the refined protein structure is divided into three regions: the main or core region(s), the buffer region(s), and the stereochemistry restraint region(s) (Figure 7.2). The core region(s) contains one or more ligands of interest as well as the selection of the target residues or other species such as water molecules, metal ions, cofactors, and so on within the given radius (e.g. 5 Å) from any ligand atom. The buffer region(s) are a second set of selection residues beyond each core region, which are the residues located at a second given distance (e.g. 3 Å) from any atom of the core region. Finally, the balance of the protein is treated as a pure stereochemical restraint region. The core regions, if there are more than one in the structure, do not need to be contiguous (and neither do buffer regions). The entire core and buffer regions are computed at the QM level of theory, but only QM gradients of the core region are employed in the refinement. Thus, each buffer region chemically insulates its core region to limit errors that may occur in the gradients due to capping or other artifacts in the surrounding chemical environment. Finally, for the remainder of the protein (outside of the core region), the atomic gradients are calculated using the standard stereochemistry restraint functional as implemented in the chosen X-ray crystallography platform (PHENIX or BUSTER). Mathematically, the QM

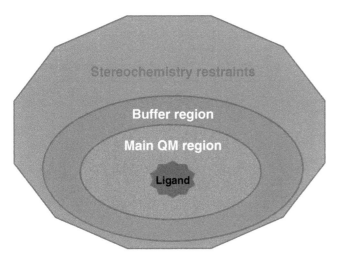

Figure 7.2 Schematic view of the QM region refinement concept.

and stereochemistry gradients on each atom with coordinates **x** are combined in the crystallography platform according to the equation,

$$(\nabla \mathbf{x}_i)_{total} = \kappa \times \Omega_{xray} \times (\nabla \mathbf{x}_i)_{xray} + \varpi_i \times (\nabla \mathbf{x}_i)_{QM} + (1 - \varpi_i) \times (\nabla \mathbf{x}_i)_{geom} \quad (7.5)$$

where the weight ϖ is set to 1 for the core QM region(s) and 0 for the rest of the atoms, including the buffer region. Ω_{xray} is a variable weight determined using an automatic procedure in PHENIX or a fixed weight in BUSTER, and κ is the additional scale factor implemented in PHENIX. It is notable that the *full* QM refinement can be performed by setting the weight ϖ of 1 for all atoms in the whole system.

Prior to this first effort, it was shown that the local chemistry of the ligand within the binding pocket could be improved with the integration of the QM methods into the X-ray refinement on the example of several crystal structures [9–13]. The Region-QM refinement approach is consistent, and we systematically demonstrated significant improvement of E_{strain} of 50, quasi-randomly chosen protein–ligand structures from the PDB. In particular, the average ligand strain energy for the set of 50 structures calculated for the deposited coordinates is 83.50 ± 9.03 kcal/mol, and the minimum and maximum values are 6.88 and 283.35 kcal/mol, respectively, or a range of 276.47 kcal/mol. After Region-QM refinement, significant improvement was observed in the E_{strain} throughout the set: the average strain energy of the re-refined set of structures is 24.60 ± 3.67 kcal/mol, or 3.5 times smaller than that of the deposited structures (Table 7.1). To validate these QM E_{strain} energies, we compared them with those calculated ab initio with the HF/6–311 + G** basis set. The change in the strain energies based on the ab initio calculations is less pronounced than that obtained with the SE-QM Hamiltonian (Table 7.1). It was expected, as the ab initio method was not used directly in the X-ray refinement. Also, the HF level of theory, despite the large basis set, does not consider electronic correlation, while it is partially incorporated into SE-QM methods such as AM1 and PM6. However, despite those factors, in all cases studied, SE-QM X-ray crystallographic refinement

Table 7.1 Average ligand strain energies over 50 crystal structures refined using region-QM refinement method.

	Deposited PDB	QM refined	Improvement, fold
StrainEnergy, AM1	83.50 ± 9.03	24.60 ± 3.67	3.4
StrainEnergy, HF/6-311G**	93.39 ± 9.00	53.76 ± 4.61	1.5

Source: Adapted from Borbulevych et al. [19].

lead to significantly improved ab initio-calculated ligand strains. It confirms the robustness of the SE-QM methods for X-ray refinement [19].

As an example, PDB 2X7T at 2.8 Å resolution [53] is the structure of the enzyme carbonic anhydrase inhibited by the ligand WZB in an active site that includes tetrahedral zinc coordinated by the three histidine residues. The WZB is bound to the zinc via an amino group to complete the coordination sphere of the metal (Figure 7.3). In conventional X-ray crystallographic refinement protocols, structures involving ligands coordinated with metals usually require tedious work to create library files to account for these interactions. For instance, in the deposited structure 2X7T, all coordination distances involving Zn are in the range of 2.14–2.25 Å, which is longer than the average length of 2.00(2) Å for Zn···N and Zn···O coordination bonds [54]. Such discrepancies result in a distortion of the coordination sphere of the metal (Figure 7.3), which also affects the ligand geometry. In particular, the phenyl ring of the ligand is heavily distorted, including a bond angle of 147 degree and significant deviations of bond lengths from the average C_{ar}–C_{ar} bond length of 1.398 Å [55]. These anomalies contribute to a high ligand E_{strain} of 96.71 kcal/mol for the deposited conformation. The QM region refinement was performed with none of these geometry assumptions concerning the coordination, yielding a model in which the coordination distances with the zinc are in the range of 1.98–2.02 Å. When these coordination distances are compared to the literature, we find them to be very close to the average values mentioned above [54]. Furthermore, the aforementioned ligand distortions evaporate, and the ligand E_{strain} drops to 14.5 kcal/mol indicating a significant improvement in the geometry of the bound ligand WZB.

We argued [19] that the main source of observed ligand E_{strain} improvement in the structures refined using the region-QM protocol is the elimination of errors

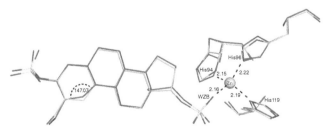

Figure 7.3 Superimposition of the residues in the coordination sphere of zinc in the structure 2X7T from the region-QM (*green*) refinements and the original PDB (*magenta*).

in the ligand geometry. On the one hand, because it completely disregards any bond angle/length parameters provided by the CIF, QM X-ray crystallographic refinement *automatically* resolves the "garbage in/garbage out" problem [5], which results from the inaccurate or imprecise ligand descriptions found in these standard ligand libraries. On the other hand, the QM potential also influences the geometry of the ligand through electrostatic, polarization, and charge transfer interactions observed in situ that are not available in the rudimentary conventional restraints (especially those built from ex situ states).

7.5 ONIOM Refinement

Despite the success of the region-QM X-ray crystallographic refinement, this approach has two fundamental drawbacks. First, most protein atoms are still treated using simple conventional restraints. Second, nonbonded and electrostatic interactions between the QM region and the rest of the macromolecular system (outside of the core + buffer regions) are not considered. To address these weaknesses, we developed a more holistic approach to conduct QM refinement based on the all-atom QM/MM scheme [29]. Subtractive QM/MM or ONIOM [28], allows for the straightforward computation of the system's energy (7.6),

$$E_{ONIOM}^{QM/MM} = E_{region}^{QM} + E_{all}^{MM} - E_{region}^{MM} \tag{7.6}$$

where the E_{all}^{MM} term is the MM energy calculated for the entire system, the E_{region}^{MM} term is the MM energy for the QM region, and E_{region}^{QM} is the energy of the QM region computed with the chosen SE-QM Hamiltonian. QM/MM gradients in the subtractive scheme are calculated as follows,

$$\nabla x_{ONIOM}^{QM/MM} = \nabla x_{region}^{QM} + \nabla x_{all}^{MM} - \nabla x_{region}^{MM} \tag{7.7}$$

in which the gradients of the QM region(s) include terms from both the QM and the MM functionals, and the electrostatics and van der Waals interactions between the QM and MM regions are explicitly included in the energy and gradient calculations. Generally, the ONIOM approach leads to faster and more convergent calculations vs. other QM/MM methods (like additive QM/MM), and the approach readily supports models with multiple QM regions such as those with multiple active site/ligand centers commonly found in crystal structures. Using the ONIOM formalism for the whole system, X-ray refinement with PHENIX and BUSTER is expressed as follows,

$$(\nabla x_i)_{total} = \kappa \times \Omega_{Xray} \times (\nabla x_i)_{Xray} + \nabla x_{ONIOM}^{QM/MM} \tag{7.8}$$

where $\nabla x_{ONIOM}^{QM/MM}$ corresponds to the ONIOM gradients determined using expression (7.7). Therefore, any ligand(s) and surrounding binding pocket(s) is (are) defined as the QM region(s), and the remainder of the protein–ligand model is designated as the MM region(s) and characterized using the AMBER force field such as amberff14sb [37]. When this calculation is performed in the PHENIX platform, the Ω_{xray} term is a variable weight determined using an automatic procedure in the platform [56], and κ is the additional scale factor implemented in the platform [57]. Within this ONIOM X-ray crystallographic protocol, *all* stereochemical restraint gradients (including those for the ligand(s), the waters, the cofactor(s), any metals, and all of the protein/DNA/RNA residues) are replaced with high-quality QM/MM gradients.

Overall, we consider ONIOM particularly well suited to fast, routine, high-throughput, and user-friendly QM/MM-based crystallographic refinement, as demonstrated in [29]. In that work, we show that ONIOM refinement exhibits superior performance as judged by four different metrics, including strain, ZDD, MolProbity, and Clashscore (Figure 7.4), when validated against the 80 structures and 141 discrete ligand poses of the Astex Diverse Set [58]. In this validation, three different types of X-ray refinements were performed and compared: ONIOM QM/MM, region-QM, and conventional PHENIX [29]. The strain energy distributions are similar for the 2 QM-driven refinements in which the average strain energies calculated over 141 ligands equal 9.95 ± 3.77 kcal/mol for ONIOM and 10.49 ± 4.52 kcal/mol for region-QM refinements. Furthermore, the ligand strain histograms (Figure 7.4a) for both QM refinements have peaks around 3.0 kcal/mol, accounting for approximately 75% of the ligand poses in the set. This congruence makes sense when one considers that in both cases, the same SE-QM Hamiltonian was brought to bear on approximately the same residues in each case. Therefore, the difference between the two values is likely due to the impact of the more complete Hamiltonian in ONIOM vs. the presence of stereochemical restraints in the region-QM method. Contrasting with the QM results, the conventional refinement yields strain energy data that are mostly evenly distributed across a broad range from 10 to 40 kcal/mol of which ~30% of data are in the last bin of 50+ kcal/mol. As a result, the average ligand strain energy after the Conventional refinement of the Astex set is 35.64 ± 9.35 kcal/mol, or about 3.5-fold higher than in the QM-based refinements.

The histogram for ZDD (Figure 7.4b) shows a broad peak at 1.4 units, a similar feature of all three distributions. Nevertheless, the number of ligands that fall into the bin range from 0 to 1.2 ZDD units is higher for ONIOM and region-QM refined structures than that of conventionally refined ones. Thus, the average ZDD for the ligands in ONIOM-refined structures (2.3 ± 0.8) is slightly lower (better) than that after the conventional refinements (2.9 ± 1.1). As expected, the region-QM refinement results (2.6 ± 0.9) fall in the middle. Interestingly, ligand strain and ZDD arrays are uncorrelated, as concluded from the Pearson correlation coefficient between those two metrics being close to zero for all refinement methods.

Where ONIOM refinement shows significant impact vs. both region-QM and conventional refinement is the improvement of the overall structure quality measured using Clashscore. Across the 80 refined Astex structures, the average Clashscores for conventional (4.83 ± 1.2 units) and region-QM (5.54 ± 1.6 units) refined models are similar. In contrast, the average Clashscore for the ONIOM structures is 1.10 ± 0.41 units, thus exhibiting a dramatic improvement of 4.5–5.0-fold. Furthermore, the Clashscore histogram (Figure 7.4c) reveals a sharp peak located at the 0.5 unit mark, which comprises 90% of the ONIOM models compared to the peak for both conventional and region-QM model data at around 3.5 units. About 50% of those data in the conventional and region-QM histograms are found in the tail of the respective peaks and distributed in histogram bins 4.5+ units, and above

7.5 ONIOM Refinement | 167

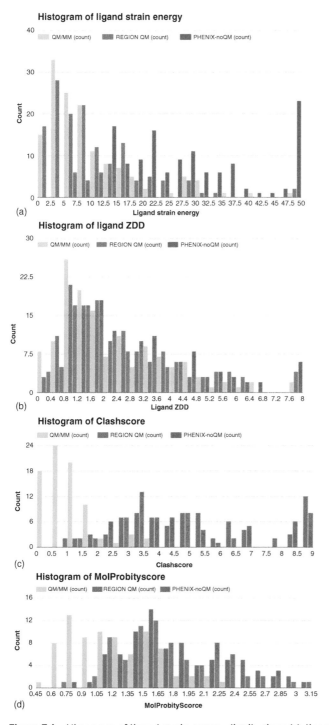

Figure 7.4 Histogram of ligand strain energy distributions (a), ligand ZDD distributions (b), MolProbity Clashscore distributions (c), and MolProbity score distributions (d) for 80 Astex structures refined with QM/MM (ONIOM), region-QM, and conventional methods. Histograms (a) and (b) include data for 141 ligand instances.

while ONIOM data are not even observed in that range. The similarity between region-QM and conventional refinements result indicates that the significant improvement in Clashscore for the bulk of the protein structure arises from using the QM/MM Hamiltonian on the entire structure. Notably, a similar improvement upon ONIOM refinement is observed for MPScore (Figure 7.4d).

7.6 XModeScore: Distinguish Protomers, Tautomers, Flip States, and Docked Ligand Poses

Given that X-rays are scattered by electrons, this leads to a fundamental limitation of X-ray crystallography in which the experiment is unable to (easily) directly detect the positions of hydrogen atoms. H atoms not only have just one electron, but their electron cloud is shifted toward the heavy atoms to which they are bound. Thus, the hydrogen atom has the weakest scattering power for X-rays among all elements [59]. Hence, with the rare exception of ultrahigh-resolution X-ray data, it is generally impossible to experimentally determine the protonation or tautomeric state of both the ligand and the surrounding active site in macromolecular crystallography at the resolutions often used in SBDD. Thus, a possible application of the QM methods to facilitate the determination of the protonation states of molecules or protein residues (e.g. GLU, ASP, or HIS) in the crystal structure and this has attracted attention in the past [18, 60]. Previous work has also shown that distinguishing among possible ligand tautomers is often pivotal in steering new drug discovery campaigns in the right direction [61].

Although X-ray crystallography does not permit direct observation of hydrogen atoms, QM/MM does not suffer from this problem, and therefore, using the QM/MM functional, it should be possible to detect the **influence** of hydrogen atoms on the heavy atoms (carbon, nitrogen, oxygen) to which they are bound and compare that influence to the experimental density to determine the protomer/tautomer state in the crystal. This idea is behind the XModeScore method developed by QuantumBio [50]. Using automatic enumeration of all likely protonation/tautomer ligand (and active site) states, followed by the QM/MM refinement of each protein–ligand state, this method is able to determine which protonation/tautomer form is most closely in agreement with the experimental X-ray data. The impact of the different protonation configurations is measured using two indicators of the ligand: E_{strain} and ZDD. In the XModeScore scoring procedure, both indicators are scaled according to the Z-score formula and then combined to produce the overall score of the i-tautomer form,

$$XModeScore_i = -\left\{ \frac{ZDD_i - \mu_{ZDD}}{\sigma_{ZDD}} + \frac{E_{strain_i} - \mu_{E_{strain}}}{\sigma_{E_{strain}}} \right\} \quad (7.9)$$

where μ is the mean value and σ is the standard deviation of the corresponding array of data (ZDD or E_{strain}). For example, the E_{strain} array contains E_{strain} values for all protomers/tautomers included in the calculations. $E_{strain,i}$ and ZDD_i are corresponding values of the i-tautomer. The highest $XModeScore_i$ corresponds to the protomeric/tautomeric form "i" that best fits both E_{strain} and ZDD criteria.

XModeScore was challenged using a Human carbonic anhydrase II (HCA II) structure bound to a high-affinity inhibitor [62, 63] acetazolamide (AZM). HCA II carbonate hydration/dehydration are involved in numerous metabolic processes, including CO_2 transport and pH regulation, and AZM was approved as a drug known as "Diamox" [64, 65]. AZM binds to the Zn atom within the active site of the enzyme via the nitrogen atom of the sulfonamide group, which completes its tetrahedral coordination by making coordination bonds with nitrogen atoms of His94, His96, and His119. In this configuration, as depicted in Scheme 7.1, AZM can exist in the three tautomeric/protonation forms. However, even high-resolution X-ray diffraction studies failed to determine which state of AZM exists in the crystal [66, 67]. It was only when the community obtained a neutron diffraction model of this enzyme [67] that it was proven that AZM exists in form **3**, and thus binds to zinc via the negatively charged sulfonamide SO_2NH group in the crystal form.

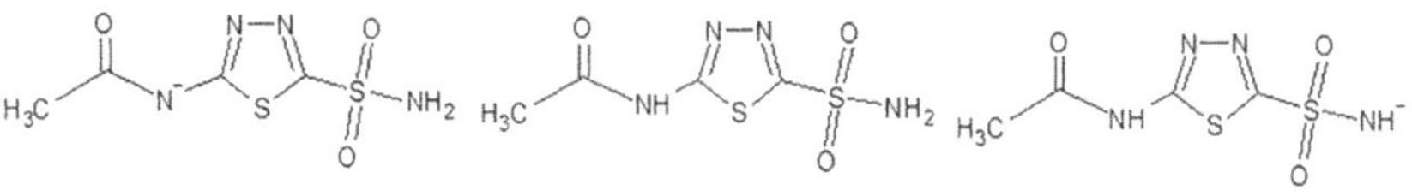

Scheme 7.1 Potential binding states of the compound AZM.

XModeScore results [50] based on the region-QM refinements of the three considered forms of AZM using the X-ray data from PDB 3HS4 reveal that form **3** is the superior form and is the best (lowest) in both E_{strain} and ZDD scoring components (Table 7.2, Figure 7.5). This finding is entirely consistent with the neutron diffraction results [67] regarding the protonation state of AZM in the crystal phase. Further examination of XModeScore results revealed that the ZDD of form **3** is twice as low as that of the other two forms, suggesting that protomer **3**, with the negatively charged N1 atom coordinated with zinc, is much more consistent with the experimental X-ray data than are the other two tautomers with the amino group at this position. Furthermore, the difference density maps of tautomers **1** and **2**, obtained after the QM refinement (Figure 7.5a,b), show prominent negative/positive difference density peaks around the nitrogen atom N1, which also support this conclusion. Notably, a series of refinements using incremental truncation of the original high-resolution data set 3HS4 demonstrates that XModeScore remains robust and predictive up to at least 3.0 Å resolution (Table 7.2).

7.7 Impact of the QM-Driven Refinement on Protein–Ligand Affinity Prediction

Given structural errors in crystal structures negatively impact ligand binding affinity predictions [2] and QM/MM crystallographic refinement improves the quality of protein–ligand structures, we have demonstrated [68] the impact of QM/MM refinement (and XModeScore) on protein–ligand binding affinity prediction. In

Table 7.2 XModeScore results for three forms of ligand AZM in PDB 3HS4 at different resolutions.

Pose	SE	RSCC	ZDD	XModeScore
Structure-3HS4				
3	5.55	0.989	12.8	2.72
2	8.89	0.978	24.9	−0.74
1	10.8	0.975	27.2	−1.98
Resolution 1.6 Å				
3	6.01	0.987	7.87	2.72
2	8.71	0.98	14.3	−0.70
1	9.75	0.978	16.8	−2.02
Resolution 2.0 Å				
3	5.58	0.989	6.56	2.68
2	8.74	0.982	12.3	−1.24
1	7.86	0.975	15.6	−1.45
Resolution 2.2 Å				
3	5.77	0.989	6.17	2.77
2	7.73	0.981	10.8	−1.31
1	8.35	0.984	10	−1.47
Resolution 2.5 Å				
3	5.4	0.989	7.65	2.47
2	8.2	0.986	8.62	−0.04
1	11.1	0.984	9.48	−2.43
Resolution 2.8 Å				
3	5.45	0.984	9.67	2.8
2	8.25	0.984	10.2	−1.39
1	8.74	0.982	10.1	−1.41

this work, QM/MM X-ray crystallographic refinement was applied to the set of structures from the community structure activity resource (CSAR) dataset released in 2012 [69]. This well-curated CSAR set is available with experimental binding affinities and is intended to be used as a benchmark for developing and testing docking and scoring functions. The CSAR set consists of the following subsets: cyclin-dependent kinase 2 (CDK2) with 15 ligands, checkpoint kinase 1 (CHK1) with 16 ligands, mitogen-activated protein kinase 1 (ERK2) with 12 ligands, and urokinase-type plasminogen activator (uPA) with 7 ligands.

As a baseline, we explored the impact of QM/MM refinement alone by measuring the R^2 correlation coefficients between experimental binding affinities and precited

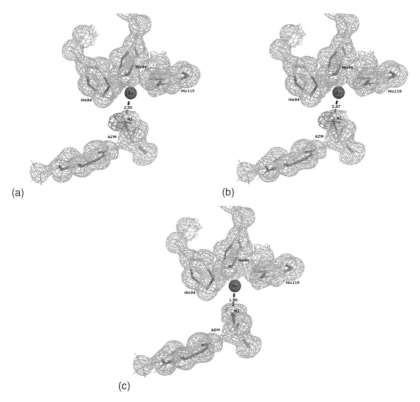

Figure 7.5 The coordination sphere of Zn in the catalytic center of HCA II with bound AZM molecules at three alternative binding modes **1** (a), **2** (b), and **3** (c) after the QM refinement of the PDB structure 3HS4. The difference density around the key nitrogen atom N1 of the sulfonamide group of AZM is contoured at 3.5σ level.

binding affinities as computed with a conventional, mainstream GBVI/WSA score function [70] available in the molecular operating environment (MOE) from Chemical Computing Group ULC. These results are depicted in Figure 7.6. QM/MM refinement improves the correlation within each subset, and the most significant magnitude of the transformation is observed for the CDK2 group, which comprises 15 structures [68]. Notably, the predicted affinities based on conventionally refined structures of this subset exhibit virtually no correlation to the experimental affinity with R^2 of 0.25 (Figures 7.6 and 7.7a), while QM/MM refined structures yield GBVI/WSA scores which correlate with the experimental binding affinity with R^2 of 0.60. Such a dramatic improvement is attributable to a much better fit of the crystal structures to the observed density in the QM/MM X-ray refinement, as these results show a twofold improvement of the average ZDD for the 15 CDK2 structures after the QM/MM refinement compared to the conventionally refined structure [68]. For example, consider the structure 4FKS, the worst outlier on the correlation graph based on the conventional refinement (Figure 7.7a). As a result of the QM/MM refinement ZDD around the ligand dramatically decreased from 16.63 to 3.86, mainly due to a different orientation of the ligand's benzyl moiety, as

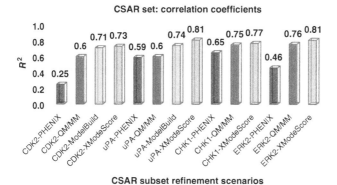

Figure 7.6 Correlation coefficients R^2 between experimental and predicted binding affinities for CSAR sets for various refinement scenarios: PHENIX (*black*), QM/MM (*red*), model-built QM/MM (*green*), and QM/MM with XModeScore chosen tautomers (*blue*).

seen in Figure 7.8. Interestingly, such an improvement model gives rise to a shift of the predicted GBVI/WSA score from −5.70 to −7.50 kcal/mol, bringing the new predicted value almost precisely on the trendline.

7.7.1 Impact of Structure Inspection and Modification

Even the most advanced refinements will ultimately only reach the nearest local minimum. Therefore, any structural changes and improvements resulting from the X-ray refinement are relatively limited in scope and achieved within a limited radius of convergence. Thus, the refinement cannot fix significant structural defects such as misplaced side chains or functional groups, including incorrect flip states, missing or extra water, etc. Visual inspection of the electron density maps usually allows one to easily spot problematic regions by the presence of prominent positive and negative peaks of the difference density, and manual model building is then required to fix such regions. Notably, these structure errors can even occur in well-curated crystal structures and could significantly impact the predicted binding scores.

For example, the correlation analysis of the uPA subset [68] revealed that the structure 4FU9 is one of the worst outliers, and the visual inspection of the ligand binding pocket in the structure revealed several problems, as shown in Figure 7.9a. First, the small peak of the positive difference density around the atom N18 indicates the presence of an alternative protonation state of the ligand 675 in the crystal. Second, there is a prominent peak of the negative electron density around the water molecule (WAT526) in the vicinity of the ligand, suggesting that this water is to be removed from the model. Third, the succinate molecule, SIN304, which is a crystallization buffer compound, was initially added to the model with an occupancy of 0.5. But a large blob of positive electron density observed around the molecule, suggests full occupancy of SIN304. With these changes made, the new QM/MM refinement leads to a better difference density distribution in the binding pocket (Figure 7.9b), and the GBVI/WSA score for the ligand 675 decreases from −7.11 to −6.40 kcal/mol. This shift leads to a considerable improvement in correlation for the

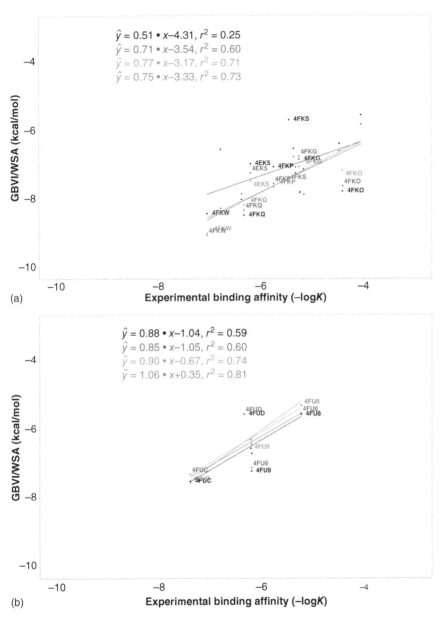

Figure 7.7 The regression lines of the correlation between experimental affinity (−log K) and computationally predicted GBVI/WSA scores for the 15 protein–ligand CDK2 complexes (a) and the 7 protein–ligand uPA complexes (b) for PHENIX structures (*black*), QM/MM structures (*red*), model-built QM/MM structures (*green*), and QM/MM refined structures with XModeScore chosen tautomers (*blue*).

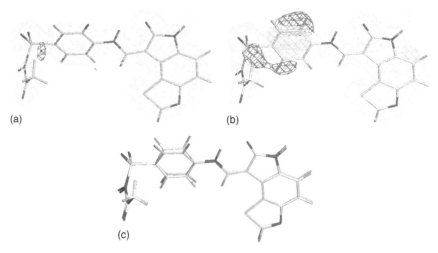

Figure 7.8 The σA-weighted mFo-DFc difference electron density map drawn at 3σ level around the ligand (ligand ID 46K) in the PDB structure 4FKS was refined with QM/MM (*green*) (a) and conventional (*yellow*) (b), as well as the superimposition of the two structures (c). The σA-weighted 2mFo-DFc electron density map is contoured at 1σ.

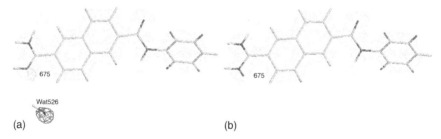

Figure 7.9 Positive (*green*) and negative (*red*) peaks of the σA-weighted mFo-DFc difference electron density map around the ligand (ligand ID 675) and Wat526 in the binding pocket of the protein target uPA in the PDB structure 4FU9 refined with QM/MM before (a) and after (b) the manual fit. The σA-weighted 2mFo-DFc electron density map is contoured at 1σ.

uPA set in which increases from R^2 0.6 to 0.74. A similar improvement in correlation was also achieved for the CDK2 subset by manual model correction (removing unjustified waters and choosing alternative side-chain positions of certain residues) of the worst outlier structures [68].

7.7.2 Impact of Selecting Protomer States: Implications of XModeScore on SBDD

As discussed above, the correct selection of the tautomeric/protonation state is often critical to the success of SBDD campaigns. In this work, we demonstrated that the choice of the correct ligand state significantly impacts the GBVI/WSA score and therefore the binding affinity of the method [68]. The application of

XModeScore to the CSAR set demonstrated that while often the default ligand protonation states chosen using default protonation are correct, alternative ligand protonation states are found to be correct in a significant plurality of the cases explored. Those structures are distributed across the subsets as follows: CDK2–4 structures, uPA – 3 structures, CHK1 2 structures, and one structure is from ERK2. Notably, including those alternative forms in the QM/MM refinement improves – to various degrees – the overall correlation between predicted and experimental affinities for all subsets, as shown in Figure 7.6.

When exploring the uPA subset, it was mentioned above that the protonation state of the ligand 675 in the structure 4FU9 was adjusted based on the study of the electron density map (Figure 7.9a). The XModeScore procedure also confirms that this ligand form – with a fully protonated amino(imino)methyl group – is the most favorable state (Figure 7.9b). The same alternative state with the fully protonated amino(imino)methyl group was also established by XModeScore for another ligand 2UP (PDB 4FU8) from the uPA subset, and using the alternative protomer in the QM/MM refinement resulted in the change of GBVI/WSA score from −5.52 to −5.29 for 4FU8. It was also found that in structure 4FUC that the default protonation state of the ligand 239 with the charged NH_3^+ group has a *worse* XModeScore score than that of the protomer the NH_2 group. Despite weaker H-bond interactions between the ammonia group and neighboring Asp50, the new-QM/MM refinement with NH_2 state reveals much better agreement with the experimental density as proven by a smaller ZDD value (1.97 units) compared to its magnitude of 4.42 units in the original QM/MM structure [68]. Overall results of incorporation of the correct protonation states according to XModeScore lead to a significant improvement in correlation for the uPA subset in which the R^2 increases from 0.74 to 0.81.

7.8 Conclusion

X-ray crystallography has become an integral tool in SBDD, and it provides the primary data used in new method development in CADD (including docking/scoring/sampling algorithm innovation, force field parameterization, and even training for artificial intelligence/machine learning (AI/ML)-based methods like AlphaFold [71]). While traditionally, computational chemists and medicinal chemists will begin with the X-ray structure, add protons, and complete an optimization process prior to using the model, QM/MM refinement is able to strike the right balance and provide insights into SBDD while still staying true to the experimental data. And given that QM/MM is built on QM for critical areas of the structure, it can account for interactions that are not properly represented in conventional, CIF-based methods like hydrogen bonds, electrostatics, charge transfer, polarization, and even metal coordination. We have been able to deploy these methods and make them available for routine use by also implementing mature molecular perception and preparation protocols, tautomer/protomer (and flip-state and chirality) enumeration methods, and density-based statistical analyses. QM/MM refinement allows us to not only better understand what target:ligand interactions a particular ligand satisfies (and perhaps more importantly, what interactions a candidate molecule does not), but with XModeScore, these methods also give us a more accurate picture of more subtle effects due to protonation state, binding mode, rotamer position, and bound-isomer character.

Going forward, cryogenic electron microscopy (Cryo-EM) is another approach to structure solution, which has several strengths vs. X-ray crystallography [72–74]. For example, cryo-EM does not require protein–ligand crystallization, and therefore it is applicable to target classes that poorly crystallize (such as certain globular proteins, many membrane-bound proteins, and so on). This method has opened opportunities for SBDD for additional campaigns, but cryo-EM does suffer from several limitations. These include mixed resolution characteristics in which different parts of the structure (e.g. pocket vs. ligand vs. protein-bulk) may resolve at vastly different resolutions [72, 73], EM is often unable to obtain atomic-scale resolutions required for SBDD, greater disorder and heterogeneity since the sample is frozen instead of crystallized, and so on. Because QM and MM functionals include more information than stereochemical restraints alone, they are applicable to the mixed or lower quality data available in cryo-EM: often in X-ray, for example, as shown in Table 7.2, we obtain similar results regardless of resolution.

To explore the question of whether or not ZDD will also work with cryo-EM data and an estimated difference map [75], we implemented a proof-of-concept cryo-EM real-space refinement protocol using DivCon combined with the opensource Clipper v. 2.1.20201109 library [76] for real-space map and density gradient calculation. During this proof-of-concept study, DivCon was joined with `phenix.real_space_diff_map` tool within the PHENIX package [4, 75] to estimate a real-space difference density map between both the initial input model and map downloaded from the PDB and the final model and map resulting from 50 steps of real-space QM/MM cryo-EM refinement. These estimated maps were supplied to our built-in ZDD calculator, and as hoped, ZDD does indeed improve upon refinement. In fact, as shown in Table 7.3, we observe consistent improvement in each case in all considered metrics. We, therefore, expect that, given the improvement in both strain and ZDD that XModeScore will be applicable to Cryo-EM

Table 7.3 Cryo-EM RS-DivCon refinement preliminary results.

Cryo-EM PDBid	Res (Å)	Ligand	$Strain_{Initial}$	$Strain_{Final}$	$ZDD_{Initial}$	ZDD_{Final}	$MP_{Initial}$	MP_{Final}
7efc	1.7	BTN	39.65	15.02	7.89	6.27	1.49	0.84
7jsy	1.8	I3C	38.75	14.09	7.4	5.47	1.3	0.7
7sq6	2.3	AQV	32.71	11.97	17.57	13.38	1.4	1.17
7vdf	2.6	BMA	13.22	7.64	17.71	15.61	1.65	1.25
7p02	2.9	CLR	24.73	11.18	11.96	11.05	1.8	1.25

ZDD = Z score of the *estimated* difference density calculated with phenix.real_space_diff_map.
Strain = all-atom ligand strain (E_{strain}) of the Ligand calculated at the QM level of theory.
MP = MolProbity score calculated with the phenix.molprobity tool.
Initial = published structure/final = QM(ligand+3.0 Å pocket)/MM real-space refined structure.

refinement as we have already shown it to be with X-ray crystallographic refinement. Going forward, we have implemented a built-in real space difference map estimator in DivCon, and we will continue to explore the impact of local (pocket:ligand) resolution, bound-ligand disorder, and ensemble heterogeneity in XModeScore in order to provide actionable intelligence in cryo-EM-enabled SBDD campaigns.

QM/MM refinement – both with X-ray and cryo-EM data – is a logical next step in routine structure solution, and as these methods have become faster and more highly convergent, they have shown themselves to be extremely useful for SBDD and FBDD efforts in the pharmaceutical space both for the determination of better, more accurate models and for the characterization of states – like protonation states – that would be difficult if not impossible to characterize without access to the more accurate models. As we move forward, routine QM/MM X-ray and cryo-EM refinement will no doubt become as indispensable for structural biology teams as QM, QM/MM, and MM sampling and characterization methods have already become for computer-aided drug design campaigns.

Acknowledgments

The authors wish to acknowledge the support of our clients and users, who have provided valuable feedback. We also thank the continued support of the PHENIX Consortium, in particular Drs. Nigel Moriarty, Pavel Afonine, and Paul Adams, for maintaining the application programming interface (API) "hooks" to our software within PHENIX. Likewise, we thank Global Phasing Limited, in particular Drs. Clemens Vonrhein and Gerard Bricogne, for supporting our development of analogous hooks with BUSTER. We would also like to thank Chemical Computing Group (in particular Alain Deschenes, Chris Williams, Paul Labute, and the entire CCG support team) for their continued support with MOE best practices and with the scientific vector language (SVL). Finally, we thank the National Institutes of Health (NIH) through SBIRs #R44GM121162 and #R44GM134781 for funding the research and development effort. The DivCon plugin to PHENIX and BUSTER along with XModeScore are provided by QuantumBio Inc. and they are available at the following: https://www.quantumbioinc.com/products/software_licensing.

References

1 Chilingaryan, Z., Yin, Z., and Oakley, A.J. (2012). Fragment-based screening by protein crystallography: successes and pitfalls. *Int J Mol Sci* 13: 12857.
2 Davis, A.M., Teague, S.J., and Kleywegt, G.J. (2003). Application and limitations of X-ray crystallographic data in structure-based ligand and drug design. *Angew Chem Int Ed* 42: 2718–2736.
3 Davis, I.W., Leaver-Fay, A., Chen, V.B. et al. (2007). MolProbity: all-atom contacts and structure validation for proteins and nucleic acids. *Nucleic Acids Res* 35: W375–W383.

4 Adams, P.D., Afonine, P.V., Bunkoczi, G. et al. (2010). PHENIX: a comprehensive python-based system for macromolecular structure solution. *Acta Cryst Sect D* 66: 213–221.

5 Kleywegt, G.J. (2007). Crystallographic refinement of ligand complexes. *Acta Cryst Sect D* 63: 94–100.

6 Kleywegt, G.J., Henrick, K., Dodson, E.J., and van Aalten, D.M.F. (2003). Pound-wise but penny-foolish: how well do micromolecules fare in macromolecular refinement? *Structure* 11: 1051–1059.

7 Read, R.J., Adams, P.D., Arendall, W.B. III, et al. (2011). A new generation of crystallographic validation tools for the protein data bank. *Structure* 19: 1395–1412.

8 Ryde, U., Olsen, L., and Nilsson, K. (2002). Quantum chemical geometry optimizations in proteins using crystallographic raw data. *J Comput Chem* 23: 1058–1070.

9 Caldararu, O., Manzoni, F., Oksanen, E. et al. (2020). Refinement of protein structures using a combination of quantum-mechanical calculations with neutron and X-ray crystallographic data. Corrigendum. *Acta Crystallogr D Struct Biol* 76: 85–86.

10 Nilsson, K. and Ryde, U. (2004). Protonation status of metal-bound ligands can be determined by quantum refinement. *J Inorg Biochem* 98: 1539–1546.

11 Rulíšek, L. and Ryde, U. (2006). Structure of reduced and oxidized manganese superoxide dismutase: a combined computational and experimental approach. *J Phys Chem B* 110: 11511–11518.

12 Ryde, U. and Nilsson, K. (2003). Quantum chemistry can locally improve protein crystal structures. *J Am Chem Soc* 125: 14232–14233.

13 Yu, N., Yennawar, H.P., and Merz, K.M. (2005). Refinement of protein crystal structures using energy restraints derived from linear-scaling quantum mechanics. *Acta Cryst Sect D* 61: 322–332.

14 Bergmann, J., Oksanen, E., and Ryde, U. (2022). Combining crystallography with quantum mechanics. *Curr Opin Struct Biol* 72: 18–26.

15 Fu, Z., Li, X., and Merz, K.M. (2011). Accurate assessment of the strain energy in a protein-bound drug using QM/MM X-ray refinement and converged quantum chemistry. *J Comput Chem* 32: 2587–2597.

16 Li, X., Hayik, S.A., and Merz, K.M. (2010). QM/MM X-ray refinement of zinc metalloenzymes. *J Inorg Biochem* 104: 512–522.

17 Yu, N., Li, X., Cui, G. et al. (2006). Critical assessment of quantum mechanics based energy restraints in protein crystal structure refinement. *Protein Sci* 15: 2773–2784.

18 Yu, N., Hayik, S.A., Wang, B. et al. (2006). Assigning the protonation states of the key aspartates in β-secretase using QM/MM X-ray structure refinement. *J Chem Theory Comput* 2: 1057–1069.

19 Borbulevych, O.Y., Plumley, J.A., Martin, R.I. et al. (2014). Accurate macromolecular crystallographic refinement: incorporation of the linear scaling, semiempirical quantum-mechanics program DivCon into the PHENIX refinement package. *Acta Cryst Sect D* 70: 1233–1247.

20 QuantumBio Inc. (2015) LibQB. (Inc, Q., ed., 6.0, www.quantumbioinc.com (Ed.), QuantumBio Inc.

21 Dixon, S.L. and Merz, K.M. (1996). Semiempirical molecular orbital calculations with linear system size scaling. *J Chem Phys* 104: 6643–6649.

22 Dixon, S.L. and Merz, K.M. (1997). Fast, accurate semiempirical molecular orbital calculations for macromolecules. *J Chem Phys* 107: 879–893.

23 Diller, D.J., Humblet, C., Zhang, X.H., and Westerhoff, L.M. (2010). Computational alanine scanning with linear scaling semiempirical quantum mechanical methods. *Proteins* 78: 2329–2337.

24 Raha, K., Peters, M.B., Wang, B. et al. (2007). The role of quantum mechanics in structure-based drug design. *Drug Discov Today* 12: 725–731.

25 Raha, K., van der Vaart, A.J., Riley, K.E. et al. (2005). Pairwise decomposition of residue interaction energies using semiempirical quantum mechanical methods in studies of protein-ligand interaction. *J Am Chem Soc* 127: 6583–6594.

26 van der Vaart, A. and Merz, K.M. (1999). Divide and conquer interaction energy decomposition. *J Phys Chem A* 103: 3321–3329.

27 Zhang, X.H., Gibbs, A.C., Reynolds, C.H. et al. (2010). Quantum mechanical pairwise decomposition analysis of protein kinase B inhibitors: validating a new tool for guiding drug design. *J Chem Inf Model* 50: 651–661.

28 Vreven, T., Morokuma, K., Farkas, Ö. et al. (2003). Geometry optimization with QM/MM, ONIOM, and other combined methods. I. Microiterations and constraints. *J Comput Chem* 24: 760–769.

29 Borbulevych, O., Martin, R.I., and Westerhoff, L.M. (2018). High-throughput quantum-mechanics/molecular-mechanics (ONIOM) macromolecular crystallographic refinement with PHENIX/DivCon: the impact of mixed Hamiltonian methods on ligand and protein structure. *Acta Cryst Sect D* 74: 1063–1077.

30 Brünger, A.T., Adams, P.D., Clore, G.M. et al. (1998). Crystallography & NMR system: a new software suite for macromolecular structure determination. *Acta Crystallogr D Biol Crystallogr* 54: 905–921.

31 Van der Vaart, A., Gogonea, V., Dixon, S.L., and Merz, K.M. (2000). Linear scaling molecular orbital calculations of biological systems using the semiempirical divide and conquer method. *J Comput Chem* 21: 1494–1504.

32 Van der Vaart, A., Suarez, D., and Merz, K.M. (2000). Critical assessment of the performance of the semiempirical divide and conquer method for single point calculations and geometry optimizations of large chemical systems. *J Chem Phys* 113: 10512–10523.

33 Wang, B., Westerhoff, L.M., and Merz, K.M. (2007). A critical assessment of the performance of protein-ligand scoring functions based on NMR chemical shift perturbations. *J Med Chem* 50: 5128–5134.

34 Dewar, M.J.S., Zoebisch, E.G., Healy, E.F., and Stewart, J.J.P. (1985). The development and use of quantum-mechanical molecular-models. 76. Am1 – a new general-purpose quantum-mechanical molecular-model. *J Am Chem Soc* 107: 3902–3909.

35 Rezac, J., Fanfrlik, J., Salahub, D., and Hobza, P. (2009). Semiempirical quantum chemical PM6 method augmented by dispersion and H-bonding correction terms reliably describes various types of noncovalent complexes. *J Chem Theory Comput* 5: 1749–1760.

36 Stewart, J.J.P. (2009). Application of the PM6 method to modeling proteins. *J Mol Model* 15: 765–805.

37 Case, D.A., Babin, V., Berryman, J.T. et al. (2014). *AMBER 14*. San Francisco: University of California.

38 Bricogne, G., Blanc, E., Brandl, M. et al. (2017). *BUSTER*. Cambridge, United Kingdom: Global Phasing Ltd.

39 QuantumBio. (2022) LibQB. (Inc, Q. ed., 7.0, www.quantumbioinc.com (Ed.), QuantumBio Inc.

40 Bietz, S., Urbaczek, S., Schulz, B., and Rarey, M. (2014). Protoss: a holistic approach to predict tautomers and protonation states in protein-ligand complexes. *J Chem* 6: 12.

41 Labute, P. (2005). On the perception of molecules from 3D atomic coordinates. *J Chem Inf Model* 45: 215–221.

42 Mukhopadhyay, A.K. and Mukherjee, N.G. (1981). Self-consistent methods in Hückel and extended Hückel theories. *Int J Quant Chem* 19: 515–519.

43 Tickle, I. (2012). Statistical quality indicators for electron-density maps. *Acta Cryst Sect D* 68: 454–467.

44 Chen, V.B., Arendall, W.B., Headd, J.J. et al. (2010). MolProbity: all-atom structure validation for macromolecular crystallography. *Acta Cryst Sect D* 66: 12–21.

45 Janowski, P.A., Moriarty, N.W., Kelley, B.P. et al. (2016). Improved ligand geometries in crystallographic refinement using AFITT in PHENIX. *Acta Cryst Sect D* 72: 1062–1072.

46 Mobley, D.L. and Dill, K.A. (2009). Binding of small-molecule ligands to proteins: "what you see"; is not always "what you get". *Structure* 17: 489–498.

47 Perola, E. and Charifson, P.S. (2004). Conformational analysis of drug-like molecules bound to proteins: an extensive study of ligand reorganization upon binding. *J Med Chem* 47: 2499–2510.

48 Fu, Z., Li, X., and Merz, K.M. (2012). Conformational analysis of free and bound retinoic acid. *J Chem Theory Comput* 8: 1436–1448.

49 Borbulevych, O.Y., Plumley, J.A., and Westerhoff, L.M. (2012). Systematic study of the ligand strain energy derived from the quantum mechanics crystallographic refinement using the linear scaling program DivCon integrated into the PHENIX package. *Abstr Pap Am Chem Soc* 478.

50 Borbulevych, O., Martin, R.I., Tickle, I.J., and Westerhoff, L.M. (2016). XModeScore: a novel method for accurate protonation/tautomer-state determination using quantum-mechanically driven macromolecular X-ray crystallographic refinement. *Acta Cryst Sect D* 72: 586–598.

51 MacCallum, J.L., Hua, L., Schnieders, M.J. et al. (2009). Assessment of the protein-structure refinement category in CASP8. *Proteins* 77: 66–80.

52 Word, J.M., Lovell, S.C., LaBean, T.H. et al. (1999). Visualizing and quantifying molecular goodness-of-fit: small-probe contact dots with explicit hydrogen atoms. *J Mol Biol* 285: 1711–1733.

53 Cozier, G.E., Leese, M.P., Lloyd, M.D. et al. (2010). Structures of human carbonic anhydrase II/inhibitor complexes reveal a second binding site for steroidal and nonsteroidal inhibitors. *Biochemistry* 49: 3464–3476.

54 Harding, M.M. (1999). The geometry of metal-ligand interactions relevant to proteins. *Acta Cryst Sect D* 55: 1432–1443.

55 Allen, F.H., Kennard, O., Watson, D.G. et al. (1987). Tables of bond lengths determined by X-ray and neutron-diffraction. 1. Bond lengths in organic-compounds. *J Chem Soc Perkin Trans* 2: S1–S19.

56 Adams, P.D., Pannu, N.S., Read, R.J., and Brunger, A.T. (1997). Cross-validated maximum likelihood enhances crystallographic simulated annealing refinement. *Proc Natl Acad Sci U S A* 94: 5018–5023.

57 Afonine, P.V., Grosse-Kunstleve, R.W., Echols, N. et al. (2012). Towards automated crystallographic structure refinement with phenix.refine. *Acta Cryst Sect D* 68: 352–367.

58 Hartshorn, M.J., Verdonk, M.L., Chessari, G. et al. (2007). Diverse, high-quality test set for the validation of protein-ligand docking performance. *J Med Chem* 50: 726–741.

59 Rupp, B. (2009). *Biomolecular crystallography: principles, practice, and application to structural biology*. Garland Science.

60 Ryde, U. and Nilsson, K. (2003). Quantum refinement—a combination of quantum chemistry and protein crystallography. *J Mol Struct THEOCHEM* 632: 259–275.

61 Martin, Y.C. (2009). Let's not forget tautomers. *J Comput Aided Mol Des* 23: 693–704.

62 USP-DI (1995). *United States pharmacopeia*, 15the, 659. Rockville, MD: The United States Pharmacopeial Convention Inc.

63 Moldow, B., Sander, B., Larsen, M., and Lund-Andersen, H. (1999). Effects of acetazolamide on passive and active transport of fluorescein across the normal BRB. *Invest Ophthalmol Vis Sci* 40: 1770–1775.

64 Krishnamurthy, V.M., Kaufman, G.K., Urbach, A.R. et al. (2008). Carbonic anhydrase as a model for biophysical and physical-organic studies of proteins and protein-ligand binding. *Chem Rev* 108: 946–1051.

65 Merz, K.M. and Banci, L. (1997). Binding of bicarbonate to human carbonic anhydrase II: a continuum of binding states. *J Am Chem Soc* 119: 863–871.

66 Sippel, K.H., Robbins, A.H., Domsic, J. et al. (2009). High-resolution structure of human carbonic anhydrase II complexed with acetazolamide reveals insights into inhibitor drug design. *Acta Cryst SectF* 65: 992–995.

67 Fisher, S.Z., Aggarwal, M., Kovalevsky, A.Y. et al. (2012). Neutron diffraction of acetazolamide-bound human carbonic anhydrase II reveals atomic details of drug binding. *J Am Chem Soc* 134: 14726–14729.

68 Borbulevych, O.Y., Martin, R.I., and Westerhoff, L.M. (2021). The critical role of QM/MM X-ray refinement and accurate tautomer/protomer determination in structure-based drug design. *J Comput Aided Mol Des* 35: 433–451.

69 Dunbar, J.B. Jr., Smith, R.D., Damm-Ganamet, K.L. et al. (2013). CSAR data set release 2012: ligands, affinities, complexes, and docking decoys. *J Chem Inf Model* 53: 1842–1852.

70 Corbeil, C.R., Williams, C.I., and Labute, P. (2012). Variability in docking success rates due to dataset preparation. *J Comput Aided Mol Des* 26: 775–786.

71 Jumper, J., Evans, R., Pritzel, A. et al. (2021). Highly accurate protein structure prediction with AlphaFold. *Nature* 596: 583–589.

72 Wang, H.W. and Wang, J.W. (2017). How cryo-electron microscopy and X-ray crystallography complement each other. *Protein Sci* 26: 32–39.

73 Merino, F. and Raunser, S. (2017). Electron cryo-microscopy as a tool for structure-based drug development. *Angew Chem Int Ed* 56: 2846–2860.

74 Shoemaker, S.C. and Ando, N. (2018). X-rays in the cryo-electron microscopy era: structural biology's dynamic future. *Biochemistry* 57: 277–285.

75 Afonine, P.V., Klaholz, B.P., Moriarty, N.W. et al. (2018). New tools for the analysis and validation of cryo-EM maps and atomic models. *Acta Crystallogr D Struct Biol* 74: 814–840.

76 McNicholas, S., Croll, T., Burnley, T. et al. (2018). Automating tasks in protein structure determination with the clipper python module. *Protein Sci* 27: 207–216.

8

Quantum-Chemical Analyses of Interactions for Biochemical Applications

Dmitri G. Fedorov

Research Center for Computational Design of Advanced Functional Materials (CD-FMat), National Institute of Advanced Industrial Science and Technology (AIST), Central 2, Umezono 1-1-1, Tsukuba 305-8568, Japan

8.1 Introduction

Interactions [1] in molecular systems determine their behavior. Building reliable models of interactions is the pivotal task of theoretical physics and chemistry. Quantum-mechanical (QM) methods are suitable for building these models because they are capable of describing all kinds of interactions in chemistry, including those that are difficult to obtain in faster models such as force fields.

The application of QM methods to biomolecules is challenging, given the high computational cost of these simulations. Biomolecules are not only large, but they also have the additional complexity of flexibility, which needs to be taken into account by sampling the conformational space at a given temperature and computing free energy. For biochemical processes, it may be necessary to describe a solvent, and other constituents of solutions, such as counterions.

Driven by the progress in method development and computational hardware, QM methods are increasingly used in computational drug discovery [2, 3]. There are a variety of fragment-based methods [4, 5], which not only reduce the computational cost but also deliver the properties of molecular parts, which can be useful for fragment-based drug discovery (FBDD) [6]. Interaction energies are used in structure- and interaction-based drug design (SIBDD) [7] for the rational enhancement of binding affinities.

In practice, some compromises have to be made in the application of QM methods to biochemical studies. Atomic structures are often refined in molecular mechanics (MM) simulations, although there are fast parametrized QM methods [8], such as density-functional tight-binding (DFTB) [9] that are adequate for full geometry optimizations of biochemical systems. These parametrized methods can also be used for molecular dynamics (MD) simulations, taking into account temperature effects.

Dividing biochemical systems into small parts (fragments) is very attractive because not only is the computational cost drastically reduced, but detailed properties of fragments can also be obtained. Among them, interactions between

Computational Drug Discovery: Methods and Applications, First Edition.
Edited by Vasanthanathan Poongavanam and Vijayan Ramaswamy.
© 2024 WILEY-VCH GmbH. Published 2024 by WILEY-VCH GmbH.

fragments are very useful, providing quantitative information on the role of residues, or individual functional groups, for molecular recognition in protein-ligand, protein-DNA, and protein–protein binding. Likewise, enzymes can be treated gaining valuable insight into the contributions of residues in lowering a reaction barrier.

The energy decomposition analysis (EDA) [10, 11] has been an important conceptual starting point for the fragment molecular orbital (FMO) method [12–15]. FMO-based analyses suitable for biochemical studies are presented here, with a brief methodological description and a review of their applications.

8.2 Introduction to FMO

In FMO, a molecular system is divided into fragments. There are well-defined patterns for fragmenting all major types of biochemical systems, which are accomplished automatically using graphical user interfaces, such as Facio [16]. Polypeptides (proteins) can be automatically divided into amino acid residues (Figure 8.1), polynucleotides (DNA and RNA) into nucleotides, polysaccharides (cellulose, heparin, etc.) into saccharides, and explicit solvents into individual molecules.

It is possible either to calculate whole proteins [17] or truncate [18] them, keeping only residues within a suitable threshold (~5 Å) from the ligand [19], although it was argued [20] that for charged ligands the threshold has to be larger (by ~1 Å when optimizing geometry).

A pattern for fragmentation is chosen to balance two factors: fragments should be as close as possible to commonly used biochemical units, and the error of fragmentation (the difference between FMO energies and the values without fragmentation) should be as small as possible. In FMO, fragments are similar but not identical to the commonly used biochemical units. Namely, for proteins, a residue fragment in FMO has the same chemical composition as a conventional residue, but the border between two fragments is shifted so that the fragment for residue i includes the carbonyl of residue i-1 rather than its own carbonyl (Figure 8.1). For polynucleotides, fragments have the same composition as conventional nucleotides, but phosphate

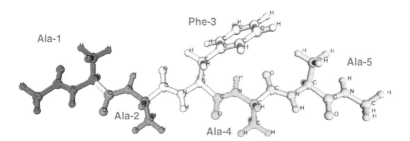

Figure 8.1 Automatic fragmentation of the AAFAA polypeptide into five residue fragments, whose names by convention include a dash to distinguish them from conventional residues. Terminal fragments include caps.

groups are likewise assigned to adjacent fragments. This is done to avoid having a fragment boundary at C—N (peptide) and P—O (nucleotide) bonds, which have a strong delocalized character, ill-suited for a QM fragmentation.

After N fragments are defined, FMO calculations can be conducted for a chosen QM level: wave function, basis set, and solvent model. First, individual fragments (monomers) are calculated in the electrostatic (ES) embedding, followed by calculations of pairs (dimers), and, optionally, triples (trimers). Combining these results, one obtains the total energy E. In a three-body expansion (FMO3), the energy is

$$E = \sum_{I}^{N} E'_I + \sum_{I>J}^{N} \Delta E_{IJ} + \sum_{I>J>K}^{N} \Delta E_{IJK} \qquad (8.1)$$

where E'_I is the internal energy of polarized fragment I, ΔE_{IJ} is the pair interaction energy (PIE) between fragments I and J, and ΔE_{IJK} is the coupling of pair interactions in trimer IJK. The most commonly used method is FMO2, in which the last term in Eq. (8.1) is omitted.

Polarization energies [21] can be obtained by computing fragments with and without the electrostatic embedding. In most FMO analyses, polarization is contained implicitly without an explicit separation of a polarization contribution.

It is possible to compute analytic gradients of E, optimize geometry [22], and perform MD simulations using FMO [23]. Molecular structures can be refined using the frozen domain FMO [20, 24] and FMO-DFTB [25] methods. The latter approach can be used for FMO/MD simulations [26]. Partial geometry optimizations with density functional theory (DFT) and full optimizations with DFTB can be conducted with FMO for realistic atomic models containing thousands of atoms. Infrared (IR) and Raman spectra of proteins can be computed [27, 28].

Solvent can be treated both as explicit molecules or implicitly as a continuum in the polarizable continuum model (PCM) [29] and the solvent model density (SMD) method [30]. Analyses with explicit solvents are complicated by the conformational aspect, whereas implicit continuum models are easy to use. Because biochemical processes typically involve charged species in solution, solvent effects are of paramount importance.

Periodic boundary conditions (PBC) can be used for FMO-DFTB [31], making it possible to compute liquids and solutions [32], molecular crystals (e.g. of ice [33] and proteins [34]), and solid state of inorganic materials [35]. Some analyses can be combined with PBC, as described below.

QM calculations with FMO-DFTB can be conducted for molecular systems containing more than 1 million atoms [36, 37], whereas ab initio methods (second-order Møller–Plesset perturbation theory, MP2) for thousands of atoms can be routinely done [17].

Analyses described in this chapter can be performed with FMO implemented [38–40] in GAMESS [41]. Molecular electrostatic potential (MEP) [42], taking into account polarization and charge transfer, can be computed using FMO to guide ligand docking. MEP can be used to visualize electrostatic complementarity, for example, in protein–protein complexes [43].

8.3 Pair Energy Decomposition Analysis (PIEDA)

Pair interaction energies ΔE_{IJ} in Eq. (8.1) quantify the strength of interactions, often taken as a measure of binding, although binding and interaction energies are not the same. Binding energies are defined with respect to the reference state of isolated, noninteracting species, whereas interaction energies in FMO are defined with respect to polarized fragments [40].

One can argue that for molecules, the isolated state might be a better reference, whereas for covalently bound fragments (as residues in a protein), the naturally embedded polarized state is more suitable.

For example, a PIE between two polarized water molecules in a water droplet can be taken as a measure of the hydrogen bond energy. In a vacuum, a pair interaction may be slightly more attractive than the corresponding binding energy because it includes the stabilization part of the polarization [21]. In solution, interaction and binding energies differ more, as explained below.

PIEs between two fragments connected by a covalent bond reflect the effects of separating electrons in the bond detached atom (BDA) on the interfragment border between two fragments. Such PIEs are quite large, and they are seldom used in analyses, although it is possible to split the artificial component using BDA corrections [44].

PIEs can be decomposed into components for the purpose of gaining deeper physicochemical insight. A decomposition has to be designed for each QM method and solvent model, yielding quantitative information about the role of different physical interactions as described next.

8.3.1 Formulation of PIEDA

The pair interaction energy decomposition analysis (PIEDA) [44–46] is based on ideas from EDA [10]. The original PIEDA was formulated for FMO2 [44], later generalized to FMO3 [47]. These decompositions can be thought of as FMOn combined with EDA, so they are referred to as FMOn-EDA (PIEDA is FMO2-EDA). FMO1-EDA is used for defining polarization energies.

PIEDA can be done for DFTB, Hartree–Fock (HF), DFT, second-order Møller–Plesset perturbation theory (MP2), and coupled cluster, e.g. CCSD(T). The components in PIEDA differ depending on the QM method.

Taking as an example PIEDA for MP2, a commonly used method in biochemical applications, the decomposition of a PIE is

$$\Delta E_{IJ} = \Delta E_{IJ}^{ES} + \Delta E_{IJ}^{EX} + \Delta E_{IJ}^{CT+MIX} + \Delta E_{IJ}^{DI+RC} + \Delta E_{IJ}^{solv} \qquad (8.2)$$

with the following components: electrostatic (ES), exchange-repulsion (EX), charge transfer (CT) and mix coupling terms (CT + MIX), dispersion (DI) and remainder correlation (RC), and solvent screening (solv). The convention is to use capital and small letters for solute and solvent terms, respectively.

The ES term is the Coulomb interaction between two polarized fragments, including the stabilization component of the polarization. ES describes the interaction

8.3 Pair Energy Decomposition Analysis (PIEDA)

between charge distributions (electron density clouds and point charges of protons in the nuclei) in a vacuum. This interaction can be very strong for charged fragments. In solution, there is a solvent screening solv term, which typically reduces the Coulomb (ES) interaction. The solv term is computed as

$$\Delta E_{IJ}^{solv} = \Delta E_{IJ}^{es} + \Delta E_{IJ}^{non\text{-}es} \tag{8.3}$$

where es and non-es are the solute-solvent electrostatic and non-electrostatic screening interactions, respectively.

The es term is defined in continuum models, PCM or SMD, whereas the non-es term is present in PCM only. The es term is computed as the interaction of the solute charge distribution with induced solvent charges of the solvent. There are two models for the es term: local [45] and partial [48]. They differ in the definition of solvent charges. In the local model, solvent charges induced by the combined potential of all fragments are divided among fragments geometrically. In the partial model, solvent charges are induced by the partial potential of individual fragments. The charge quenching effect [45] (the cancelation of the solvent charges due to the partial potentials of oppositely charged fragments) is responsible for a large underestimation of the solvent screening in the local model. So the partial model, which has some extra cost, is the preferred way of defining solvent screening.

It is useful to add mutually canceling ES and es terms, producing the solute–solute electrostatic interaction screened in solution (ES + es). The ES and ES + es terms are long-ranged interactions, slowly decaying with interfragment separation. If two fragments I and J are sufficiently separated, the interaction energy can be computed as

$$\Delta E_{IJ} \approx \Delta E_{IJ}^{ES} + \Delta E_{IJ}^{solv} \tag{8.4}$$

which reduces the cost of FMO calculations very considerably.

The exchange-repulsion (EX) interaction arises due to the Pauli exclusion principle, describing the repulsion of two fermions (electrons). It corresponds to the repulsive part of the Lennard-Jones potential. In QM methods, the EX term is rigorously computed based on the wave function. Without this repulsion, two ions of the opposite charge would stick to each other. EX is a short-ranged interaction, which arises whenever two fragments are strongly attracted to each other. Thus, EX is an inevitable companion of a strong attraction. However, EX may be substantial without a strong attraction due to a steric repulsion in a poorly optimized structure. If a large EX term is found without other attractive terms, it is an indication of a need to refine the structure, although it may be inevitable that for one pair of fragments to be strongly attracted, another pair may be forced into repulsion.

The RC + DI term is the contribution of the electron correlation, some part of which is the dispersion (DI) interaction, and the rest is called the RC. For DFT with empirical dispersion, the RC and DI terms are separable. The DI term corresponds to the attractive term of the Lennard-Jones potential describing the van-der-Waals interactions, as pertinent to hydrophobic contacts in biochemical systems. The RC term describes non-dispersive interaction due to the electron correlation [49].

Basis set superposition error (BSSE) can be corrected using the auxiliary polarization (AP) method [50] or HF-3c [51], with a basis set (BS) term ΔE_{IJ}^{BS} added to

Eq. (8.2). HF-3c is Hartree-Fock with three corrections (3c): empirical dispersion, short- and long-ranged BSSE corrections. For PIEDA/AP, each interaction term (see Eq. (8.2)) can be BSSE-corrected [40].

In DFTB, the RC and EX terms are combined in the so-called 0-order term (0-order refers to the Taylor expansion of the electron density in DFTB), and due to the parametrization, the two components cannot be separated.

In FMO3/EDA, a three-body interaction ΔE_{IJK} in Eq. (8.1) can be decomposed for MP2 as

$$\Delta E_{IJK} = \Delta E_{IJK}^{EX} + \Delta E_{IJK}^{CT+MIX} + \Delta E_{IJK}^{RC+DI} + \Delta E_{IJK}^{solv} \qquad (8.5)$$

Comparing Eqs. (8.2) and (8.5), it can be seen that an ES term is absent in the latter. It is because ES is purely two-body without three-body corrections. The same applies to DI and BS in empirical models (HF-3c).

Sometimes, FMO/EDA3 terms are compressed [52, 53] into three-body corrected effective two-body terms. This can be done for the total PIE and for individual components. The contracted PIE is defined as,

$$\Delta \tilde{E}_{IJ} = \Delta E_{IJ} + \frac{1}{3} \sum_{K \neq I,J}^{N} \Delta E_{IJK} \qquad (8.6)$$

The total interaction energy (TIE) can be computed for the binding of a ligand I to a protein via summing over residues J in the protein as

$$\Delta E_I = \sum_{J \in protein} \Delta E_{IJ} \qquad (8.7)$$

For drug design, it may be useful to split a large ligand into several fragments. Then, for each ligand fragment I, its fragment efficiency [54] can be defined using the number of heavy (non-hydrogen) atoms N_I as

$$\Delta \overline{E}_I = \frac{\Delta E_I}{N_I} \qquad (8.8)$$

By comparing $\Delta \overline{E}_I$ for different ligand fragments I, a decision can be made to replace or remove those parts that contribute little, guiding drug design.

Repulsions can be excluded from analyses by defining $\Delta \tilde{E}_I$ as ΔE_I minus the EX interaction. It can be useful if the structure optimization is not done with the same QM method as the analysis.

For a clear labeling of residue contributions, it was suggested [55] to define the fraction of a component A to binding,

$$f_I^A = \frac{\Delta E_I^A}{\Delta \tilde{E}_I} \qquad (8.9)$$

where A can be ES, DI, or CT + MIX (in the original scheme, the solvent screening was not considered). In other words, for each residue I, its ligand binding is represented by a composite "color" assigned in an RGB-like scheme, with three primary colors (A = electrostatic, dispersion, or charge transfer) mixed with fractions f_I^A. The fractions are normalized to 1 for each I,

$$\sum_A f_I^A = 1 \qquad (8.10)$$

Using the fractions, signature ratios were defined [55],

$$1 : \frac{f_I^{DI}}{f_I^{ES}} : \frac{f_I^{CT+MIX}}{f_I^{ES}} \quad (8.11)$$

with which the physical type of the contribution of residue I can be easily identified. Using the fractions in Eq. (8.11), it was shown [55] that dispersion and electrostatics contribute comparably to a wide range of representative GPCR proteins. As a further simplification, fractions can be grouped into two, rather than three, categories: electrostatic (ES plus CT + MIX) and dispersion (DI) [56, 57].

PIEs can be used as descriptors in structure–activity relationships (SAR) [58], in particular, in quantitative SAR (QSAR) [59]. There is an ongoing effort to use PIE [60] and TIE [61] in combination with machine learning, to predict free energies of binding based on FMO.

8.3.2 Applications of PIEs and PIEDA

For cross-validation, PIEs were compared for two different QM methods, and TIEs were compared to the experiment, as summarized in Table 8.1. There are two aspects of the validation: verify that QM results have a good correlation to the experiment (of course, not perfect, because some effects such as entropy are not accounted for in QM calculations) and test if very fast QM methods like DFTB or HF-3c can be reliably used for a high-throughput drug design pipelines.

Questions that can be answered by considering the signatures of PIE components include: should CH...O bonds be classified as weak hydrogen bonds or do they have a different physical nature? Answer: the latter [67]. What is the nature of CH-π bonding that is frequently found in biochemical systems [68]? A singular value decomposition can be applied to PIEs [69] for a clear identification of the principal factors in protein–ligand binding. Some representative applications of PIEs are summarized in Table 8.2.

8.3.3 Example of PIEDA

To show the kind of data that can be obtained, PIEDA is applied to a protein-ligand complex of Trp-cage (PDB: 1L2Y) with deprotonated p-phenolic acid (an anion) [20] (Figure 8.2). The structures of the complex and isolated proteins and ligands were optimized at the level of FMO2-DFTB3/D3(BJ)/C-PCM<1> using 3ob parameters [81]. The protein was divided into 1 residue per fragment, and the ligand was a separate fragment.

PIEDA calculations for the complex were done at the level of MP2/PCM/cc-pVDZ. PIEs for fragments are shown in Figure 8.3. According to this plot, there are strong residue–ligand interactions involving residue fragments Gln-5, Pro-18, and Trp-6. The ligand interacts mainly with the two terminal sections of the protein, as can be seen in Figure 8.2.

Individual components describing the nature of interactions are shown in Figure 8.4. For Gln-5, a strong attraction is found in the CT + MIX, RC + DI, and

Table 8.1 Validation of PIEs and TIEs to other computational and experimental results.

Methods	System (PDB ID)	Correlation	Reference
MP2 vs experiment	OX$_2$ orexin (4S0V)	TIE vs pEC$_{50}$ ($R^2 = 0.872$).	[62]
MP2 vs experiment	OX$_2$ orexin receptor (4S0V) β$_2$ adrenergic receptor (3SN6) κ opioid receptor (4DJH) P2Y$_{12}$ receptor (4NTJ)	TIE vs pK$_i$ ($R^2 = 0.748$), pEC$_{50}$ ($R^2 = 0.729$), pK$_e$ ($R^2 = 0.576$), and pIC$_{50}$ ($R^2 = 0.763$).[a]	[55]
DFTB vs experiment	β$_2$ adrenergic receptor (3SN6) κ opioid receptor (4DJH) P2Y$_{12}$ receptor (4NTJ)	TIE vs pK$_i$ ($R^2 = 0.783$), pEC$_{50}$ ($R^2 = 0.662$), and pIC$_{50}$ ($R^2 = 0.812$).[a]	[63]
MP2 vs experiment	Cyclin-dependent kinase-2 inhibitor (4FKL, etc.)	TIE vs ΔG ($R^2 = 0.99$),	[64]
MP2 vs experiment	Estrogen receptor β (7XVY, etc.)	TIE vs ΔH ($R^2 = 0.870$)	[65]
DFTB vs MP2	β$_2$ adrenergic receptor (3SN6) κ opioid receptor (4DJH) P2Y$_{12}$ receptor (4NTJ)	TIEs ($R^2 = 0.943$, $R^2 = 0.913$, and $R^2 = 0.959$).[a]	[63]
DFTB vs MP2	Trp-cage (1L2Y)	PIEs ($R^2 = 0.990$ and 0.988).[b]	[66]
HF-3c vs MP2	Trp-cage (1L2Y)	PIEs ($R^2 = 0.999$ and 0.983).[b]	[51]

a) Each of these R^2 values is for multiple ligands bound to a single protein.
b) Each of these R^2 values is for multiple residues bound to a single ligand.

ES + solv terms, compensated partially by repulsion in the EX term, as is usually the case. RC + DI has substantial contributions, mainly because of the phenyl ring, and the hydrophobic RC + DI term is roughly on par with the screened electrostatics ES + solv, although the latter is somewhat larger on average (the ligand has a charge of −1).

8.4 Partition Analysis (PA)

Pair interaction energies between fragments are very useful, but there are three problems with them: (1) fragments differ (albeit slightly) from conventional units (residues or nucleotides), (2) fragment pairs with a covalent boundary between the two fragments have a large artificial interaction, and (3) it is not feasible to get functional group contributions. All of these problems are solved by the use of

Table 8.2 Representative applications of PIEs.

Level	Task	References
ROMP2	Prediction of the hydrogen abstraction site for radical damage in lipids.	[70]
MP2	Hit-to-lead phase in drug design using SAR-by-FMO (CDK2 receptor).	[39, 54]
MP2	Signatures of interaction types in GPCR proteins.	[55, 57]
MP2	Lead optimization for kinase inhibition.	[19]
MP2	Discovery of a preclinical candidate for the treatment of type 2 diabetes.	[71]
MP2	Formation of an amorphous formulation in pharmaceutics.	[72]
MP2	Protein-residue topological network models of protein–protein complexes.	[73]
MP2	Dissipative particle dynamics using PIE-based parametrization.	[74]
MP2	Hot spot identification in protein–protein interfaces in relation to site-directed mutagenesis.	[75]
MP2	Protonation and pK_a in polynucleotides	[76]
RI-MP2 and DFTB	Virtual screening for metalloprotease.	[77]
RI-MP2	Virtual screening for SARS-CoV-2	[78]
RI-MP2	Combination of a molecular interaction field (MIF) with PIEs for ligand docking.	[79]
DFTB	Machine learning for predicting binding affinities to the SARS-CoV-2 spike glycoprotein.	[61]
DFTB	Ensemble docking drug discovery pipeline for COVID-19.	[80]

segments in the partition analysis (PA) [82] for electronic energies and the partition analysis of vibrational energies (PAVEs) [83].

What are segments? Segments are sets of atoms, like fragments. The most fundamental difference is that QM calculations are done for fragments, but not for segments. The QM results of fragments are post-processed (repartitioned) into segments, somewhat analogously to computing atomic charges for a converged wave function. Capital (I,J) and small (i,j) indices are used for fragments and segments, respectively.

There is no limit to the definition of segments. They can be conventional residues, functional groups, or even individual atoms. There is no accuracy loss due to the use of small segments because the partitioning of properties is exact.

FMO fragments are used for a fast computation of QM properties, which are at the end partitioned into segments in PA. It can be done for the electronic energy in DFTB or the vibrational energy in any QM method.

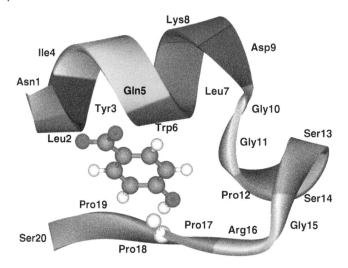

Figure 8.2 Protein–ligand complex for Trp-cage (1L2Y). The two yellow atoms are the carbonyl group of Pro-17 assigned to the Pro-18 fragment.

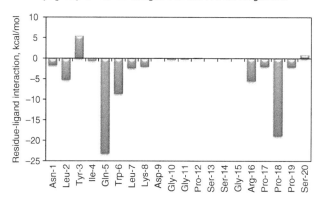

Figure 8.3 Total residue–ligand interactions ΔE_{IJ} (PIEDA) in the protein–ligand complex at the level of MP2/PCM/cc-pVDZ.

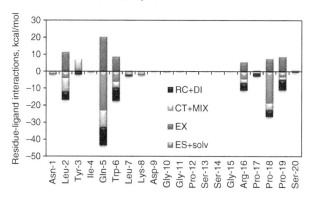

Figure 8.4 Components A of residue-ligand interactions ΔE_{IJ}^A (PIEDA) in the protein-ligand complex at the level of MP2/PCM/cc-pVDZ ($A =$ RC + DI, CT + MIX, EX, and ES + solv).

8.4 Partition Analysis (PA)

Table 8.3 Fragments vs segments as units for analyses.

Property	Fragments	Segments
QM method	Any	DFTB(elec), any (vib)[a]
Analyses	PIEDA	PA, PAVE
Bonds between units	Single	Any
Functional groups as units	No[b]	Yes
Residues/nucleotides	Shifted	Conventional[c]
Unit charges	Integer	Fractional
Invariance[d]	No	Yes
Use without FMO	No	Yes[e]

a) For electronic (elec) or vibrational (vib) energies.
b) Technically possible but not recommended due to artificial results.
c) Shifted residues can also be used, with segments defined identically to fragments.
d) Invariance of the total energies with respect to a different division into units.
e) Segments can be used in unfragmented calculations (technically accomplished as FMO with 1 fragment).

On a fragment boundary, one atom (BDA) is shared between two fragments [40], whereas no atom is shared between segments. Bonds of any order can be on a segment boundary, and PIEs for segment pairs with a covalent bond are in the same order as non-covalent PIEs.

Another major difference is that for segments, charge transfer is treated at a higher order than for fragments. It is accomplished by using FMOn atomic charges for defining the charges of segments, which are in general fractional. A summary of the differences is shown in Table 8.3.

8.4.1 Formulation of PA

The energy of M segments in PA [82] is (compare to Eq. (8.1)),

$$E = \sum_{i=1}^{M} E'_i + \sum_{i>j} \Delta E_{ij} \qquad (8.12)$$

where E'_i is the internal energy of segment i and the PIE for two segments is

$$\Delta E_{ij} = \Delta E_{ij}^{ES} + \Delta E_{ij}^{DI} + \Delta E_{ij}^{solv} \qquad (8.13)$$

PA does not have trimer terms, even if PA is conducted for post-processing FMO3 results. It is because of the nature of DFTB interaction terms, which involve at most two particles (two atoms).

There are three components, electrostatic (ES), dispersion (DI) and solvent screening (solv). They have the same meaning (but, in general, different values) as fragments. 0-order (i.e. EX+RC) and CT-related terms are absent in Eq. (8.13), and these interactions are incorporated into monomer values E'_i. As a result, ΔE_{ij} in PA has no repulsive term corresponding to EX in PIEDA.

PA may be used with DFTB only, possibly combined with PCM, SMD, or PBC. A PA calculation can use a PDB file, reading all important biochemical information, such as atomic and residue names. Conventional residues can be used exactly as defined in a PDB. Side chains can be automatically split from amino acid residues, whereas bases can be split from nucleotides. This produces two segments for each residue or nucleotide. Functional groups can be treated as separate segments by adjusting the residue ID in the PDB, which is used to index atoms in segments [40].

8.4.2 Applications and an Example of PA

PA/PBC was applied [34] to calculate the interactions between solvent molecules and proteins in protein crystals, where some superbinding water molecules were identified that formed 3 hydrogen bonds and 2 charge–dipole interactions with the protein. PA/PBC was used [33] to analyze a self-assembled monolayer of organic molecules on an ice surface, where the importance of dispersion for guest-guest binding in competition with guest-surface binding was shown. PA/PBC was applied [35] to analyze the catalytic activity of faujasite zeolite, where the importance of the charge delocalization and interactions of multiple zeolite segments in the transition state stabilization was shown.

As a demonstration of PA, it is applied to the protein–ligand complex used in Section 7.3.3, at the level of FMO2-DFTB3/3ob/D3(BJ)/PCM. In the Trp-cage protein, each conventional residue out of 20 is defined as a segment. The ligand is split into 3 segments: phenyl (Ph), hydroxyl (OH), and carboxylate (COO$^-$).

The interaction energies of residues with the three ligand segments are shown in Figure 8.5 (here, conventional residues are used for analysis, so that their names have no dash). The phenyl interacts with Gln5, Trp6, Pro17, and Pro19. The hydroxyl interacts with Trp6, Arg16, and Pro17. The carboxylate interacts with Ile5 and Leu2. Such quantitative information can be useful for FBDD, where segment efficiency $\Delta \overline{E}_i$ can be defined similarly to Eq. (8.8).

Essentially no repulsion is observed in PA (Figure 8.5), whereas some is seen in PIEDA (Figure 8.3). It is attributed to (a) the consistent use of the same method (DFTB) in the structure optimization and analysis (PA in Figure 8.5) and (b) the assignment of EX repulsion to monomer terms in PA without a corresponding PIE contribution.

In Figure 8.3, a strong interaction of the ligand is observed for Pro-18 fragment. However, according to Figure 8.5, Pro17 segment has a strong interaction. A close inspection reveals that it is because of the carbonyl of Pro17 that is assigned to Pro-18 in the fragment-based analysis. This carbonyl is shown in Figure 8.2.

Because solvent screening is taken into account, the binding of the anionic carboxylate segment is on par with the binding of neutral functional groups. Segment charges are fractional; for phenyl, carboxylate, and hydroxyl, the computed segment charges in the complex are 0.026, −0.868, and −0.131, respectively (a.u.). The sum of

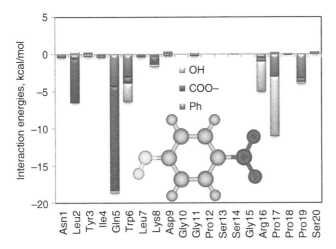

Figure 8.5 Interaction energies ΔE_{ij} (PA) of functional groups of the ligand with residues in the protein-ligand complex at the level of DFTB3/PCM.

them, −0.973, reflects the protein-ligand charge transfer (the formal ligand charge is −1).

PA offers a unique feature of evaluating the energy of individual bonds between functional groups. In the application of PA to a protein-DNA complex [84], nucleotides in the DNA were divided into phosphates, ring remainders, and tiny segments made of 2–3 atoms, which constitute functional groups (such as amides or carbonyls). By doing this, the energies of individual hydrogen bonds and CH...O interactions were defined in PA, explaining the relative strength of the nucleotide-nucleotide binding in the natural pairs C...G and A...T, as well as in a mutant pair G...T.

8.5 Partition Analysis of Vibrational Energy (PAVE)

PAVE [83] is used for computing the vibrational enthalpy, entropy, and free energy of individual segments. It is accomplished by computing a Hessian matrix of second derivatives of E with respect to nuclear coordinates either analytically or semi-numerically (by a numerical differentiation of analytic gradients), evaluating partition functions, and hence thermodynamical quantities. PAVE is computed at a given temperature for a single minimum (it does not take into account conformational entropy).

PAVE is done for a manually chosen selection of active atoms by constructing a partial Hessian for it. PAVE can be used for PBC (semi-numerical Hessians only). The decomposition of vibrational energies is done based on the localization of the eigenvectors of the Hessian (the squared coefficients summed for each atom are taken as the weight of the contribution of that atom, and atomic values are summed into segments).

8.5.1 Formulation of PAVE

The vibrational zero-point energy (ZPE) is decomposed into segment contributions as

$$E^{ZPE} = \sum_{i=1}^{M} E_i^{ZPE} \tag{8.14}$$

The vibrational (vib) enthalpy is

$$H^{vib} = E^{ZPE} + H^T \tag{8.15}$$

where the temperature (T) dependent enthalpy contribution is

$$H^T = \sum_{i=1}^{M} H_i^T \tag{8.16}$$

The vibrational Gibbs free energy is

$$G^{vib} = H^{vib} - TS^{vib} \tag{8.17}$$

where the entropy is

$$S^{vib} = \sum_{i=1}^{M} S_i^{vib} \tag{8.18}$$

By combining terms, the free energy is decomposed as

$$G^{vib} = \sum_{i=1}^{M} G_i^{vib} = \sum_{i=1}^{M} \left(E_i^{ZPE} + H_i^T - TS_i^{vib} \right) \tag{8.19}$$

PAVE has no pairwise contributions. Instead, only monomer vibrational properties are defined (e.g. G_i^{vib}), which correspond to electronic energies E_i' in PA. Segment contributions to a binding or activation energy can be computed by subtracting appropriate values of G_i^{vib} (e.g. for transition states and reactants). PAVE results can be combined with PA, for a composite electronic+vibrational analysis.

Like PA, PAVE can be used for unfragmented calculations (technically done as FMO with 1 fragment), although typically PAVE is used with FMO. PAVE is a post-processing of Hessian results, like PA is a post-processing of DFTB energies.

PAVE is sensitive to the accuracy of the Hessian, especially for H^T and S contributions, which are determined by low-frequency vibrations. The most accurate gradient (Hessian) should be used for a minimum located with a tight threshold (10^{-5} Eh/bohr or smaller) in geometry optimizations.

8.5.2 Applications of PAVE

PAVE was applied to evaluate the enthalpy and entropy of the guest binding on an ice surface, with a good agreement between computed and experimental enthalpies but a less favorable comparison of entropies [33].

PAVE was applied [35] to study guest binding to two types of zeolite crystals and the full production cycle of p-xylene catalyzed by faujasite zeolite. In both of these studies (adsorption and solid-state catalysis), the PAVE engine was used to get the total values of enthalpy H, entropy S, and free energy G, but individual segment contributions were not discussed. A demonstrative example of an application of PAVE is given below.

8.6 Subsystem Analysis (SA)

The analyses described above, PIEDA and PA, can be conducted for a single structure, for example, for a protein-ligand complex. Such a calculation can reveal insight into the complex stability. Interaction and binding energies are inherently different, and subsystem analysis (SA) [85] can be used to analyze the binding energies, which are more relevant to many biochemical applications.

8.6.1 Formulation of SA

For an FMO-based analysis of binding, the starting point is the equation that describes the process (a complex formation or a chemical reaction). Taking as an example a protein (A) – ligand (B) complex formation,

$$A + B \to AB \tag{8.20}$$

its binding (bind) energy is simply

$$\Delta E^{bind} = E^{AB} - E^A - E^B \tag{8.21}$$

using the FMO energies E in Eq. (8.1) computed for A, B, and AB. One can take electronic energies only, or add vibrational contributions in Eq. (8.19). It is possible to consider the effects of a structure's deformation (optimizing each system separately) or, as an approximation, neglect the deformation effects, optimizing only the complex and using its geometry for computing A and B separately.

The binding energy can be studied with SA, which requires at least 3 separate calculations (of AB, A, and B), and the arithmetic burden of subtracting numbers lies with the user. SA can be performed with either fragments or segments.

The energy decomposition in SA can be written for fragments as

$$\Delta E^{bind} = \sum_{I \in A} \Delta E_I^{part} + \sum_{I \in B} \Delta E_I^{part} + \sum_{I \in A} \sum_{J \in B} \Delta E_{IJ} \tag{8.22}$$

where ΔE_I^{part} is the difference in the partial (part) energies E_I^{part} of fragment I in the complex and isolated states; ΔE_I^{part} describes deformation, polarization, and desolvation (it is possible to decouple these three contributions and define them separately [40]). For residues I and J, the partial energy of I in the protein is

$$E_I^{part} = E_I' + \frac{1}{2} \sum_{J \neq I} \Delta E_{IJ} \tag{8.23}$$

The reason for using partial energies is to reduce the complexity of data. For example, when a ligand binds to a protein, it can affect residue–residue interactions (the ΔE_{IJ} term in Eq. (8.23)) via polarization. Usually, these effects are small, and they can be conveniently compressed into more manageable partial energies of residues E_I^{part} in the protein, so that the decomposition in Eq. (8.22) is formulated using differential partial energies ΔE_I^{part} of residues ($I \in A$) and ligand ($I \in B$), and residue-ligand interactions ΔE_{IJ}. The number of terms in Eq. (8.22) is linear with respect to the number of residues N^{res} (if the ligand is not fragmented, there are $2N^{res}+1$ terms). In contrast, the number of symmetric residue–residue interactions ΔE_{IJ} is quadratic $\sim N^{res2}/2$.

The difference in residue–residue interactions ΔE_{IJ} (differential $\Delta\Delta E_{IJ}$) contributes to the binding. The values of $\Delta\Delta E_{IJ}$ may be pertinent to rationalize allosteric regulation in proteins. Heat maps of ΔE_{IJ} in the complex and $\Delta\Delta E_{IJ}$ are shown in Figure 8.6 (fragment pairs connected by a covalent bond are excluded). Differential interactions reflect two effects: deformation (the structure is separately optimized for the bound and isolated states) and polarization of residue–residue interactions by the ligand. In other words, they explain how the ligand changes the protein. For absolute values in the complex, Lys-8 and Gln-5 are the happiest pair, and Pro-17 and Pro-12 are the unhappiest couple.

An isolated ligand (protein) is fully immersed in the solution, but in a complex, some solute-solvent interaction energy is lost (the desolvation penalty). Charged ligands can have a large attractive interaction energy ΔE_{IJ} with the protein and a large repulsive desolvation penalty in ΔE_I. The values of TIE (the last term in Eq. (8.22)) can be a large overestimate of the binding energy.

If there is just one ligand fragment J, then it is possible to simplify Eq. (8.22) as

$$\Delta E^{bind} = \sum_{I \in A,B} \Delta E_I^{bind} \tag{8.24}$$

where the binding energies ΔE_I^{bind} of fragments I include all relevant effects (deformation, polarization, desolvation, and interaction). For residue I and ligand J, $\Delta E_I^{bind} = \Delta E_I^{part} + \Delta E_{IJ}$ and $\Delta E_J^{bind} = \Delta E_J^{part}$. The values of ΔE_I^{bind} are better descriptors of binding than PIEs ΔE_{IJ}.

SA can be performed for segments. By combining PA, PAVE, and SA, the free energy of binding can be decomposed as [83]

$$\Delta G^{bind} = \Delta E^{bind} + \Delta G^{vib} = \sum_{i \in A,B} \Delta G_i^{bind} \tag{8.25}$$

where the free binding energy contribution of residue i is

$$\Delta G_i^{bind} = \Delta E_i^{part} + \Delta G_i^{vib} + \sum_{j \in ligand} \Delta E_{ij} \tag{8.26}$$

and the contribution of functional group i of ligand is

$$\Delta G_i^{bind} = \Delta E_i^{part} + \Delta G_i^{vib} \tag{8.27}$$

where ΔG_i^{vib} is obtained by subtracting G_i^{vib} for the complex and isolated species.

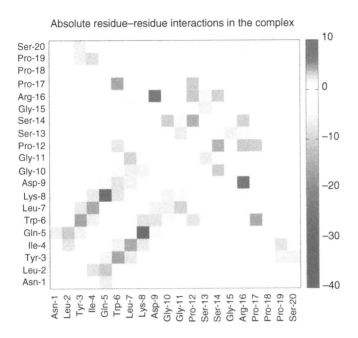
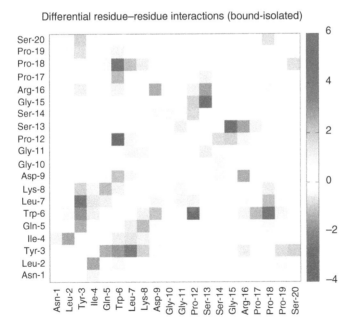

Figure 8.6 Residue–residue pair interactions ΔE_{IJ} in the complex (top) and differential values of $\Delta\Delta E_{IJ}$ (bound minus isolated protein, bottom) for unconnected dimers in the complex of Trp-cage with its ligand (MP2/PCM/cc-pVDZ), in kcal/mol.

8.6.2 Examples of SA and PAVE

The same system as in Section 7.4.3 is analyzed using a combination of SA, PA, and PAVE. The protein is divided into 20 residue segments, and the ligand into 3 functional group segments. Two subsystems are defined in SA, the protein and the ligand. The structures of AB, A, and B are separately optimized, so that deformation energies are implicitly included in each term. PAVE was conducted at 298 K. To obtain the vibrational contribution to the carboxylate binding, the atoms Leu2, Gln5, and COO^- are defined to be active (for them a partial Hessian is computed). The results are shown in Figure 8.7.

The binding energy ΔE^{bind} is obtained by subtracting three total energies in PA (Eq. (8.12)). The result is -24.7 kcal/mol. In contrast, TIE (the sum of all residue–ligand PIEs, the last term in Eq. (8.26)) is -54.3 kcal/mol, so that TIE is an overestimate of binding mainly due to desolvation effects. ΔG^{vib} is -1.5 kcal/mol. ΔG^{bind} is -26.2 kcal/mol.

The polarization, deformation, and desolvation are described by the partial energy, ΔE_i^{part}. Two ligand segments have a substantial value: phenyl, because of the desolvation loss of the dispersion interaction with solvent (the non-es term in Eq. (8.3)), and carboxylate, because some of its electrostatic interaction with solvent (the es term in Eq. (8.3)) is lost in the complex. The cavitation energy (included in E_i) also makes a contribution to binding, describing the entropy loss of the solvent. After all contributions are added, the following final binding analysis is obtained (Figure 8.8).

Ligand segments (Ph and COO^-) lose quite a bit of energy due to complexation (ΔG_i^{bind} is repulsive). On the other hand, there is a gain due to the protein-ligand binding, which overweighs the loss. Thus, in analyzing protein-ligand binding, the desolvation penalty (the main difference between interaction and binding energies) should be considered.

SA has been applied to biscarbene-Gold(I)/DNA G-quadruplex complex [86], stabilized in part by π-cation interactions involving by Au(I).

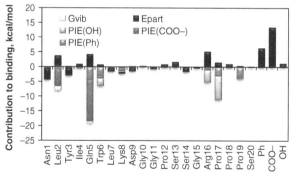

Figure 8.7 Contributions of partial energies (Epart, ΔE_i^{part}), vibrational free energies (Gvib, ΔG_i^{vib}), and pair interaction energies of residue i with functional group j of the ligand (PIE(j), ΔE_{ij}) to the protein-ligand binding energy (DFTB3/PCM).

Figure 8.8 Free energies of binding ΔG_i^{bind} at 298 K for residues and functional groups in the ligand (DFTB3/PCM).

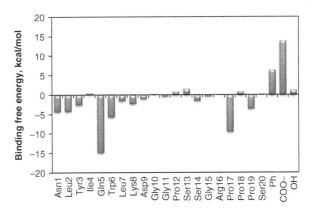

8.7 Fluctuation Analysis (FA)

Temperature and conformational flexibility are important for soft materials, such as proteins or sugars. In the fluctuation analysis (FA) [66], a conformational averaging in FMO/MD is combined with the many-body expansion in Eq. (8.1).

In their most common form, energy fluctuations are measured relative to a reference value E_0, typically chosen to be the minimal energy in MD. FMO2/FA can be written as

$$\langle E \rangle = E_0 + \langle E - E_0 \rangle = \sum_{I=1}^{N} E_I^0 + \sum_{I>J}^{N} \Delta E_{IJ}^0 + \sum_{I=1}^{N} \langle \Delta E_I \rangle + \sum_{I>J}^{N} \langle \Delta \Delta E_{IJ} \rangle \quad (8.28)$$

where brackets indicate averaging over an MD trajectory. Thus, the total QM energy (the internal energy in the thermodynamical sense) is a sum of the reference energy and fluctuations from it, decomposed into monomer and dimer values as

$$\langle \Delta E_I \rangle = \langle E_I \rangle - E_I^0 \quad (8.29)$$

$$\langle \Delta \Delta E_{IJ} \rangle = \langle \Delta E_{IJ} \rangle - \Delta E_{IJ}^0 \quad (8.30)$$

The total kinetic (kin) energy, normalized in NVT or NPT simulations to a given temperature T, can be decomposed into fragment values E_I^{kin}. Using the number of atoms N_I^{at}, the effective temperature of each fragment I is

$$T_I = \frac{2 \langle E_I^{kin} \rangle}{3 N_I^{at} k_B} \quad (8.31)$$

where k_B is the Boltzmann's constant. Some fragments are hotter than others, although not necessarily hot spots.

FA was applied to a protein-ligand binding complex [85], where the weakening of interactions at room temperature was attributed to the kinetic pressure, forcing the molecular structure to stray away from the minimum due to thermal fluctuations. Dynamical aspects of protein–ligand binding can be very important for designing new drugs, in particular because they affect the drug residence time [87]. Averaging over multiple structures helps to increase the correlation of QM predictions to experimental binding energies [64].

8.8 Free Energy Decomposition Analysis (FEDA)

In the free energy decomposition analysis (FEDA) [32], FA is combined with constrained MD (umbrella sampling MD), used to obtain the potential of mean force (PMF). PMF is taken to be equal to the free energy.

So far, FEDA has been applied to chemical reactions, for which a reaction coordinate ζ can be designed a priori. In FMO, all reactants are usually assigned to one fragment. By doing a series of FMO/MD simulations for a set of values of ζ_0, adding the constraining potential

$$U(\zeta) = \frac{k}{2}(\zeta - \zeta_0)^2 \tag{8.32}$$

a PMF $F(\zeta)$ is obtained. By plotting it, one can identify three values of the reaction coordinate, describing reactants ζ^A, transition state ζ^B, and products ζ^C.

To analyze the reaction barrier, the points ζ^A and ζ^B are important. Two more MD simulations are performed with ζ_0 equal to ζ^A and ζ^B, with a very large value of k, strongly constraining the system to be near the desirable points. The free energy of the reaction barrier is then decomposed as

$$\Delta F = F(\zeta^B) - F(\zeta^A) = \Delta E - T\Delta S \tag{8.33}$$

The QM energy ΔE is obtained using FA,

$$\Delta E = \langle E(\zeta^B) \rangle - \langle E(\zeta^A) \rangle = \sum_{I=1}^{N} \Delta E_I + \sum_{I>J}^{N} \Delta \Delta E_{IJ} \tag{8.34}$$

where monomer terms describe the polarization and deformation of fragments

$$\Delta E_I = \langle E_I(\zeta^B) \rangle - \langle E_I(\zeta^A) \rangle \tag{8.35}$$

and differential PIEs describe the change in PIEs due to the transition state formation,

$$\Delta \Delta E_{IJ} = \langle \Delta E_{IJ}(\zeta^B) \rangle - \langle \Delta E_{IJ}(\zeta^A) \rangle \tag{8.36}$$

FEDA is performed by first computing ΔF (from MD trajectories for a chosen set of ζ_0) using PMF, then ΔE is calculated from two more MD trajectories (for ζ^A and ζ^B), and, finally, the entropy ΔS is computed from Eq. (8.33).

FEDA was applied [32] to study $S_N 2$ reactions in explicit solvent using FMO/PBC/MD, where the mechanism of the intrasolute and solute-solvent charge transfer during the reaction was elucidated.

8.9 Other Analyses of Chemical Reactions

Prior to the development of FEDA, there were other applications of FMO/MD to find reaction pathways for organic reactions in solution [88, 89].

FMO can be used to locate a transition state by computing the Hessian and using standard engines for a saddle point search. A reaction path can be mapped using the intrinsic reaction coordinate (IRC) combined with FMO [90]. For enzymes, a full structure relaxation may be time-consuming, and a feasible approach is to use the frozen domain (FD) formulation of FMO [20] to optimize only a part of the system while treating the whole enzyme quantum-mechanically. There are several examples [90–92] of mapping a reaction path for enzymatic catalysis in this way using FMO for systems up to about 9000 atoms.

Alternatively, a reaction path can be mapped with QM/MM [93]. For some representative structures, PIEs can be computed, and the roles of different residues can be identified [94–96].

8.10 Conclusions

The FMO method is feasible for applications to biochemical systems on a realistic time scale. Both traditional methods like DFT and MP2, and parametrized approaches like DFTB can be used with FMO. The latter is suitable for full geometry optimizations and MD simulations, with a conformational sampling (although rather short by modern MM standards, on the order of 1 ns for fully QM MD simulations).

For a balanced description of binding, the contributions of desolvation, polarization, and deformation should be included, whereas interaction energies overestimate binding.

Free binding energies in PA + PAVE can be decomposed into segment values, although they include only electronic and vibrational contributions for a single minimum. There are important conformational entropic effects not considered in such analysis.

An important role of theoretical chemistry is to explain chemical phenomena and enhance our understanding of them, with the goal of being able to improve the materials, such as drugs or enzymes, by designing them in silico. FMO delivers very useful quantitative data, in particular pair interaction energies, which can pinpoint hot spots in proteins and other macromolecules.

A molecular design in silico is by no means easy, because of the complexity of molecular objects, including conformational sampling. However, pathways to solve typical problems in biochemistry using QM methods are set out.

Machine learning and artificial intelligence can be useful in building predictive models based on some training data, but it remains to be seen if such models can explain the physics of a given problem rather than simply make a black-box prediction (though useful it may be).

References

1 Phipps, M.J.S., Fox, T., Tautermann, C.S., and Skylaris, C.-K. (2015). Energy decomposition analysis approaches and their evaluation on prototypical protein–drug interaction patterns. *Chem. Soc. Rev.* 44: 3177–3211.

2 Merz, K.M. (2014). Using quantum mechanical approaches to study biological systems. *Acc. Chem. Res.* 47: 2804–2811.

3 Ryde, U. and Söderhjelm, P. (2016). Ligand-binding affinity estimates supported by quantum-mechanical methods. *Chem. Rev.* 116: 5520–5566.

4 Gordon, M.S., Pruitt, S.R., Fedorov, D.G., and Slipchenko, L.V. (2012). Fragmentation methods: a route to accurate calculations on large systems. *Chem. Rev.* 112: 632–672.

5 Raghavachari, K. and Saha, A. (2015). Accurate composite and fragment-based quantum chemical models for large molecules. *Chem. Rev.* 115: 5643–5677.

6 Erlanson, D.A., Fesik, S.W., Hubbard, R.E. et al. (2016). Twenty years on: the impact of fragments on drug discovery. *Nat. Rev. Drug Disc.* 15: 605–619.

7 Mironov, V., Shchugoreva, I.A., Artyushenko, P.V. et al. (2022). Structure- and interaction-based design of anti-SARS-CoV-2 aptamers. *Chem. Eur. J.* 28: e202104481.

8 Fedorov, D.G. (2022). Parametrized quantum-mechanical approaches combined with the fragment molecular orbital method. *J. Phys. Chem.* in press.

9 Christensen, A.S., Kubar, T., Cui, Q., and Elstner, M. (2016). Semiempirical quantum mechanical methods for noncovalent interactions for chemical and biochemical applications. *Chem. Rev.* 116: 5301–5337.

10 Kitaura, K. and Morokuma, K. (1976). A new energy decomposition scheme for molecular interactions within the Hartree-Fock approximation. *Int. J. Quant. Chem.* 10: 325–340.

11 Chen, W. and Gordon, M.S. (1996). Energy decomposition analyses for many-body interaction and applications to water complexes. *J. Phys. Chem.* 100: 14316–14328.

12 Kitaura, K., Ikeo, E., Asada, T. et al. (1999). Fragment molecular orbital method: an approximate computational method for large molecules. *Chem. Phys. Lett.* 313: 701–706.

13 Fedorov, D.G. (2017). The fragment molecular orbital method: theoretical development, implementation in GAMESS, and applications. *WIREs: Comput. Mol. Sc.* 7: e1322.

14 Fukuzawa, K. and Tanaka, S. (2022). Fragment molecular orbital calculations for biomolecules. *Curr. Opin. Struct. Biol.* 72: 127–134.

15 Mochizuki, Y., Tanaka, S., and Fukuzawa, K. (ed.) (2021). *Recent Advances of the Fragment Molecular Orbital Method*. Singapore: Springer.

16 Suenaga, M. (2008). Development of GUI for GAMESS/FMO calculation. *J. Comput. Chem. Jap.* 7: 33–54. (in Japanese).

17 Sawada, T., Fedorov, D.G., and Kitaura, K. (2010). Binding of influenza a virus hemagglutinin to the sialoside receptor is not controlled by the homotropic allosteric effect. *J. Phys. Chem. B* 114: 15700–15705.

18 Nakamura, S., Akaki, T., Nishiwaki, K. et al. (2023). System truncation accelerates binding affinity calculations with the fragment molecular orbital method: a benchmark study. *J. Comput. Chem.* 44:824–831.

19 Heifetz, A., Trani, G., Aldeghi, M. et al. (2016). Fragment molecular orbital method applied to lead optimization of novel interleukin-2 inducible T-cell kinase (ITK) inhibitors. *J. Med. Chem.* 59: 4352–4363.

20 Fedorov, D.G., Alexeev, Y., and Kitaura, K. (2011). Geometry optimization of the active site of a large system with the fragment molecular orbital method. *J. Phys. Chem. Lett.* 2: 282–288.

21 Fedorov, D.G. (2022). Polarization energies in the fragment molecular orbital method. *J. Comput. Chem.* 43: 1094–1103.

22 Nakata, H. and Fedorov, D.G. (2020). Geometry optimization, transition state search, and reaction path mapping accomplished with the fragment molecular orbital method. In: *Quantum Mechanics in Drug Discovery, A. Heifetz (Ed.), Methods in Molecular Biology*, vol. Vol. 2114, 87–104. New York: Springer.

23 Komeiji, Y., Mochizuki, Y., Nakano, T., and Fedorov, D.G. (2009). Fragment molecular orbital-based molecular dynamics (FMO-MD), a quantum simulation tool for large molecular systems. *J. Mol. Str. (THEOCHEM)* 898: 2–7.

24 Nakata, H. and Fedorov, D.G. (2016). Efficient geometry optimization of large molecular systems in solution using the fragment molecular orbital method. *J. Phys. Chem. A* 120: 9794–9804.

25 Nishimoto, Y. and Fedorov, D.G. (2016). The fragment molecular orbital method combined with density-functional tight-binding and the polarizable continuum model. *Phys. Chem. Chem. Phys.* 18: 22047–22061.

26 Nishimoto, Y., Nakata, H., Fedorov, D.G., and Irle, S. (2015). Large-scale quantum-mechanical molecular dynamics simulations using density-functional tight-binding combined with the fragment molecular orbital method. *J. Phys. Chem. Lett.* 6: 5034–5039.

27 Nakata, H., Fedorov, D.G., Yokojima, S. et al. (2014). Simulations of Raman spectra using the fragment molecular orbital method. *J. Chem. Theory Comput.* 10: 3689–3698.

28 Nakata, H. and Fedorov, D.G. (2020). Analytic first and second derivatives of the energy in the fragment molecular orbital method combined with molecular mechanics. *Int. J. Quantum Chem.* 120: e26414.

29 Fedorov, D.G., Kitaura, K., Li, H. et al. (2006). The polarizable continuum model (PCM) interfaced with the fragment molecular orbital method (FMO). *J. Comput. Chem.* 27: 976–985.

30 Fedorov, D.G. (2018). Analysis of solute-solvent interactions using the solvation model density combined with the fragment molecular orbital method. *Chem. Phys. Lett.* 702: 111–116.

31 Nishimoto, Y. and Fedorov, D.G. (2021). The fragment molecular orbital method combined with density-functional tight-binding and periodic boundary conditions. *J. Chem. Phys.* 154: 111102.

32 Fedorov, D.G. and Nakamura, T. (2022). Free energy decomposition analysis based on the fragment molecular orbital method. *J. Phys. Chem. Lett.* 13: 1596–1601.

33 Nakamura, T., Yokaichiya, T., and Fedorov, D.G. (2022). Analysis of guest adsorption on crystal surfaces based on the fragment molecular orbital method. *J. Phys. Chem. A* 126: 957–969.

34 Nakamura, T., Yokaichiya, T., and Fedorov, D.G. (2021). Quantum-mechanical structure optimization of protein crystals and analysis of interactions in periodic systems. *J. Phys. Chem. Lett.* 12: 8757–8762.

35 Nakamura, T. and Fedorov, D.G. (2022). The catalytic activity and adsorption in faujasite and ZSM-5 zeolites: the role of differential stabilization and charge delocalization. *Phys. Chem. Chem. Phys.* 24: 7739–7747.

36 Nishimoto, Y., Fedorov, D.G., and Irle, S. (2014). Density-functional tight-binding combined with the fragment molecular orbital method. *J. Chem. Theory Comput.* 10: 4801–4812.

37 Nishimoto, Y. and Fedorov, D.G. (2018). Adaptive frozen orbital treatment for the fragment molecular orbital method combined with density-functional tight-binding. *J. Chem. Phys.* 148: 064115.

38 Fedorov, D.G. and Kitaura, K. (2004). The importance of three-body terms in the fragment molecular orbital method. *J. Chem. Phys.* 120: 6832–6840.

39 Alexeev, Y., Mazanetz, M.P., Ichihara, O., and Fedorov, D.G. (2012). GAMESS as a free quantum-mechanical platform for drug research. *Curr. Top. Med. Chem.* 12: 2013–2033.

40 Fedorov, D.G. (2023). *Complete Guide to the Fragment Molecular Orbital Method in GAMESS*. Singapore: World Scientific.

41 Barca, G.M.J., Bertoni, C., Carrington, L. et al. (2020). Recent developments in the general atomic and molecular electronic structure system. *J. Chem. Phys.* 152: 154102.

42 Fedorov, D.G., Brekhov, A., Mironov, V., and Alexeev, Y. (2019). Molecular electrostatic potential and electron density of large systems in solution computed with the fragment molecular orbital method. *J. Phys. Chem. A* 123: 6281–6290.

43 Ozono, H., Mimoto, K., and Ishikawa, T. (2022). Quantification and neutralization of the interfacial electrostatic potential and visualization of the dispersion interaction in visualization of the interfacial electrostatic complementarity. *J. Phys. Chem. B* 126: 8415–8426.

44 Fedorov, D.G. and Kitaura, K. (2007). Pair interaction energy decomposition analysis. *J. Comput. Chem.* 28: 222–237.

45 Fedorov, D.G. and Kitaura, K. (2012). Energy decomposition analysis in solution based on the fragment molecular orbital method. *J. Phys. Chem. A* 116: 704–719.

46 Green, M.C., Fedorov, D.G., Kitaura, K. et al. (2013). Open-shell pair interaction energy decomposition analysis (PIEDA): formulation and application to the hydrogen abstraction in tripeptides. *J. Chem. Phys.* 138: 074111.

47 Fedorov, D.G. (2020). Three-body energy decomposition analysis based on the fragment molecular orbital method. *J. Phys. Chem. A* 124: 4956–4971.

48 Fedorov, D.G. (2019). Solvent screening in zwitterions analyzed with the fragment molecular orbital method. *J. Chem. Theory Comput.* 15: 5404–5416.

49 Thirman, J. and Head-Gordon, M. (2015). An energy decomposition analysis for second-order Møller-Plesset perturbation theory based on absolutely localized molecular orbitals. *J. Chem. Phys.* 143: 084124.

50 Fedorov, D.G. and Kitaura, K. (2014). Use of an auxiliary basis set to describe the polarization in the fragment molecular orbital method. *Chem. Phys. Lett.* 597: 99–105.

51 Fedorov, D.G., Kromann, J.C., and Jensen, J.H. (2018). Empirical corrections and pair interaction energies in the fragment molecular orbital method. *Chem. Phys. Lett.* 702: 111–116.

52 Nakano, T., Mochizuki, Y., Yamashita, K. et al. (2012). Development of the four-body corrected fragment molecular orbital (FMO4) method. *Chem. Phys. Lett.* 523: 128–133.

53 Watanabe, C., Fukuzawa, K., Okiyama, Y. et al. (2013). Three- and four-body corrected fragment molecular orbital calculations with a novel subdividing fragmentation method applicable to structure-based drug design. *J. Mol. Graphics Modell.* 41: 31–42.

54 Mazanetz, M.P., Chudyk, E., Fedorov, D.G., and Alexeev, Y. (2016). Applications of the fragment molecular orbital method to drug research. In: *Computer Aided Drug Discovery* (ed. W. Zhang), 217–255. York: Springer, New.

55 Heifetz, A., Chudyk, E.I., Gleave, L. et al. (2016). The fragment molecular orbital method reveals new insight into the chemical nature of GPCR-ligand interactions. *J. Chem. Inf. Model.* 56: 159–172.

56 Chudyk, E.I., Sarrat, L., Aldeghi, M. et al. (2018). Exploring GPCR-ligand interactions with the fragment molecular orbital (FMO) method. In: *Computational Methods for GPCR Drug Discovery* (ed. A. Heifetz), 179–195. New York: Humana Press.

57 Heifetz, A., Morao, I., Babu, M.M. et al. (2020). Characterizing interhelical interactions of G-protein coupled receptors with the fragment molecular orbital method. *J. Chem. Theory Comput.* 16: 2814–2824.

58 Mazanetz, M.P., Ichihara, O., Law, R.J., and Whittaker, M. (2011). Prediction of cyclin- dependent kinase 2 inhibitor potency using the fragment molecular orbital method. *J. Cheminf.* 3: 2.

59 Yoshida, T. and Hirono, S. (2019). A 3D-QSAR analysis of CDK2 inhibitors using FMO calculations and PLS regression. *Chem. Pharm. Bull.* 67: 546–555.

60 Tokutomi, S., Shimamura, K., Fukuzawa, K., and Tanaka, S. (2020). Machine learning prediction of inter-fragment interaction energies between ligand and amino-acid residues on the fragment molecular orbital calculations for Janus kinase-inhibitor complex. *Chem. Phys. Lett.* 757: 137883.

61 Lim, H., Jeon, H.-N., Lim, S. et al. (2022). Evaluation of protein descriptors in computer-aided rational protein engineering tasks and its application in property prediction in SARS-CoV-2 spike glycoprotein. *Comp. Str. Biotechn. J.* 20: 788–798.

62 Heifetz, A., Aldeghi, M., Chudyk, E.I. et al. (2016). Using the fragment molecular orbital method to investigate agonist-orexin-2 receptor interactions. *Biochem. Soc. Trans.* 44: 574–581.

63 Morao, I., Fedorov, D.G., Robinson, R. et al. (2017). Rapid and accurate assessment of GPCR-ligand interactions using the fragment molecular orbital-based density-functional tight-binding method. *J. Comput. Chem.* 38: 1987–1990.

64 Takaba, K., Watanabe, C., Tokuhisa, A. et al. (2022). Protein-ligand binding affinity prediction of cyclin-dependent kinase-2 inhibitors by dynamically averaged fragment molecular orbital-based interaction energy. *J. Comput. Chem.* 43: 1362–1371.

65 Handa, C., Yamazaki, Y., Yonekubo, S. et al. (2022). Evaluating the correlation of binding affinities between isothermal titration calorimetry and fragment molecular orbital method of estrogen receptor beta with diarylpropionitrile (DPN) or DPN derivatives. *J. Ster. Biochem. Mol. Biol.* 222: 106152.

66 Fedorov, D.G. and Kitaura, K. (2018). Pair interaction energy decomposition analysis for density functional theory and density-functional tight-binding with an evaluation of energy fluctuations in molecular dynamics. *J. Phys. Chem. A* 122: 1781–1795.

67 Nakanishi, I., Fedorov, D.G., and Kitaura, K. (2007). Molecular recognition mechanism of FK506 binding protein: an all-electron fragment molecular orbital study. *Proteins: Struct., Funct. Bioinf.* 68: 145–158.

68 Ozawa, M., Ozawa, T., Nishio, M., and Ueda, K. (2017). The role of CH/π interactions in the high affinity binding of streptavidin and biotin. *J. Mol. Graph. Model.* 75: 117–124.

69 Maruyama, K., Sheng, Y., Watanabe, H. et al. (2018). Application of singular value decomposition to the inter-fragment interaction energy analysis for ligand screening. *Comp. Theor. Chem.* 1132: 23–34.

70 Green, M.C., Nakata, H., Fedorov, D.G., and Slipchenko, L.V. (2016). Radical damage in lipids investigated with the fragment molecular orbital method. *Chem. Phys. Lett.* 651: 56–61.

71 Li, S., Qin, C., Cui, S. et al. (2019). Discovery of a natural-product-derived preclinical candidate for once-weekly treatment of type 2 diabetes. *J. Med. Chem.* 62: 2348–2361.

72 Mai, X., Higashi, K., Fukuzawa, K. et al. (2022). Computational approach to elucidate the formation and stabilization mechanism of amorphous formulation using molecular dynamics simulation and fragment molecular orbital calculation. *Int. J. Pharmaceutics* 615: 121477.

73 Sladek, V., Tokiwa, H., Shimano, H., and Shigeta, Y. (2018). Protein residue networks from energetic and geometric data: are they identical? *J. Chem. Theory. Comput.* 14: 6623–6631.

74 Doi, H., Okuwaki, K., Mochizuki, Y. et al. (2017). Dissipative particle dynamics (DPD) simulations with fragment molecular orbital (FMO) based effective parameters for 1-palmitoyl-2-oleoyl phosphatidyl choline (POPC) membrane. *Chem. Phys. Lett.* 684: 427–432.

75 Monteleone, S., Fedorov, D.G., Townsend-Nicholson, A. et al. (2022). Hotspot identification and drug design of protein-protein interaction modulators using the fragment molecular orbital method. *J. Chem. Info. Model.* 62: 3784–3799.

76 González-Olvera, J.C., Zamorano-Carrillo, A., Arreola-Jardón, G., and Pless, R.C. (2022). Residue interactions affecting the deprotonation of internal guanine moieties in oligodeoxyribonucleotides, calculated by FMO methods. *J. Mol. Model.* 28: 43.

77 Lim, H., Hong, H., Hwang, S. et al. (2022). Identification of novel natural product inhibitors against matrix metalloproteinase 9 using quantum mechanical fragment molecular orbital-based virtual screening methods. *Int. J. Mol. Sci.* 23: 4438.

78 Hengphasatporn, K., Wilasluck, P., Deetanya, P. et al. (2022). Halogenated baicalein as a promising antiviral agent toward SARS-CoV-2 main protease. *J. Chem. Inf. Model.* 62: 1498–1509.

79 Paciotti, R., Agamennone, M., Coletti, C., and Storchi, L. (2020). Characterization of PD-L1 binding sites by a combined FMO/GRID-DRY approach. *J. Comput.-aided Mol. Des.* 34: 897–914.

80 Acharya, A., Agarwal, R., Baker, M.B. et al. (2020). Supercomputer-based ensemble docking drug discovery pipeline with application to covid-19. *J. Chem. Inf. Model.* 60: 5832–5852.

81 Gaus, M., Goez, A., and Elstner, M. (2013). Parametrization and benchmark of DFTB3 for organic molecules. *J. Chem. Theory Comput.* 9: 338–354.

82 Fedorov, D.G. (2020). Partition analysis for density-functional tight-binding. *J. Phys. Chem. A* 124: 10346–10358.

83 Fedorov, D.G. (2021). Partitioning of the vibrational free energy. *J. Phys. Chem. Lett.* 21: 6628–6633.

84 Sladek, V. and Fedorov, D.G. (2022). The importance of charge transfer and solvent screening in the interactions of backbones and functional groups in amino acid residues and nucleotides. *Int. J. Mol. Sci.* 23: 13514.

85 Fedorov, D.G. and Kitaura, K. (2016). Subsystem analysis for the fragment molecular orbital method and its application to protein-ligand binding in solution. *J. Phys. Chem. A* 120: 2218–2231.

86 Paciotti, R., Coletti, C., Marrone, A., and Re, N. (2022). The FMO2 analysis of the ligand- receptor binding energy: the biscarbene-gold(I)/DNA G-quadruplex case study. *J. Comput. Aided Mol. Des.* 36: 851–866.

87 Zhang, Q., Zhao, N., Meng, X. et al. (2022). The prediction of protein-ligand unbinding for modern drug discovery. *Exp. Op. Drug Disc.* 17: 191–205.

88 Sato, M., Yamataka, H., Komeiji, Y. et al. (2008). How does an S_N2 reaction take place in solution? Full ab initio MD simulations for the hydrolysis of the methyl diazonium ion. *J. Am. Chem. Soc.* 130: 2396–2397.

89 Sato, M., Yamataka, H., Komeiji, Y. et al. (2010). Does amination of formaldehyde proceed through a zwitterionic intermediate in water? Fragment molecular orbital molecular dynamics simulations by using constraint dynamics. *Chem. Eur. J.* 16: 6430–6433.

90 Nakata, H., Fedorov, D.G., Nagata, T. et al. (2015). Simulations of chemical reactions with the frozen domain formulation of the fragment molecular orbital method. *J. Chem. Theory Comput.* 11: 3053–3064.

91 Steinmann, C., Fedorov, D.G., and Jensen, J.H. (2013). Mapping enzymatic catalysis using the effective fragment molecular orbital method: towards all ab initio biochemistry. *PLoS ONE* 8: e60602.

92 Pruitt, S.R. and Steinmann, C. (2017). Mapping interaction energies in chorismate mutase with the fragment molecular orbital method. *J. Phys. Chem A* 121: 1798–1808.

93 Ishida, T., Fedorov, D.G., and Kitaura, K. (2006). All electron quantum chemical calculation of the entire enzyme system confirms a collective catalytic device in the chorismate mutase reaction. *J. Phys. Chem. B* 110: 1457–1463.

94 Ito, M. and Brinck, T. (2014). Novel approach for identifying key residues in enzymatic reactions: proton abstraction in ketosteroid isomerase. *J. Phys. Chem. B* 118: 13050–13058.

95 Abe, Y., Shoji, M., Nishiya, Y. et al. (2017). The reaction mechanism of sarcosine oxidase elucidated using FMO and QM/MM methods. *Phys. Chem. Chem. Phys.* 19: 9811–9822.

96 Tribedi, S., Kitaura, K., Nakajima, T., and Sunoj, R.B. (2021). On the question of steric repulsion versus noncovalent attractive interactions in chiral phosphoric acid catalyzed asymmetric reactions. *Phys. Chem. Chem. Phys.* 23: 18936–18950.

Part III

Artificial Intelligence in Pre-clinical Drug Discovery

9

The Role of Computer-Aided Drug Design in Drug Discovery

Storm van der Voort[1], Andreas Bender[2], and Bart A. Westerman[1]

[1] Department of Neurosurgery, Amsterdam UMC, location VUMC, Cancer Center, Amsterdam, the Netherlands
[2] Yusuf Hamied Department of Chemistry, University of Cambridge, Lensfield Rd, Cambridge, United Kingdom

9.1 Introduction to Drug–Target Interactions, Hit Identification

The therapeutic effect of drugs is dependent on their interactions with their target molecules, such as kinases, GPCRs, phosphodiesterases, nuclear receptors, or ion channels [1]. Drug-target interactions (DTIs) are not only responsible for therapeutic efficacy but could also lead to adverse events that might conflict with clinical benefits [2]. Therefore, accurate assessment of DTIs is an important step in the drug discovery process, allowing researchers to probe the target properties, efficacy, and safety of a drug, thereby propelling it into various stages of the drug development process.

Hit identification is commonly based on phenotypic assays based on high-throughput screening (HTS) using compounds or fragment-based phenotypic assays, which can subsequently be linked to structural information such as X-ray crystallography or NMR structural information. Subsequently, assessment of DTIs is often done through *in vitro* methods, although these methods have practical limitations when considering the enormous number of potential small-molecule-to-target interactions. Virtual screening is a computational method used to identify potential drug candidates by screening large databases of compounds against a target protein. Therefore, high throughput *in silico* DTI prediction can facilitate the matching of a wide variety of compounds against an array of targets – after which the most promising drug–target combinations can be verified experimentally [3]. Recently, AlphaFold, a neural network (NN)-based predictor of 3D protein structure from the sequence [1], won the 14th Critical Assessment of Protein Structure Prediction [1–4]. This was followed by the publication of 350,000 models of protein structures generated by AlphaFold, showing the potential that NN-based methods have within drug discovery [5], although only a single, *holo*, structure is generated for each protein [6]. Here we will describe how DTI predictors, as well as other complementary predictive models, can assist in the development of new drugs. Figure 9.1 shows that DTI prediction is used at various stages in the clinical development

Computational Drug Discovery: Methods and Applications, First Edition.
Edited by Vasanthanathan Poongavanam and Vijayan Ramaswamy.
© 2024 WILEY-VCH GmbH. Published 2024 by WILEY-VCH GmbH.

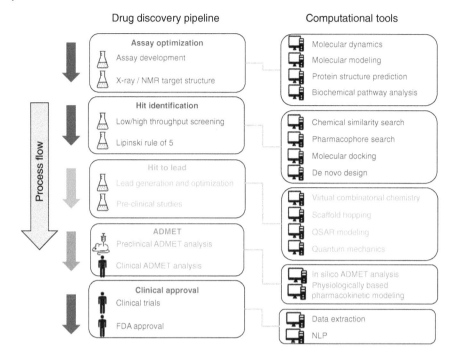

Figure 9.1 Overview of drug discovery pipeline vs. computer-aided design. The figure is partially based on Schaduangrat et al. [7].

pipeline, where each drug discovery stage (left) can be informed by their respective bioinformatic tools (right).

Box 9.1 Hit identification, single or more targets?

An interesting concept in drug discovery is the use of multi-target approaches, relevant for many diseases, such as cancer. A specific multi-target approach is the utilization of polypharmacology, or the use of a single drug to target multiple proteins at once, thereby potentially enhancing its efficacy [4]. This concept can for instance be applied to kinase inhibitors since kinase mutations often drive the process of carcinogenesis [8]. Most of the kinase inhibitors prevent downstream signal transduction by binding to the highly conserved ATP binding site of kinase to block the binding of ATP [9]. Consequently, the binding of a kinase inhibitor to kinases is not highly selective. Given that only approximately 1–2% of kinase inhibitors' targets are known [10–12], proper DTI prediction could uncover this as yet undisclosed target space.

One of the popular techniques in virtual screening is fragment-based screening, which involves breaking down larger molecules into smaller fragments to search for potential binding sites on the target protein. Binding affinity prediction is also a crucial step in drug discovery, where various computational tools are used to predict the strength of binding between a potential drug candidate and the target protein. The success of drug development depends on understanding protein–ligand interactions, which determine the stability and specificity of the DTI. Therefore, a thorough understanding of these methods is essential for efficient and effective drug discovery.

9.2 Lead Identification and Optimization: QSAR and Docking-Based Approaches

The most established methods for *in silico* DTI prediction are *ligand-based* and *docking-based* approaches [1]. In ligand-based methods like quantitative structure-activity relationships (QSAR), a large collection of confirmed binders to a certain target are collected, and linear (regression) models are built to correlate certain structural features to biological activity [5–7]. This model can then be used to predict the activity of untested molecules. An advantage of QSAR methods is that they do not require structural information about the target protein, making them suitable for protein classes where structural information is scarce, such as GPCRs [8]. However, QSAR modeling has pitfalls such as data overfitting, poor generalizability, and inadequate model validation [13, 14].

In contrast, docking-based approaches use structural information of the protein target to "fit" a molecule into the active site [15, 16]. Here, affinity is typically predicted by assessing the free energy gain upon placing the ligand in the active site using a "scoring function." However, docking has pitfalls too, as it requires structural information [17], model accuracy is highly dependent on the scoring function used [18–21], and it often fails to incorporate receptor structural flexibility [22]. Furthermore, it is computationally expensive compared to ligand-based methods and therefore less suitable for probing polypharmacology [23].

Matched molecular pair analysis is a common tool used to identify and analyze structural modifications in drug molecules [24]. Knowledge of molecules with similar physical and chemical properties can be used to improve the pharmacokinetics and pharmacodynamics of a drug. Solubility issues are an important issue in drug development, as poorly soluble drugs may have limited bioavailability. Features such as LogD, i.e. the logarithm of the partition coefficient between a drug and water/octanol, allow one to predict a drug's solubility and distribution [25]. Furthermore, pK_a, the negative logarithm of the acid dissociation constant, affects the drug's solubility and permeability [26]. A thorough understanding of these parameters can help identify potential drug candidates and optimize their properties for clinical use.

9.3 DTI Machine Learning Methods

Machine learning (ML) methods have recently been aimed to address the shortcomings of traditional ligand-based and docking-based approaches. ML methods in DTI prediction learn from a set of known data points to predict whether a compound

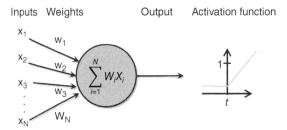

Figure 9.2 (a) An overview of the McCulloch-Pitts model for a neuron, which receives N inputs x with weights w. The neuron sums the inputs and weights to obtain the total input value and activates if the input value exceeds a certain threshold. (b) An example of a neural network containing 7 inputs, one hidden layer, and one output layer. Adapted from Krogh [35].

binds to a target [27]. Generally, a model (a set of rules) is "trained" by making predictions about labeled data. One of the simplest ML methods is linear regression (also used in QSAR), where a line is fit through a set of known data points, and predictions are made by extrapolating from this fitted line [28]. ML approaches have received increased interest over the last years because of their low computational cost, high performance, and applicability to proteins without structural information [29–32].

One of the more recent methods used in ML is deep learning [33, 34] (DL). DL makes use of artificial NNs, thereby mimicking the neural structure of the human brain to generate complex predictive models [35]. NNs are built from neurons, which are connected via links (Figure 9.2, a). NNs are trained by making predictions about a labeled training set, after which the "loss" is defined (the error between the predictions and the actual data labels). The weights of all the neurons are optimized so that the loss function, and therefore the prediction error, is minimized.

9.4 Supervised, Non-supervised and Semi-supervised Learning Methods

ML methods require "training," i.e. they must be taught how to make predictions based on the input data. There are two main ways of training ML models, supervised and unsupervised learning [36]. The main difference between these training methods is the presence of labeled data [37]. Supervised learning uses labeled training data: data has been pre-labeled (e.g. whether a specific molecule binds to the protein target or not). The model is then trained by making predictions about the input data and comparing the predictions to the label. The variables within the model (connections, thresholds, and weights) are subsequently optimized such that they increase the accuracy of the prediction. A limitation of supervised learning is the need for high-quality labeled data, which can be time-consuming to generate as this is often done through manual curation [37, 38]. Furthermore, since labels are defined in advance, this method has limited capacity to discover new patterns outside the realm of this predefined angle [37].

Unsupervised methods do not use labeled training data and can be used to find patterns or commonalities in data [39]. Unsupervised learning can facilitate to find unique patterns in relation to common patterns within datasets [39]. Unsupervised learning can also be used to preprocess a dataset and drastically cut down on the effort required to label data for supervised learning. This type of combined unsupervised and supervised learning is a form of semi-supervised learning, which is positioned between supervised and unsupervised learning and aims to address the shortcomings of both [39]. Semi-supervised learning can also be applied by training a model on partially labeled data, letting the model infer labels based on partial training during the training process [40].

9.5 Graph-Based Methods to Label Data for DTI Prediction

Typical ML methods can be applied to two- or three-dimensional data sources that can be mapped to Euclidian coordinates and can therefore be referred to as Euclidian data [41, 42]. In contrast, graphs have a non-Euclidian structure, featuring nodes (points on the graph) and edges (connections between nodes, Figure 9.3).

Graphs make it possible to examine inputs that are not well represented in Euclidian space. For example, the DTI network is naturally represented as a graph. Drugs and targets make up nodes, and the observed interactions between them are represented as edges (Figure 9.4a). Chemical and protein structures can also be represented as graphs, with atoms or amino acids acting as nodes and chemical bonds or interactions between molecules represented as edges (Figure 9.4b). This more "natural" translation of the DTI network or chemical structures compared to string representations has the potential to make graph-based methods for DTI prediction superior over traditional Euclidian methods.

Still, graphs present additional challenges for computation. Traditional NNs use string representations to perform calculations. However, the irregular nature of a graph makes it hard to perform these calculations. GNNs typically learn by generating representations of the graphs, which are lower-order strings that incorporate information on nodes, edges, or a graph. An example is shown in Figure 9.4c. Here, a representation of target node A is generated by taking its connectivity information, showing it is connected to nodes B, D, and C. An extra layer of the GNN then considers the connectivity information of nodes B, D, and C. This connectivity information is embedded into a lower-dimension representation that can be used as the input for NN-based learning [42, 43].

Figure 9.3 A general overview of the structure of a graph. The nodes (points on the graph) can be connected to each other via edges (connections).

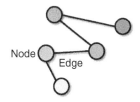

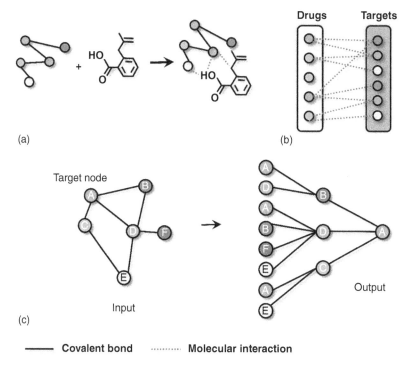

Figure 9.4 (a) Low-order graph methods model the protein and the drug as graphs, using atoms/residues as nodes. (b) High-order methods model the DTI network as a graph, using drugs and targets as nodes. (c) An example of how calculations can be performed on graph-based inputs. Adapted from Zhang et al. [42].

9.6 The Importance of Explainable ML Methods: Linking Molecular Properties to Effects

The main criticism of using ML methods, and especially NNs, for predicting DTIs is their "black-box" nature [44, 45]. Often, the intricate nature of the hidden layer(s) of the NN, combined with the automated nature in which the weights and connections within an NN are optimized, means they are not understood and are therefore not transparent [44]. Since the predictions that NNs make on DTIs are used for predicting drug efficacies and safeties, the models used must be understood by humans. This has been stressed by the recent debate surrounding the "right to explain" language used in the European General Data Protection Regulation, which could imply that "black-box" NNs will not be allowed for important automated decisions [46, 47]. Therefore, it is necessary to have human-understandable or "explainable" ML methods in DTI predictions. Furthermore, explainable ML methods for DTI prediction can offer more insights beyond simply predicting whether a particular compound will bind well. This has the potential to improve the rational design of drugs for a particular protein target, advancing the field of medicinal chemistry. However, this will require a shift away from optimizing ML methods for predictive accuracy and toward ML methods that can be explained.

In order to explain the DTI molecular mechanisms, one would like to pinpoint individual atoms that explain the interaction. However, this molecular interaction assessment is performed in the context of general drug properties required for the solubility and membrane permeability of compounds, hence, this aspect cannot be uncoupled from features that explain the molecular specificity of the compound. Optimal general features consist of physicochemical properties that have values according to Lipinski and others [48, 49], consisting of restrictions of logD, pKa, the number of hydrogen bond acceptors (HBA), intramolecular hydrogen bonds (IHMB) [50], hydrogen bond donors (HBD) [51], and rotational bonds, as well as the polar surface area (PSA). Rather than physicochemical properties, molecular properties are more suited to explain molecular interactions. Different methods have been developed to describe small-molecule compounds, ranging from relatively simple to more extended descriptors. Simple descriptors such as MACCS fingerprints (166 features) have the advantage that the results are relatively easy to understand, at the cost of limited overlap with the enormous amount of possibilities in the chemical space. More complex graph-based descriptors, such as ECFP fingerprints (Extended Connectivity Fingerprint) [52], describe molecules with a fixed number of bits, commonly between 1024 and 4096 bits, as a binary vector representation. Each bit corresponds to a certain molecular feature where the total number of bits is used to describe the molecular structure. A recent development is the use of NNs, including graph convolutional networks (GCNs), that can go beyond the complexity of ECFP fingerprints. Physical-chemical properties and molecular descriptors are therefore valuable to make DTI models explainable, and this is currently a field of innovation [53].

9.7 Predicting Therapeutic Responses

When possible, the identified therapeutics should be customized to patients' molecular profiles to achieve better therapeutic effects [54]. Many studies have shown the value of precision medicine in a clinical setting, especially in the field of cancer [55, 56], especially for kinase inhibitors, which are commonly mutated and therefore patient-specific targets for cancer treatment. With the increasing availability of HTS data and rapid development in the field of genomics, researchers are able to employ computational methods to build drug response prediction models with good performance and interpretability for monotherapy [57, 58] as well as combination therapy [59, 60]. Therefore, the ability to make links between drug effects and patient-specific molecular profiles, and drug sensitivity data required for certain patients to a certain drug is both scientifically as well as clinically relevant. Molecular markers that predict a therapeutic response in the clinic commonly consist of single or multiple defined mutations [61]. In addition, mutations are considered the primary cause of abnormal growth in cancer cells and are associated with drug responses [61, 62]. In spite of that, drug sensitivity predictions based on mutation as well as based on other molecular data such as mRNA expression profiles generally do not perform well [57, 63–66].

9.8 ADMET-tox Prediction

Computer-based predictions of a drug's absorption, distribution, metabolism, excretion, and toxicity (ADMET) form a significant step in drug discovery. The bioavailability of a drug, or its ability to reach the bloodstream, is dependent on its absorption and metabolism. Highly specific compounds should have good evolvability (e.g. favorable molecular and physical properties) to ensure oral bioavailability, half-life, metabolism, transporters, cell permeability, distribution, and solubility. In silico models can predict how well a drug is absorbed in the gut, how quickly it is metabolized, and whether it can cross cell membranes. Cell permeability is also a critical factor in drug discovery, as drugs must be able to penetrate cells to reach their targets. Transporters also play a significant role in drug distribution, and in silico models can predict how a drug interacts with these proteins.

The suitability of a compound for further clinical implementation depends in large part on the potential adverse events. Without proper insight into adverse events, it is possible that therapy could do more harm than good for the patient. The explainability of adverse events based on drug targets would particularly help to provide insight into this balance. So, for a drug therapy to be used safely and effectively, models that can predict the toxicity to the patient of a given therapy would be highly valuable. To create models that can predict the toxicity of compounds or combinations thereof, several data sources can be used, among which are clinicaltrials.gov and the FDA's Adverse Event Reporting System (FAERS) database. Both databases contain millions of patient observations for adverse events linked to drug therapies. Using Natural Language Processing (NLP), structured data can be extracted from these data sources and converted to a structured Common Terminology Criteria for Adverse Events (CTCAE) format [67]. hERG, a potassium ion channel, is a common target for drug-induced cardiotoxicity, and in silico models can predict a drug's risk of causing hERG inhibition [68].

9.9 Challenging Aspects of Using Computational Methods in Drug Discovery

Over the last decades, computational methods have been employed at least since the 1980s, with the first docking programs such as DOCK [69] becoming available. Another peak in the utilization of computational methods happened around 2000, coinciding with the publication of the draft sequence of the human genome [70], which has led to high (and probably also inflated) expectations regarding the impact on drug discovery in subsequent years. As an article from 2001 [71] states, there was the expectation that there would be "3,000–10,000 targets compared with 483," which were targeted at the time of writing – however, an article from 2017 [72] put this number only at or around 667. Hence, it is important to keep in mind also the limitations of novel technologies, such as computational methods in drug discovery.

9.9.1 What are Those Limitations?

Firstly, most computational methods in drug discovery are based on data – which has the advantage that the data that has been generated is used in models (such as HTS data, measurements of and physicochemical properties). On the other hand, the use of data for algorithms reinforces a focus on areas that have been explored already, hence potentially slowing down explorations of novel areas of chemical and biological target space (if such algorithms are based on historic data only). On the other hand, computational approaches can be based on modeling, so on understanding the underlying system, e.g. when modeling receptors and their interaction with a ligand or their activation mechanisms, or, on a scale probably more relevant for physiology, cellular, and higher-level systems. The crux is, though, that biological systems are often poorly understood, rendering this engineering-style approach not immediately feasible in many cases [73].

Secondly, there is the question of *in vivo* relevance of the resulting predictions (an area that has been extensively discussed in a set of two recent review articles [74, 75]. Drug discovery at its core is not about preclinical assays or computational methods – rather, all that matters is the efficacy and safety of a compound in a clinic, for a patient cohort we are able to identify via diagnosis. However, much of the data that is available for generating predictive models at the current point in time is of preclinical nature, such as bioactivity data against isolated targets and physicochemical characterization of compounds. This data can be labeled, and hence predictive models can be built – however, the question is whether those predictions translate to clinically relevant endpoints. For example, we are able to predict protein targets of small molecules reasonably well [31] – but whether this translates to in vivo relevant effects also depends on the dose and compound pharmacokinetics (reaching the target tissue, etc.). Hence, just predicting unconditional compound properties is not sufficient to generate in vivo-relevant computational models. On the other hand, *in vivo*-relevant data, such as clinical readouts, is much more difficult to label, and hence to generate models for – effects obtained depend on the dose, patient genotype, disease endotype, sex, age, co-medication, etc. In addition, such data is available on a much smaller scale, usually only in the number of hundreds or in the small thousands of compounds. So, when it comes to *in vivo*-relevant endpoints, we are dealing with small datasets, which are difficult to label – and hence, it is difficult to generate meaningful models for them. Therefore, there is a disconnect between the data we need to translate into the clinic (*in vivo* relevant data) and the labelable, large-scale data (which is largely *in vitro* data, far removed from clinical relevance).

Thirdly, there are problems with performing meaningful validation of computational models in the context of drug discovery. Unlike biological sequence space, where we are able to assign transition likelihoods for mutations for, e.g. BLOSUM matrices, chemical space is large (on the order of 10^{60} possible compounds) and difficult to characterize. This has severe consequences for all subsequent steps of model generation – since we are unable to identify biases in the data or perform any type of meaningful statistics on the results obtained. Model validation in itself is also insufficient to impact drug discovery as a whole – computational models do not take any project context into account, and neither do they take into account downstream assays that can (or cannot) be performed for experimental validation of the results obtained. However, those are key factors that matter in drug discovery – when a model predicts toxicity of a certain type; so what does this mean in the context of disease (lifestyle disease or terminal cancer?), dose, reaching the target tissue, etc.? Is the prediction of the model *relevant* in the given context, and *how can the output be confirmed (or refuted)*? It is less likely that projects are stopped just because of a prediction, so integration with the *process* is key here, as illustrated in Figure 9.5 below. Just establishing model performance metrics by themselves, even if they are better than preceding metrics, does not address this aspect of how a model translates to decision-making in a project in practice.

Finally, the generic performance metrics of models, which are frequently found in publications are often irrelevant in practical settings. Performance such as AUC and class-averaged accuracy. are generic – but in a practical setting there is a context of to what extent follow-up assays can be performed, and whether the model operates in an "abundance of options" setting (discovery), or in a "scarcity of options" setting (e.g. in later stage optimization). In the former case, usually sufficient recall in the top few percentages of any ranked library is what is needed (but in absolute terms, this recall may be small, given we are in a situation with many options, say finding active compounds); while in the latter case, a much bigger attention needs to be paid to avoiding false-positive predictions in say a toxicity prediction setting (since losing compounds is very costly and needs to be avoided). All of this depends on the context of how the model is used and what the local experimental follow-up capabilities of a model are – and generic performance metrics are generally not sufficient to be able to translate to this real-world situation.

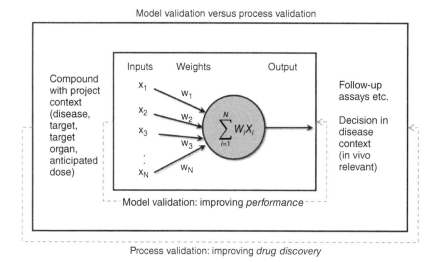

Figure 9.5 Model validation (center) in itself is insufficient to improve drug discovery as a process, since the project context, available follow-up assays, and use of situation-relevant performance metrics also need to be taken into account to have a real-world impact.

References

1. Jumper, J. et al. (2021). Highly accurate protein structure prediction with AlphaFold. *Nature* 596: 583.
2. Ozden, B., Structure, A.K. et al. (2021). Function, undefined, and, undefined & 2021, undefined. Assessment of the CASP14 assembly predictions. *Wiley Online Libr.* 89: 1787–1799.
3. Pereira, J. et al. (2021). High-accuracy protein structure prediction in CASP14. *Wiley Online Libr.* 89: 1687–1699.
4. Jumper, J. et al. (2021). Applying and improving AlphaFold at CASP14. *Proteins Struct. Funct. Bioinform.* 89: 1711–1721.
5. Thornton, J.M., Laskowski, R.A., and Borkakoti, N. (2021). AlphaFold heralds a data-driven revolution in biology and medicine. *Nat. Med.* 2710 (27): 1666–1669.
6. Perrakis, A. and Sixma, T.K. (2021). AI revolutions in biology. *EMBO Rep.* 22: e54046.
7. Schaduangrat, N. et al. (2020). Towards reproducible computational drug discovery. *J. Cheminform.* 12: 9.
8. Bhullar, K.S. et al. (2018). Kinase-targeted cancer therapies: progress, challenges and future directions. *Mol. Cancer* 17: 48.
9. Ahn, N.G. and Resing, K.A. (2005). Lessons in rational drug design for protein kinases. *Science (80-)* 308: 1266–1267.
10. Fabian, M.A. et al. (2005). A small molecule–kinase interaction map for clinical kinase inhibitors. *Nat. Biotechnol.* 23: 329–336.
11. Kanev, G.K. et al. (2019). The landscape of atypical and eukaryotic protein kinases. *Trends Pharmacol. Sci.* 40: 818–832.
12. Roskoski, R. (2019). Properties of FDA-approved small molecule protein kinase inhibitors. *Pharmacol. Res.* 144: 19–50.
13. Kubinyi, H. (2004). Validation and predictivity of QSAR models. In: *European Symposium on QSAR & Molecular Modelling*.
14. Cherkasov, A. et al. (2014). QSAR modeling: Where have you been? Where are you going to? *J. Med. Chem.* 57: 4977–5010.
15. Pinzi, L. and Rastelli, G. (2019). Molecular docking: shifting paradigms in drug discovery. *Int. J. Mol. Sci.* 20: 4331.
16. Ferreira, L.G., Dos Santos, R.N., Oliva, G., and Andricopulo, A.D. (2015). Molecular docking and structure-based drug design strategies. *Molecules* 20: 13384–13421.
17. Lavecchia, A. and Di Giovanni, C. (2013). Virtual screening strategies in drug discovery: a critical review. *Curr. Med. Chem.* 20 (23): 2839–2860.
18. Neves, M. A. C., Totrov, M., Llc, M. & Abagyan, R. Docking and scoring with ICM: the benchmarking results and strategies for improvement. doi:https://doi.org/10.1007/s10822-012-9547-0
19. Park, H., Lee, J., and Lee, S. (2006). Critical assessment of the automated AutoDock as a new docking tool for virtual screening. *Proteins Struct. Funct. Bioinform.* 65: 549–554.

20 Ha, S., Andreani, R., Robbins, A., and Muegge, I. (2000). Evaluation of docking/scoring approaches: A comparative study based on MMP3 inhibitors. *J. Comput. Mol. Des.* 145 (14): 435–448.

21 Mysinger, M.M., Carchia, M., Irwin, J.J., and Shoichet, B.K. (2012). Directory of useful decoys, enhanced (DUD-E): better ligands and decoys for better benchmarking. *J. Med. Chem.* 55: 6582–6594.

22 De Vivo, M. and Cavalli, A. (2017). Recent advances in dynamic docking for drug discovery. *Wiley Interdiscip. Rev. Comput. Mol. Sci.* 7: e1320.

23 Jakhar, R., Dangi, M., Khichi, A., and Chhillar, A.K. (2019). Relevance of molecular docking studies in drug designing. *Curr. Bioinform.* 15: 270–278.

24 Griffen, E., Leach, A.G., Robb, G.R., and Warner, D.J. (2011). Matched molecular pairs as a medicinal chemistry tool. *J. Med. Chem.* 54: 7739–7750.

25 Hsieh, C.-M., Wang, S., Lin, S.-T., and Sandler, S.I. (2011). A predictive model for the solubility and octanol–water partition coefficient of pharmaceuticals. *J. Chem. Eng. Data* 56: 936–945.

26 Navo, C.D. and Jiménez-Osés, G. (2021). Computer prediction of pKa values in small molecules and proteins. *ACS Med. Chem. Lett.* 12: 1624–1628.

27 Vamathevan, J. et al. (2019). Applications of machine learning in drug discovery and development. *Nat. Rev. Drug Discov.* 186 (18): 463–477.

28 Freedman, D.A. (2009). Statistical models: theory and practice answers to selected exercises the labs. *Statistics (Ber).* 442.

29 Ru, X. et al. (2021). Current status and future prospects of drug–target interaction prediction. *Brief. Funct. Genomics* 20: 312–322.

30 Mousavian, Z. and Masoudi-Nejad, A. (2014). Drug-target interaction prediction via chemogenomic space: Learning-based methods. *Expert Opin. Drug Metab. Toxicol.* 10: 1273–1287.

31 Mayr, A., Klambauer, G., Unterthiner, T., et al. (2018). Large-scale comparison of machine learning methods for drug target prediction on ChEMBL, undefined. pubs.rsc.org

32 Bagherian, M., Sabeti, E., Wang, K. et al. (2021). Machine learning approaches and databases for prediction of drug–target interaction: a survey paper, undefined. academic.oup.com

33 Carpenter, K.A., Cohen, D.S., Jarrell, J.T., and Huang, X. (2018). Deep learning and virtual drug screening. *Future Med. Chem.* 10: 2557–2567.

34 D'Souza, S., Prema, K.V., and Balaji, S. (2020). Machine learning models for drug–target interactions: current knowledge and future directions. *Drug Discov. Today* 25: 748–756.

35 Krogh, A. (2008). What are artificial neural networks? *Nat. Biotechnol.* 262 (26): 195–197.

36 Alloghani, M., Al-Jumeily, D., Mustafina, J., Hussain, A. & Aljaaf, A. J. A systematic review on supervised and unsupervised machine learning algorithms for data science. 3–21 (2020). doi:https://doi.org/10.1007/978-3-030-22475-2_1

37 Rajoub, B. Supervised and unsupervised learning. *Biomed. Signal Process. Artif. Intell. Healthc.* 51–89 (2020). doi:https://doi.org/10.1016/B978-0-12-818946-7.00003-2

38 Tuia, D., Volpi, M., Copa, L. et al. (2011). A survey of active learning algorithms for supervised remote sensing image classification. *IEEE J. Sel. Top. Signal Process.* 5: 606–617.

39 Usama, M. et al. (2019). Unsupervised machine learning for networking: techniques, applications and research challenges. *IEEE Access* 7: 65579–65615.

40 Chapelle, O., Schölkopf, B., and Zien, A. (2006). *Semi-Supervised Learning*, 2. Cambridge, MA: MIT Press.

41 Bronstein, M.M., Bruna, J., Lecun, Y. et al. (2017). Geometric deep learning: going beyond Euclidean data. *IEEE Signal Process. Mag.* 34: 18–42.

42 Zhang, Z. et al. (2022). Graph neural network approaches for drug-target interactions. *Curr. Opin. Struct. Biol.* 73: 102327.

43 Zhou, J. et al. Graph neural networks: a review of methods and applications (2021). doi:https://doi.org/10.1016/j.aiopen.2021.01.001

44 Li, O., Liu, H., Chen, C., and Rudin, C. (2018). Deep learning for case-based reasoning through prototypes: a neural network that explains its predictions. *Proc. AAAI Conf. Artif. Intell.* 32.

45 Angelov, P. and Soares, E. (2020). Towards explainable deep neural networks (xDNN). *Neural Netw.* 130: 185–194.

46 Ratner, M. (2018). FDA backs clinician-free AI imaging diagnostic tools. *Nat. Biotechnol.* 36: 673–674.

47 Selbst, A. D. & Powles, J. Meaningful information and the right to explanation. doi:https://doi.org/10.1007/s13347-017-0263-5

48 Lipinski, C.A. (2004). Lead- and drug-like compounds: the rule-of-five revolution. *Drug Discov. Today Technol.* 1: 337–341.

49 Veber, D.F. et al. (2002). Molecular properties that influence the oral bioavailability of drug candidates. *J. Med. Chem.* 45: 2615–2623.

50 Kuhn, B., Mohr, P., and Stahl, M. (2010). Intramolecular hydrogen bonding in medicinal chemistry. *J. Med. Chem.* 53: 2601–2611.

51 Kenny, P.W. (2022). Hydrogen-bond donors in drug design. *J. Med. Chem.* 65: 14261–14275.

52 Thomas, M., Smith, R., Boyle, N.M.O. et al. Comparison of structure- and ligand-based scoring functions for deep generative models: a GPCR case study. In: , 1–39.

53 Askr, H., Elgeldawi, E., Aboul Ella, H. et al. Deep learning in drug discovery: an integrative review and future challenges. *Artif Intell Rev.* https://doi.org/10.1007/s10462-022-10306-1. Epub ahead of print.

54 Chiu, Y.-C. et al. (2019). Predicting drug response of tumors from integrated genomic profiles by deep neural networks. *BMC Med. Genomics* 12: 18.

55 Pishvaian, M.J. et al. (2020). Overall survival in patients with pancreatic cancer receiving matched therapies following molecular profiling: a retrospective analysis of the know your tumor registry trial. *Lancet Oncol.* 21: 508–518.

56 van der Velden, D.L. et al. (2019). The drug rediscovery protocol facilitates the expanded use of existing anticancer drugs. *Nature* 574: 127–131.

57 Iorio, F. et al. (2016). A landscape of pharmacogenomic interactions in cancer. *Cell* 166: 740–754.

58 Yang, W. et al. (2013). Genomics of drug sensitivity in cancer (GDSC): a resource for therapeutic biomarker discovery in cancer cells. *Nucleic Acids Res.* 41: D955–D961.

59 Bansal, M. et al. (2014). A community computational challenge to predict the activity of pairs of compounds. *Nat. Biotechnol.* 32: 1213–1222.

60 Narayan, R.S. et al. (2020). A cancer drug atlas enables synergistic targeting of independent drug vulnerabilities. *Nat. Commun.* 11: 2935.

61 Mina, M. et al. (2017). Conditional selection of genomic alterations dictates cancer evolution and oncogenic dependencies. *Cancer Cell* 32: 155–168.e6.

62 Martínez-Jiménez, F. et al. (2020). A compendium of mutational cancer driver genes. *Nat. Rev. Cancer* 20: 555–572.

63 Costello, J.C. et al. (2014). A community effort to assess and improve drug sensitivity prediction algorithms. *Nat. Biotechnol.* 32: 1202–1212.

64 Jang, I. N. S., Neto, E. C., Guinney, J., Friend, S. H. & Margolin, A. A. Systematic assessment of analytical methods for drug sensitivity prediction from cancer cell line data. *Biocomputing* 2014 (2013). doi:https://doi.org/10.1142/9789814583220_0007

65 Garnett, M.J. et al. (2012). Systematic identification of genomic markers of drug sensitivity in cancer cells. *Nature* 483: 570–575.

66 Liu, H., Zhao, Y., Zhang, L., and Chen, X. (2018). Anti-cancer drug response prediction using neighbor-based collaborative filtering with global effect removal. *Mol. Ther. Nucleic Acids* 13: 303–311.

67 Trotti, A., Colevas, A.D., Setser, A. et al. (2003). CTCAE v3.0: development of a comprehensive grading system for the adverse effects of cancer treatment. *Semin. Radiat. Oncol.* 13: 176–181.

68 Creanza, T.M. et al. (2021). Structure-based prediction of hERG-related cardiotoxicity: a benchmark study. *J. Chem. Inf. Model.* 61: 4758–4770.

69 Kuntz, I.D., Blaney, J.M., Oatley, S.J. et al. (1982). A geometric approach to macromolecule-ligand interactions. *J. Mol. Biol.* 161: 269–288.

70 Lander, E.S. et al. (2001). Initial sequencing and analysis of the human genome. *Nature* 409: 860–921.

71 Reiss, T. (2001). Drug discovery of the future: the implications of the human genome project. *Trends Biotechnol.* 19: 496–499.

72 Santos, R. et al. (2016). A comprehensive map of molecular drug targets. *Nat. Rev. Drug Discov.* 16: 19–34.

73 Lazebnik, Y. (2002). Can a biologist fix a radio?—Or, what I learned while studying apoptosis. *Cancer Cell* 2: 179–182.

74 Bender, A. & Cortés-Ciriano, I. Artificial intelligence in drug discovery: what is realistic, what are illusions? Part 1: Ways to make an impact, and why we are not there yet. *Drug Discov. Today* (2021). 26(2):511-524.

75 Bender, A. and Cortes-Ciriano, I. (2021 Apr). Artificial intelligence in drug discovery: what is realistic, what are illusions? Part 2: a discussion of chemical and biological data. *Drug Discov. Today* 26 (4): 1040–1052.

10

AI-Based Protein Structure Predictions and Their Implications in Drug Discovery

Tahsin F. Kellici[1], Dimitar Hristozov[2], and Inaki Morao[3]

[1] Department of Computational Drug Discovery, Evotec UK Ltd., Milton Park, Abingdon, Oxfordshire OX14 4RZ, United Kingdom
[2] Department of In Silico Research and Development, Evotec UK Ltd., Milton Park, Abingdon, Oxfordshire OX14 4RZ, United Kingdom
[3] Department of Protein Homeostasis, Evotec UK Ltd., Milton Park, Abingdon, Oxfordshire OX14 4RZ, United Kingdom

10.1 Introduction

Proteins perform many different functions within organisms, including catalyzing metabolic reactions, DNA replication, providing structure to cells, responding to stimuli, and transporting molecules. The particular function performed by a protein is determined by its three-dimensional (3D) structure, which in turn is encoded in its 1D amino acid sequence. A computational method that predicts the 3D structure from the 1D amino acid sequence has been long sought. Such a method would open up new doors for both protein function prediction and rational protein design, thus accelerating the discovery of new drugs [1].

Traditionally, knowledge-based methods or physics-based protein-structure prediction have been used. Both methods are appealing but prone to computational issues when applied to even moderate-sized proteins. On the other hand, recent advances in the field of artificial intelligence (AI) have brought the data-driven approach to the fore. Therefore, new tools and software packages (RoseTTAFold [2], AlphaFold [3], trRosetta [4] and trRosettaX [5], ESMFold [6], RGN2 [7], among others) have been recently developed and have produced in silico protein structure predictions of unrivaled accuracy.

In 2020, at CASP 14 (a community-wide blind competition to predict 3D structures of not yet publicly available proteins) [8], AlphaFold2 [3] – a deep neural network approach devised by DeepMind – took the world by storm. It was the top-ranked protein structure prediction method by a large margin, producing predictions with high accuracy and achieving a median GlobalDistanceTest_TotalScore (GDT_TS) of 92.4 (out of 100) across all the proteins to be solved in the competition. GDT_TS over 90 is comparable with the results of laborious experimental determinations of

Computational Drug Discovery: Methods and Applications, First Edition.
Edited by Vasanthanathan Poongavanam and Vijayan Ramaswamy.
© 2024 WILEY-VCH GmbH. Published 2024 by WILEY-VCH GmbH.

protein structures and is considered as a correct solution that scientists can rely on with confidence.

A detailed discussion of the technical aspects of the AlphaFold2 implementation is beyond the scope of this chapter. Briefly, the method makes use of the growing availability of public data (PDB) [9, 10] and incorporates novel neural network architectures and training procedures. This allows the simultaneous tuning of the model parameters in order to optimize the final 3D structure. More details can be found in the AlphaFold2 manuscript [3], its extensive support information, and in the GitHub repository [11].

An important feature of AlphaFold2 model is its ability to assign a confidence score per residue to its own predictions. This score is termed the "predicted local-distance difference test" (pLDDT). pLDDT estimates how well the prediction would agree with an experimental structure based on the local distance difference test Cα (lDDT-Cα). It has been shown to be well-calibrated and to be a competitive predictor of disordered regions [3].

The source code for the AlphaFold2 model, trained weights, and inference script have been publicly released [11]. This has allowed researchers to use and extend the original model. In addition, DeepMind teamed up with the European Bioinformatics Institute (EMBL-EBI) to create the AlphaFold Protein Structure Database [12]. As of August 2022, there are 214 684 311 structures available on the AlphaFold DB website, including 48 complete proteomes for bulk download.

Inspired by the AlphaFold2 success and with the goal of increasing protein structure prediction accuracy for structural biology research and advancing protein design, Baek et al. [2] developed RoseTTAFold – a 3-track deep neural network model that achieved similar performance as AlphaFold2 but with significantly lower hardware requirements. This model allows the generation of 3D protein structures on a single-GPU workstation. In addition, due to its architecture, RoseTTAFold offers the potential to predict complexes of unknown structure that possess more than three chains.

Guided by the advances in natural language processing, a few methods (ESM-Fold [6] and RGN2 [7]) that do not require multiple sequence alignment (MSE) have been recently proposed. Both methods offer comparable performances in some cases while significantly improving the inference time.

In order to render AI-generated protein structures suitable for driving structure-based drug design projects (e.g. via virtual screening, lead optimization [13], etc.), it is usually necessary to reorganize the binding site to accommodate a given ligand series or the generation of a biologically significant conformational ensemble for the protein. This chapter starts with a review of state-of-the-art methods for combining deep-learning structural models with experimental data. Such a combination allows the refinement of the models produced by deep learning alone, making them more suitable for structure-based design. An overview of the combination of AI-based methods with computational approaches follows, leading to a summary of the advances and some of the remaining challenges in using deep learning structural models for drug design and discovery.

10.2 Impact of AI-Based Protein Models in Structural Biology

The accurate prediction of protein structure achieved by the deep learning methods discussed in the introduction has opened new possibilities in integrative structural biology. Experimental models created by methodologies such as cryo-EM and X-Ray crystallography are being combined with predicted models (Figure 10.1) [14]. In X-ray crystallography, predicted models can help with the "phase problem", an issue that is frequently solved by using molecular replacement with previously solved experimental structures. In the case of nuclear magnetic resonance (NMR), combining models and experimental restraints would offer a direct path to determining the multiple conformational states a protein can assume in solution. In cryo-EM, theoretical modeling methods can be useful to either accelerate model building by providing an initial model or fill parts of the EM map that are at a lower resolution. The reminder of this section presents recent examples of the use of experimental data in combination with deep learning models to provide a set of optimized conformations of the target protein. Such optimized conformations can be further used in drug discovery projects.

10.2.1 Combination of AI-Based Predictions with Cryo-EM and X-Ray Crystallography

Dramatic advances in the technology of electron microscopy (EM) at cryogenic temperatures (cryo-EM) have resulted in the production of an abundance of structures at near-atomic or better resolution [15]. Continuous developments in applying

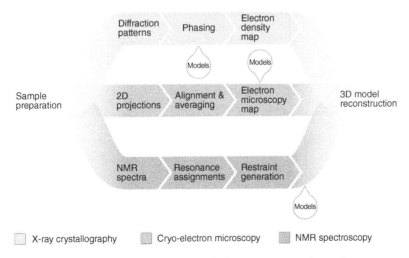

Figure 10.1 Exploiting deep learning models for accurate experimental structure determination of proteins (Figure from Ref. [14]). Deep learning models can help with the phasing of crystal structures, electron microscopy maps, and the generation and interpretation of restraints coming from NMR or other techniques.

cryo-EM to smaller proteins will further advance the power and applicability of this technology for structure-based drug design [16]. The key advantages of cryo-EM over crystallography relate to lower sample requirements, no need for crystal formation, and the fact that cryo-EM allows visualization of samples in various conformational states [17]. Due to these advantages, cryo-EM has found various uses in drug discovery [18], including:

- Handling of cases where the ligand binding site is unknown, and where induced fit phenomena bring large conformational changes to the protein [15]
- Elucidation of targets that involve one or two proteins in oligomeric structures
- Dealing with targets that are resistant to crystallization (thus hindering the establishment of the structure–activity relationship [SAR] and/or hindering the achievement of the target profile by ligand optimization alone).

Computational techniques and predictive modeling play an important role in building atomic models in cryo-EM density maps [19]. Cryo-EM images have extremely low signal-to-noise levels because biological macromolecules are highly radiation-sensitive, requiring low-dose imaging, and because the molecules are poor in contrast [20]. The improved accuracy of deep learning-based protein structure prediction has inspired fresh approaches for model building in cryo-EM density maps, some of which are summarized in Table 10.1. As can be seen from the cited examples, the combination of deep learning methods with cryo-EM data has enhanced the processing of cryo-EM density maps. Two new tools called phenix.process_predicted_model and ISOLDE further help with the integration of AlphaFold models in the experimental structure determination by X-ray crystallography and cryo-EM [32]. The phenix.process_predicted_model tool down-weights or removes low-confidence residues and can break a model into confidently predicted domains in preparation for molecular replacement or cryo-EM docking. These confidence metrics are further used in ISOLDE to weight torsion and atom–atom distance restraints, allowing the complete AlphaFold model to be interactively rearranged to match the docked fragments and reducing the need for the rebuilding of connecting regions [32].

A detailed study of the importance of theoretical models in cryo-EM was provided by Hryc and Baker [33]. The authors examined a set of community-established standard density maps representing 12 unique biological datasets ranging from 1.8 to 4.5 Å in resolution. Those targets represent a variety of macromolecular complexes, making them ideal candidates to evaluate the accuracy of predictive modeling in the context of cryo-EM density maps. It was found that AlphaFold2 produced models of superior or equal quality to the state-of-the-art methods used to model the density map for the major capsid protein gp7 of $\varepsilon15$, an infectious bacteriophage, as well as for most of the proteins of Cyanophage Syn5 [33]. The authors also found that the AlphaFold2 prediction confidence score (pLDDT) is a good guide to the overall model quality and accuracy. However, there were a number of examples where pLDDT scores were not sufficient indicators of potential model accuracy [33].

In a similar analysis, Kryshtafovych et al. [34] described the solution of four experimental structures using AlphaFold2 models submitted to CASP14 [35] (crystal structure of the inner membrane reductase FoxB; subunits of phage AR9 non-virion RNA polymerase; the crystal structure of the baseplate anchor

Table 10.1 AlphaFold2 models are being used in combination with cryo-EM density maps and crystallography data in order to provide optimized 3D structures.

Target	Description	Electron microscopy database (EMD) accession codes and PDB ID	Ref.
NALCN-FAM155A-UNC79-UNC80 channel complex	NALCN channel mediates voltage-modulated sodium leak currents, which can be blocked by extracellular calcium. The functional NALCN channel is a hetero-tetrameric channelosome. In order to predict the structure of the tetramer, the models of UNC79-UNC80 head and tail regions were predicted by AlphaFold2, docked into a cryo-EM map, and manually adjusted using Coot [21].	EMD-32344 and 7W7G	[22]
FtsH-HflKC AAA protease complex	The membrane-bound AAA protease FtsH is the key player controlling protein quality in bacteria. The predicted FtsHTM hexamer from Alphafold 2 was manually fitted to the FtsHPD+TM map. Models of FtsHPD+TM-HflKC and FtsHCD were subjected to the Phenix real-space refinement [23].	7W13	[24]
Nuclear ring (NR) and cytoplasmic ring (CR) from the *Xenopus laevis* NPC	The authors combined "sideview" particles and "tilt-view" particles to overcome the insufficient Fourier space sampling problem and used AlphaFold2 to predict all nucleoporin structures.	EMD32056, EMD-32060, EMD-32061, 7VOP	[25]
Pentameric assembly of the Kv2.1 tetramerization domain	The T1-domain sequences from Kv2.1 and Kv8.2 (in a 3:1 ratio) were submitted to the ColabFold notebook to generate the heterotetramer using AlphaFold2-Multimer [26]. The models were inspected, and the top-ranked model was used for the addition of zinc.	7RE5	[27]
Structure of the human glucose transporter GLUT4	An initial structure model for GLUT4 was generated by AlphaFold2. The structure was docked into the density map and manually adjusted and rebuilt by COOT [21].	EMD-32760; EMDB-32761; 7WSM; 7WSN	[28]
Ternary complex of insulin-like growth factor 1 (IGF1) with IGF-binding protein 3 (IGFBP3) and acid-labile subunit (ALS)	Model building of ambiguously resolved parts was aided by a protein model generated from AlphaFold2 and a post-processed map generated from DeepEMhancer [29].	EMD-32735, 7WRQ	[30]
Crystal structure of the Ars2-Red1 complex	The structure of the complex was determined by molecular replacement using the AlphaFold2 Ars2 model (AF-094326) and refined to a Rfree of 30.2% and a Rwork of 24.5%. AlphaFold was also used to model the dimeric structure of the Red1 C-terminus.	7QUU	[31]

and partner TSP assembly region of TSP4 from Bacteriophage CBA120; crystal structure of Af1503 transmembrane receptor) [34]. The authors also reported that the AlphaFold2 models helped improve the structure of an already solved target (the bacterial exo-sialidase Sia24) [34]. Although molecular replacement is a very well-established technique, high-accuracy models are needed, and until recently, this always required the availability of templates with high levels of sequence identity. As the accounts in this paper show, the models provided by deep learning methods are indeed powerful and can be used for molecular replacement [34]. The provided results for the monomeric models of subunits (the phage AR9 nvRNA polymerase, the tail spike protein TSP4-N from bacteriophage CBA120, and the Af1503 receptor) allowed the assembly of complex folds that reflect in large parts the experimentally determined oligomeric structures. Exceptions are the flexible linkers and loops without a defined secondary structure that introduce errors.

10.2.2 Combination of AI-Based Predictions with NMR Structures

NMR spectroscopy is now a well-established technique to elucidate the structure, interaction, and dynamics of molecules in solution and is also used to guide structure-based lead discovery campaigns [36]. The advantages of this technique include

- both chemical compounds and proteins of interest give NMR signals
- the binding mode between ligand and protein, such as conformational transitions upon ligand binding and interaction interface, can be determined at an atomic resolution
- NMR performs well for weak intermolecular interactions with a dissociation constant (Kd) in the µM/mM range.

Fast NMR data acquisition has led to remarkable improvements in the throughput of high-resolution and sensitive NMR methodologies. These improvements allow the identification of new fragments, thus creating a new avenue for fragment-based drug discovery and development (FBDD) [37].

In order to determine the 3D structure of a protein by NMR, a large number of distance restraints derived from nuclear Overhauser effect data together with spin–spin coupling constants are collected. The most accurate solution structures of proteins, however, are obtained when residual dipolar couplings (RDCs), preferentially for different internuclear vectors, are measured and included in the refinement of the 3D structure. RDCs improve the local backbone geometry of NMR-based solution structures of proteins and can accurately define the relative orientation of secondary structure elements and protein domains. In a recent article, Zweckstetter compared AlphaFold2 structures with NMR structures for the third IGG-binding domain from streptococcal protein G (GB3), the DNA damage-inducible protein (DinI), and ubiquitin [38]. Particularly useful for the comparison of RDCs and RDC-derived solution structures with models predicted by AlphaFold2 is GB3, because of the small rigid domain, the existence of a high-resolution crystal structure, and multiple high-resolution NMR structures using a large number of RDCs [38]. The study shows that 3D structures predicted by AlphaFold2 can be highly representative of the solution conformation of proteins. The excellent agreement of a large number of RDCs with the structures predicted

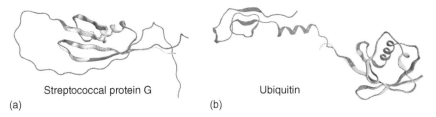

Figure 10.2 a Streptococcal protein G RDC-derived NMR structure (green; PDB id: 1P7F), 1.1 Å X-ray structure (yellow; PDB id: 1IGD), AlphaFold2-structure (red). RMSD value for the Alphafold model compared to the NMR structure is 0.42 Å; b Ubiquitin: RDC-derived NMR structure (green; PDB id: 2MJB), AlphaFold2-model (red). The structures have an RMSD of 0.65 Å.

by AlphaFold2 for GB3, DinI, and ubiquitin demonstrates the high accuracy of the predicted structures both in terms of local geometry and relative orientation of secondary structure elements (Figure 10.2), that is, the global structure [38]. These proteins provide appropriate cases for a successful AlphaFold2 prediction since they are very small and several high-resolution structures are available in the PDB. Thus were used in the training of the AlphaFold2 neural network. Problems could arise, however, for proteins that do not have these advantages. These problems are alleviated when AlphaFold2 models are combined with RDCs: either the AlphaFold2 model that best fits to the experimental RDCs can be selected (e.g. N-terminal domain of Ca2 + ligated calmodulin) or the AlphaFold2 model can be used as starting structure for RDC-based refinement calculations [38].

Similar encouraging results are reported for the 68-kDa SARS-CoV-2 M^{pro} enzyme, where measured RDCs, using a new, highly precise TROSY-AntiTROSY Encoded RDC (TATER) experiment, are compared with values derived from both high-resolution X-ray structures and AlphaFold2 models [39]. The highest pLDDT-scoring model of the full AlphaFold2 implementation fits RDCs better than 92% of all X-ray structures. Relative to the best X-ray structures, AlphaFold2 M^{pro} models agree more closely with solution RDCs for residues that are part of regular secondary structure than the remainder. This result indicates that catalytic scaffolds are well defined by AlphaFold2. The authors further hypothesize that new opportunities for combining experimentation with molecular dynamics simulations as solution RDCs provide highly precise input for QM/MM simulations of substrate binding/reaction trajectories [39].

In another study, Fowler et al. used the program ANSURR (Accuracy of NMR Structures Using RCI and Rigidity) [40], a software that measures the accuracy of solution structures, and showed that AlphaFold2 tends to be more accurate than NMR ensembles that have been calculated from chemical shifts [41]. In some cases of dynamic structures, however, like the EF-hand domain of human polycystin 2 or the transmembrane and juxtamembrane domains of the epidermal growth factor receptor in dodecylphosphocholine (DPC) micelles, the NMR ensembles are more accurate, and AlphaFold2 had low confidence. The authors found that AlphaFold2 could be used as the model for NMR-structure refinements and that AlphaFold2 structures validated by ANSURR may require no further refinement

[41]. A similar conclusion was reached by Tejero et al. [42]. The team used 12 data sets available for nine protein targets, and the results showed that the AlphaFold2 models have a remarkably good fit to the experimental NMR data. Across a wide range of structure validation methods, including both knowledge-based validations of backbone/sidechain dihedral angle distribution and packing scores, and model vs. data validation against experimental NOESY and RDC data, the AlphaFold2 models have similar, and in some cases, better structure quality scores compared with models generated using conventional structure generation methods in the hands of experts using these same NMR data [42].

The examples provided so far show that in most cases, deep learning models can predict the structures of small, relatively rigid, single-domain proteins in solution without the use of structural templates. These theoretical models could be used for construct optimization, surface analysis for buffer optimization, and site-directed mutagenesis to improve spectral quality by interpreting chemical shift perturbations. At the same time, NMR data can be used for the refining of AlphaFold2 models against RDC, sparse NOE, and chemical shift, as this data takes into account the multiple conformational states of proteins [42].

10.2.3 Combination of AI-Based Predictions with Other Experimental Restraints

Methods like mass spectrometry (MS), fluorescence resonance energy transfer (FRET), double electron–electron resonance (DEER), and electron paramagnetic resonance (EPR) can also help determine the structure of a protein by providing sets of experimental restraints. In the case of native MS, the method allows the topological investigation of intact protein complexes with high sensitivity and a theoretically unrestricted mass range. MS offers the crucial advantage of being able to provide structural data on the proteome scale. For example, proteome-wide crosslinking studies can help to filter biologically irrelevant interactions. Native MS started from a few laboratories in the 1990s, which demonstrated that noncovalent interactions could be preserved in the gas phase for analysis, enabling information on subunit stoichiometry, binding partners, protein complex topology, protein dynamics, and even binding affinities from a single mass spectrometric analysis [43]. Native MS does not yield detailed molecular structure information, but it has some major advantages over traditional structural biology methods, like speed, selectivity, sensitivity, and the ability to simultaneously measure several species present in a mixture [44]. These characteristics make native MS a method that requires just a fraction of the sample needed to solve structures by NMR spectroscopy or X-ray crystallography [44].

In a recent article, Allison et al. assessed whether native MS can be used in order to verify the plausibility of structural models generated by AlphaFold2 [45]. Three protein complexes whose interactions involve disordered regions, ligands, and point mutations were selected for evaluation and analysis: the structure of dihydroorotate dehydrogenase (DHODH), the small heat shock proteins (HSP) 17.7 and 18.1 from *Pisum sativum*, and the N-terminal domain (NT) of the spider silk protein Major ampullate Spidroin 1 (MaSp1) from *Euprosthenops australis*. In the case of DHODH, the protein contains a central cavity, which in the experimental structures is occupied by the cofactor flavin mononucleotide (FMN). Overlaying the ligand binding sites of the AlphaFold2 prediction and the X-ray structure revealed a nearly identical arrangement of the residues that coordinate FMN. Native MS, in combination with crosslinking and ion mobility (IM) measurements, showed that the human protein cannot maintain the correct conformation in the absence of FMN in MS, which strongly supports that FMN is required to adopt a stable conformation. Native MS can inform about the role of the cofactor in promoting the correct fold of DHODH, a role that is not evident from the ML-based prediction alone. In the case of HSP 17.7 and 18.1, AlphaFold2 could correctly predict both homodimers but also the hypothetical HSP 17.7–18.1 heterodimer with an equal per-residue confidence score. Native MS, however, revealed homodimer formation, while at the same time suggesting that heterodimerization is practically impossible. The last example, MASP1 is monomeric above, and dimeric below, pH 6.5. This pH sensitivity is partially due to a conserved salt bridge between Asp39/Asp40 and Lys65 on the opposing subunit. AlphaFold2 was used to predict the structure of the dimeric wild-type protein, as well as a point mutant with a weakened salt bridge, Asp40Asn. AlphaFold2 predicts with the same confidence the structure of the dimer in both the wild-type and the mutant. However, native MS analysis of both proteins at pH 6.0 showed that the Asp40Asn mutation abolished dimerization nearly completely, showing that the impact of losing this salt bridge on dimer formation requires experimental validation [45].

Other sparse experimental data like DEER restraints have been used in combination with Rosetta [46] in order to model conformational changes in proteins. In order to integrate these restraints, a new tool was created called ConfChangeMover (CCM). The performance of CCM was evaluated in both soluble and membrane proteins using simulated or experimental distance restraints, respectively. The main advantage of CCM over other methods stems from its ability to automatically identify, group, and move secondary structural elements (SSEs) as rigid bodies, a task that can be combined nicely with Rosetta [47]. More recently, del Alamo et al. [48], reported an investigation of the conformational dynamics of amino acid-polyamine-organocation transporter (GadC), a protein aiding the exchange of γ-aminobutyric acid (GABA) with extracellular Glu, using DEER spectroscopy. The analysis was assisted by generating an ensemble of structural models in multiple conformations using AlphaFold2, as described in [49]. The observed correspondence between conformational changes predicted by AlphaFold2 and distance changes observed by DEER is striking. AlphaFold2 predicted that the transmembrane helix 10 (TM10) acts as an extracellular thin gate. The ensemble of models coupled with DEER data suggested that the motion of TM10 was tightly coupled to that of TM9. Additionally, the dynamics of TM10 could distinguish between glutamate and GABA [48].

10.2.4 Impact of Deep Learning Models in Other Areas of Structural Biology

AI-based predicting methods can also help with protein construct design and engineering protein surfaces for protein crystallization. The predicted structure of a fold and the pLDDT score for each residue can be utilized to locate the less compact and disordered regions. Often, omitting less ordered regions from a protein sequence is beneficial to design well-behaved recombinant proteins for structural studies [50]. Furthermore, these methods can aid with mutational surface engineering, a method used to create patches with low conformational entropy in order to achieve an enhancement in the resolution of a crystal structure [51]. In this direction, DeepREx-WS, a webserver that assists with the identification of residues to be variated in protein surface engineering processes, provides helpful results with AI-based protein structures [52].

The examples provided in this section suggest that AlphaFold2 predictions are generally highly accurate; however, as Terwilliger et al. show, many parts of these predictions are incompatible with experimental data from corresponding crystal structures [53]. The combination of deep learning models with electron density or electron microscopy maps resulted in very high-quality models. Table 10.1 shows that theoretical models derived from AlphaFold2 are becoming quite popular in aiding the determination of complex biological macromolecules. When it comes to methods like native MS, DEER spectroscopy, etc., even though they do not provide direct structural details, they can detect a wide variety of protein interactions and aid in the generation of reliable models. In the future, methods like hydrogen/deuterium exchange (HDX) MS should be combined with ML for monitoring structural and dynamic aspects of proteins in solution. Restraints derived from NMR, EPR, or other techniques can be exploited either by defining the modeling question a priori or by employing the experimental data to identify a likely model a posteriori. In addition to the examples listed previously, restraints may be obtained from data collected with EPR and other sources of geometrical restraints such as FRET and cross-linking. Ultimately, the combination of experimental techniques and 3D protein models derived from deep learning produces more reliable models, which can be further exploited in drug discovery projects. It must be noted, however, that the given examples do not specifically address the influence of bound ligands, flexible regions, and point mutations on protein interactions and that further investigation is required in this direction.

10.3 Combination of AI-Based Methods with Computational Approaches

In the past years, physics-based refinement and force-field-based methods have served their purpose as an orthogonal approach for improving the quality of protein models that were predicted by informatics-based approaches [54]. Until recently, deep learning methods were considered to be a complement to simulation in macromolecular modeling and not a way of replacing them. That was mostly the result of the fact that deep neural networks typically have a multitude of parameters that must be optimized during training. This created the danger of overfitting: training on limited data that could result in a model perfectly tuned to reproduce the training set but unable to correctly predict new inputs [54]. The emergence of highly accurate structure prediction by machine learning is now raising questions about the limitations and future role of physics-based refinement. The machine

10.3 Combination of AI-Based Methods with Computational Approaches

learning-based models still have deficiencies, but further refinement has become much harder, even though there are still multiple issues to be solved. For example, the sampling problem continues to be a major challenge, and it is still difficult to create different conformational states from a given initial model [55, 56].

One of the first attempts to optimize deep learning models and create a conformational ensemble from AlphaFold2 structures was performed by Heo et al. [55]. The overall refinement protocol used by this group at CASP14 challenge consisted of three major components, as illustrated in Figure 10.3. The pre-sampling step consisted of oligomeric state prediction, putative binding ligands, and the possibility

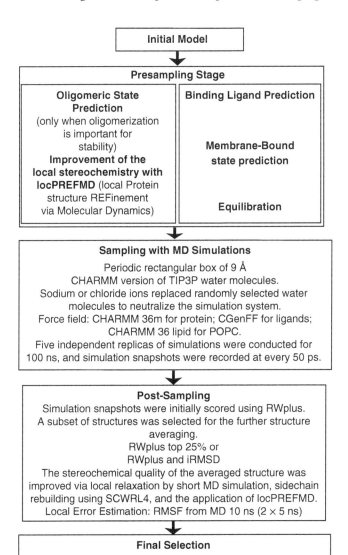

Figure 10.3 Overview of the refinement protocol as established. Source: Adapted from Heo et al. [55].

of membrane interactions. These were predicted manually based on homologous structures searched by HHsearch [57]. For the sampling step, simulation systems for non-membrane-bound proteins were constructed in an explicit water box. A periodic rectangular box was constructed with a minimal distance from any protein atom to the closest box edge of 9 Å. Empty spaces in the box were filled with the CHARMM version of TIP3P water molecules [58]. Protein conformations were sampled via molecular dynamics simulations. In principle, the protocol described by Heo et al. reached near-atomistic accuracy for many targets. However, this required long enough simulations to sample the native state with an assumption that the native state remains the lowest free energy state as other non-native conformations are being explored. This protocol was able to refine models from other predictors, including many models generated based on machine learning methods. However, when it came to improving AlphaFold2 models, the group experienced significant difficulty. A simple explanation may be that AlphaFold2 models already had very high accuracy to begin with. In terms of the global distance test – high accuracy (GDT-HA) metric [59], one of the most frequently used in literature and in CASP experiments, only 26 out of 87 TS domains had less than 70 GDT-HA units. Thus, not many models required refinement. However, even for models that had significant errors that should have been fixed, refinement was not very successful. The Markov State Model analysis and relaxation of the experimental structures revealed several issues with the current refinement protocol. First, there was still a sampling problem with the MD-refinement protocol. To reach the native state from the initial model state via MD simulations, it had to overcome several kinetic energy barriers for partial unfolding and refolding. However, the time required for the transitions was much longer than simulations for refinement, and state transitions were prohibited by the restraints. As a result, the sampled structures during the refinement simulations hardly deviated from the initial AlphaFold2 models. These were some interesting results, taking into account that a similar protocol when applied by the same group on AlphaFold models from CASP13 showed that physics-based refinement improved the accuracy of machine-learning models to exceed the accuracy of any other available method based on the targets that were assessed [56].

The applicability of AlphaFold2-generated protein conformations in virtual screening experiments – an important part of most early drug discovery projects – has been investigated in [60] with its performance being evaluated across a range of targets using the DUD-E dataset [61]. Out of the box, many AlphaFold2 structures produced low enrichment, hinting that the AlphaFold2 predictions may need further refinement before being used in the context of virtual screening [62]. These results could be the consequence of low confidence loops occluding part of the binding site, missing co-factors, and uncertainty in relative domain orientation. Where a comparison could be made, the authors found that unrefined AlphaFold2 structures deliver similar enrichments to those of an apo experimentally derived structure, significantly below the enrichments using an experimentally derived holo structure [60]. Meanwhile, the application of induced fit docking coupled with molecular dynamics (IFD-MD), a method that combines ligand-based pharmacophore docking, rigid receptor docking, and protein structure prediction with explicit solvent molecular dynamics simulations [63], can induce a binding site conformation that delivers enrichments much closer to the holo structure. This is also supported by the finding that the average binding site volume of the IFD-MD refined AlphaFold2 structure is closer to the holo structure than the raw AlphaFold2 structure [60].

Encouraged by these results [60], the authors went one step further by investigating a total of 14 protein targets, each of which consists of a congeneric set of active ligands along with a co-crystallized structure with one of those ligands [13]. Seven of the data sets are taken from the 2015 paper in which the FEP+ methodology was introduced [64], plus one homology model of PDE10A, which was used as an isolated test case; the remaining six come from internal Schrodinger drug discovery projects. In each case, the authors evaluated the performance of IFD-MD for several different homology models based on templates with differing sequence identities (roughly 30%, 40%, and 50%, although templates in all three of these categories are not available for every target). For this task, they used the ligand for which a co-crystallized structure is available for the IFD-MD calculations (so as to be able to evaluate the RMSD from the experimentally determined structure). The authors exported the top 5 poses produced by IFD-MD and carried out FEP calculations for the entire congeneric series of ligands for each pose. The final pose is selected using a scoring function, which combines several performance metrics from the FEP calculations (correlation coefficient, RMS error) as well as the absolute binding free energy. This protocol (as shown in Figure 10.4(a)) could be generalized and applied to many potential structure-based drug discovery projects, requiring experimental binding affinity data for a congeneric series obtained either from the literature (publication or patent) or in-house experiments [13]. A key aspect of this refinement protocol (Figure 10.4(a)) is the use of binding data from a ligand series to select the most appropriate protein-ligand complex structure. The ambiguity and noise that are present in a typical homology model or from deep learning methods could be addressed by differentiating proposed options with ligand-based information. These results suggest that the IFD-MD and FEP calculations provide a way to combine protein structure prediction and ligand binding information [13]. In a similar way, Beuming et al. [65] used AlphaFold2 models in order to evaluate the performance of FEP+. The authors, in order to generate the MSA, employed three databases: BFD [66], Mgnify [67], and Uniref90 [68]. The ligand was introduced into the *apo* model coming from AlphaFold2 by aligning the model with the crystal structure used for the original FEP calculations (Figure 10.4(b)). The Mean Unsigned Error of the individual perturbations for calculations done with AlphaFold2 is comparable with those performed with crystal structures, and in many cases, the R^2 values are similar to the expected values for well-behaving FEP calculations. It needs to be highlighted, however, that in this method, the introduction of the aligned ligand was performed through superposition with crystal structures. As a result, the conclusions are highly dependent on whether the initial binding poses are accurate enough.

Another approach [69] combined deep learning approaches with mechanistic modeling for a set of proteins that experimentally showed conformational changes by using trRosetta [4] as a deep learning predictive platform (Figure 10.5). By combining DeepMSA [70], with deep residual-convolutional network trRosetta

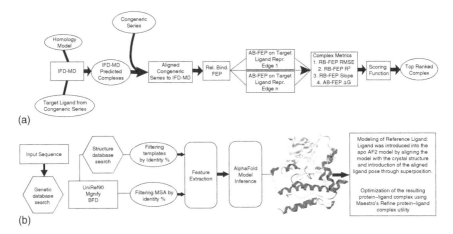

Figure 10.4 AlphaFold2 models in combination with Induced-Fit Docking or other techniques enable accurate free energy perturbation calculations [13, 60, 65].

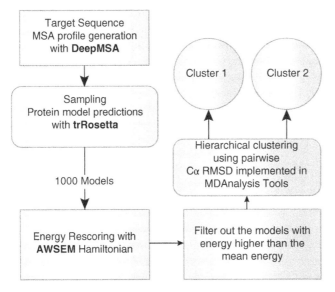

Figure 10.5 Workflow of the protein-folding pipeline used in [69]. Source: Adapted from Audagnotto et al. [69].

and the AWSEM force field [71], the authors observed that both X-ray structures of the different protein states and the similar intermediate states explored by the MD simulation were predicted. To test the ability of the pipeline to predict protein conformational ensembles, the authors investigated only X-ray structures with a maximum length of roughly 200 amino acids, a resolution equal to or less than 2.40 Å, and where more than one conformation was available in the PDB for the same sequence [69]. Four test cases were taken into consideration (Adenylate kinase, αI-domains of LFA-1, Myoglobin protein, T4 lysozyme, and Tetrahymena thermophila-BIL2). It was observed that the conformational space explored by the predicted models was similar to the one observed during the MD simulations of the experimental protein structure. In particular, for the adenylate kinase and the T4 lysozyme predictions, it was possible to predict the active and inactive structures as well as the intermediate conformations observed during the MD simulations. Although the trRosetta algorithm has been trained on static PDB structures, it was able to reproduce protein flexibility. Limitations of the technique have been observed in the case where metal ions (LFA1 test case) and cofactor (Myoglobin test case) were present influencing the conformational equilibrium. It is important to notice that the ability to predict protein flexibility is correlated to the number of available structures for the different protein conformations in the PDB, and this is the strongest limitation of the current machine-learning techniques. Another important aspect is the quality of the initial MSA profile. The authors have noted that the choice of the MSA algorithm can hamper the model prediction by favoring one conformation [69].

In a similar attempt, del Alamo et al. used a set of eight membrane proteins (transporters and GPCRs) representing different structural classes and mechanisms of action to set up an approach driving AlphaFold2 to sample their alternative conformations [49]. These included five unique transporters (LAT1, ZnT8, MCT1, STP10, and ASCT2), whose structures had been previously experimentally determined in both inward- and outward-facing conformations, and three representative G-protein-coupled receptors (CGRPR, PTH1R, and FZD7), whose structures had been solved experimentally in active and inactive states. None of these proteins were part of the original AlphaFold2 training set, which included structures located in the protein data bank (PDB) [49, 72]. The sequences of all targets were truncated at the N- and C-termini to remove large soluble and/or intrinsically disordered regions that represent a challenge for AlphaFold2. Prediction runs were executed using AlphaFold v2.0.1. The pipeline used in this study differs from the default AlphaFold2 pipeline in several aspects. First, all MSAs were obtained using the MMSeqs2 server [73]. Second, the search for a template was disabled, except when explicitly performed with specific templates of interest. Third, the number of recycles was set to one, rather than three by default. Finally, models were not refined following their prediction. This study utilized all five neural networks when predicting structures without templates, with 10 predictions per neural network per MSA size. The results indicated that AlphaFold2 could be manipulated to accurately model alternative conformations of transporters and GPCRs whose structures were not available in the training set. The use of shallow MSAs was instrumental to obtaining structurally diverse models in most proteins, and in one case (MCT1) accurate modeling of alternative conformations also required the manual curation of template structures. Thus, while the results presented here provide a blueprint for obtaining AlphaFold2 models of alternative conformations, they also argue against an optimal one-size-fits-all approach for sampling the conformational space of every protein with high accuracy. Moreover, this approach showed limited success when applied to transporters whose structures were used to train AlphaFold2, hinting at the possibility that traditional methods may still be required to capture alternative conformers [49]. The work extends the scope of AlphaFold2 beyond the structure prediction of a single state to the exploration of the conformational diversity of proteins. Even though determining the populations of alternative conformations and the interconversion pathways between them still appears to be out of reach, this work represents a crucial step toward describing the dynamic nature of proteins with modern artificial intelligence-based structure predictors [72].

The structures of membrane proteins, especially GPCRs, have been thoroughly compared with deep learning models in multiple recent articles. He et al. evaluated the performance of AlphaFold2 on GPCRs by analyzing 29 GPCR structures released after the publication of the AlphaFold2 database, thereby making sure that the prediction for these GPCR structures did not involve the experimental structural information [74]. The study was focused on subdomain assembly, ligand binding, and functional state of the receptors. Even though AlphaFold2 achieves a Cα RMSD accuracy of ~1 Å in protein structure prediction, the resulting deep learning models cannot be used directly for structure-based drug design of GPCRs. AlphaFold2 shows limitations in predicting the assembly between extracellular and transmembrane domains and the transducer-binding interface of a GPCR. Molecular docking was performed against the orthosteric sites in their predicted models and experimental structures. Different binding poses of ligands were observed between AlphaFold2 models and experimental structures with large RMSDs (especially in the case of 5-hydroxytryptamine receptor 1F) due to distinct sidechain conformations. Although Cα RMSD in the predicted model is very low, the predicted model has a narrower pocket compared with that of the experimental structure.

These findings are further confirmed by a recent report from Nicoli et al. [75], in which they use the AlphaFold2 model of OR5K1 [a member of the odorant receptors (OR)] for analysis and compare it to structures coming from traditional template-based homology modeling (HM). This allowed the authors to evaluate the use of AlphaFold2 OR structures for ligand-protein interaction studies. AlphaFold2 and homology models have differences in the backbone that unavoidably affect the binding site conformations. A difference between HM and AlphaFold2 models is the activation state. The prevalence of GPCR models in the inactive state in AlphaFold2 has been addressed in a recent paper by Heo et al. [76], and the authors found that this may also affect the accuracy of binding site predictions. The refinement process of the AlphaFold2 model included multiple consequent steps of IFD with the most potent compound. This step was needed not only to improve the performance, as for the homology model, but also to open the orthosteric binding site and allow the docking of agonists [75].

10.3.1 Combination of Structure Prediction with Other Computational Approaches

AI-based protein structures are being exploited for multiple purposes that have an impact on drug design. Binding site prediction provides important information to uncover protein functions and to direct structure-based approaches. Cavity prediction tools like PrankWeb 3 [77], CavitySpace [78], GraphSite [79], and ProBiS-Fold [80] incorporate now AlphaFold-generated models and provide data on the shape and physicochemical properties of ligand binding sites and help with druggability assessment. The ligand site prediction can lead to the comparison of pockets for drug repurposing and the prediction of off-target activities. In that direction tools like PrePCI [81], a web server that predicts the interactions between proteins and small molecules and uses, among other sources, AlphaFold structures, and DrugMAP [82] are useful resources for generating such information as well as lead candidate selection and identification of metabolites involved in mediating cellular processes. Models coming from AlphaFold have also been helpful in the analysis of cysteines in chemoproteomic datasets [83]. This reactive cysteine profiling plays an important role in covalent drug discovery.

Deep learning methods in combination with Markov Chain Monte Carlo optimization have shown great promise in protein design [84]. Anishchenko et al. showed that the trRosetta deep neural network trained using multiple sequence information could predict 3D protein structures for *de novo* designed proteins from a single sequence even in the complete absence of co-evolution information [85]. *De novo* protein design is the next frontier when it comes to drug discovery. Innovations like designing mimetics of natural immune proteins with augmented therapeutic affinity and activity but diminished immunogenicity and toxicity can be improved and expedited by using these methods [86].

AI-based methods have also been deployed in the difficult task of protein–protein docking. Protein–protein interactions are responsible for a number of key physiological processes. Modulators can target the interfaces of these interactions, called "hotspots." Structure-based design techniques can be applied to design PPI modulators once a three-dimensional structure of the protein complex is available. Protocols like Fold-and-Dock (based on trRosetta) [87], FoldDock (based on AlphaFold) [88], AlphaFold Multimer [26], and AF2Complex [89, 90] have all shown promising results when compared to methods that are based on shape complementarity and template-based docking [88].

10.4 Current Challenges and Opportunities

The achievements of AlphaFold2 and the rest of the deep learning methods that predict a 3D structure of a protein from its 1D amino acid sequence have been impressive. However, it needs to be noted that knowledge-based methods lack a fundamental energy framework. Such approaches use available data to make extrapolative predictions regarding related biological and chemical systems [91]. This limits the application of deep learning to cases in which very large numbers of training examples exist and requires careful testing of generalization error (using validation examples not used in training) and manual tuning of model hyperparameters controlling the training to minimize overfitting. The fact that the AlphaFold2 algorithm was able to create biologically-relevant conformational ensembles of proteins via workflows described in this chapter (Figures 10.3–10.5) does not help in finding the most appropriate structure for assisting in a medicinal chemistry project. The output, as shown above, will be multiple equivalent structures that need to be assessed and evaluated. One further item to be noted is that the studies that were referenced in this chapter have been retrospective in nature, and that in cases where there is not much previous knowledge on the target, these approaches will not facilitate the drug discovery process. What could help is combining the output of these deep learning methods with experimental data coming from NMR, EPR, MS, FRET, etc., as shown in the first part of this chapter.

Another question that needs to be answered is whether pLDDT and related model quality metrics are sufficient for judging the quality of the model. The accuracy of predictive models must still be evaluated based on their agreement with experimental data. With a growing training set as more and more structures become available in RCSB, predictive modeling may eventually achieve the level of accuracy needed to model dynamic protein structures and complexes. Until then, AlphaFold2 and other predictive modeling techniques, despite all their successes, cannot replace experimental methods [33].

A major issue that limits the applicability of AlphaFold2 and related deep-learning methods in drug design projects is the fact that the predicted protein conformations do not take into ligands account. In the future, these deep learning methods could be used to reliably predict the structures of protein–ligand interactions [92]. For example, AlphaFill [93, 94] is an algorithm based on sequence and structure similarity that aims to "transplant" such "missing" small molecules and ions from experimentally determined structures into predicted protein models. These publicly available structural annotations are mapped to predicted protein models, to help scientists interpret biological function and design experiments. Co-folding algorithms are also being developed, allowing the generation of protein-ligand complexes. A possible workaround is the use of binding data from a ligand series to choose between multiple options for the protein-ligand complex structure. The ambiguity and noise that are present in a typical homology model, even with the most recent advances, could be addressed by differentiating proposed options with ligand-based information. The examples given above show that an approach that combines IFD and/or MD with deep learning models could provide useful insights into the binding mode and prioritization of ligands.

Some other issues affecting the quality of deep learning structural models include:

- *Intrinsically disordered proteins/regions (IDPs/IDRs):* As noticed by Wilson et al. [95], considering a residue that pertains to a helix, strand, or β-turn in an AlphaFold2 structure, as ordered, and otherwise as disordered, results in an overestimation of disorder content and a poor prediction of disordered regions. While this may seem like a trivial observation, the abundance of AlphaFold2 structures generated for disordered proteins has made such a pitfall increasingly likely for researchers who are less familiar with IDPs and structural prediction methods. Another issue is predicting the structural dynamics and transitions (i.e. order-to-disorder, disorder-to-order, disorder-to-disorder) that an IDP may undergo [95]. The development of machine-learning methods for IDPs and polypeptide aggregation cannot be ruled out; however, at the current stage, more structural data is needed for these proteins and to link the different structures

to their stabilities [96]. Intrinsically disordered regions are not the only parts where AlphaFold2 predictions struggle. It has also been observed that modeling loops remains difficult when using neural network-based methods [97]. The poor prediction of these regions has in some cases a strong impact on the quality of the models.
- *Fold-switching proteins:* Fold-switching proteins respond to cellular stimuli by remodeling their secondary structures and changing their functions. Contrasting IDPs/IDRs, which are natively unstructured, fold-switching proteins have regions that either assume distinct stable secondary and tertiary structures under different cellular conditions or populate two stable folds at equilibrium [98]. 94% of AlphaFold2 predictions captured one experimentally determined conformation but not the other. Despite these biased results, AlphaFold2's estimated confidences were moderate-to-high for 74% of fold-switching residues [98].
- *Glycosylation:* The absence of cofactors and of co- or post-translational modifications in the models in the AlphaFold protein structure database is of particular importance when it comes to glycosylation. This issue might be remediated using sequence and structure-based comparative studies. It appears that the space where glycosylation happens is somehow preserved in AlphaFold2 models. This allows for these structural features to be directly grafted onto a model [99].
- *Folding pathways:* Outeiral et al. investigated whether state-of-the-art protein structure prediction methods can provide any insight into protein folding pathways [100]. The team generated tens of thousands of folding trajectories with seven protein structure prediction programs, obtained a set of AlphaFold2 trajectories, and used them to determine major features of folding using a simple set of statistical rules. It was found that protein structure prediction methods can in some cases distinguish the folding kinetics (two-state versus multistate) of a chain better than a random baseline, but not significantly better, and often significantly worse, than a simple, sequence-agnostic linear classifier using only the number of amino acids in the chain. In a recent opinion by Chen et al., the results of AlphaFold are compared with "interpreting a movie by fast-forwarding to the final scene without first watching the previous two hours" [101]. Scientists can see the result of the folding process but not the actual process.
- *Mutations:* Understanding the impact that missense mutations have on protein structure helps to reveal their biological effects [102, 103]. Recent papers from Sen et al. and Buel et al. showed that AlphaFold2 could not predict the full extension of the impact of a mutation. For example, alanine substitution causes the ubiquitin-associated domains (UBAs) to become intrinsically disordered; however, AlphaFold2 predicted alanine-substituted UBA1 or UBA2 to be structurally equivalent to WT UBA with only minor differences in the fold.

Although what has been listed here is a brief outline of the shortcomings of deep learning structural models, these issues also provide insights into the problems of current experimental methods. In order to build improved models, better and more training data are needed. Experiments and modeling methods are required to sample the entire conformational space of proteins. The machine learning methods themselves will also have to evolve to include ligands, post-translational modifications, and complexes of different types of molecules.

10.5 Conclusions

The ability to reliably predict the 3D structure of a protein from its amino acid sequence has potentially far-reaching consequences in many scientific fields. This is demonstrated by the rapidly growing interest in AlphaFold2 ever since the publication of the initial article [3]. It has the potential to revolutionize our understanding of biology, allowing us to derive function from structure; predict protein variants/mutations; design new proteins [85]; study the evolution of proteins and the origins of life. In traditional drug discovery, the availability of high-quality computational models, usually augmented with experimental data, has already made a big impact. However, uncertainty about the accuracy of the predictions in active sites and the inability to define the conformational state of a protein remain key limitations [92]. In addition, AlphaFold models in combination with other related methods help in enabling pocket prediction, binding site comparison for drug repurposing, off-target predictions, ligandability assessment, engineering protein surfaces for protein crystallization, protein design, and protein–protein docking.

The availability of the AlphaFold Protein Structure Database by DeepMind and the ESM Metagenomic Structure Atlas by Meta-AI as openly accessible, extensive repositories [12, 104], as well as the implementation of ColabFold [105], could support a plethora of projects, including rare diseases research programs [106]. Rare diseases in particular, are often overlooked by research investors mainly because of unfavorable costs/patient ratios, might significantly benefit from such an approach [107]. Furthermore, models coming from AlphaFold and RosettaFold are now considered trusted external resources/data content and are fully integrated with PDB data [108].

Machine learning-based fold predictions are a game changer for structural bioinformatics and experimentalists alike, with exciting possibilities ahead [109]. In the field of drug discovery, the jury on the impact of AlphaFold2 and related methods is still out. However, there is no doubt that those methods have opened up a myriad of new avenues for exciting research and have brightened up the outlook for the future of drug discovery.

References

1 Dill, K.A. and MacCallum, J.L. (2012). The protein-folding problem, 50 years on. *Science* 338 (6110): 1042–1046. https://doi.org/10.1126/science.1219021.
2 Baek, M. et al. (2021). Accurate prediction of protein structures and interactions using a three-track neural network. *Science* 373 (6557): 871–876. https://doi.org/10.1126/science.abj8754.
3 Jumper, J. et al. (2021). Highly accurate protein structure prediction with AlphaFold. *Nature* 596 (7873): 583–589. https://doi.org/10.1038/s41586-021-03819-2.

4 Yang, J. et al. (2020). Improved protein structure prediction using predicted interresidue orientations. *Proc. Natl. Acad. Sci. U. S. A.* 117 (3): 1496–1503. https://doi.org/10.1073/pnas.1914677117.

5 Su, H. et al. (2021). Improved protein structure prediction using a new multi-scale network and homologous templates. *Adv Sci (Weinh)* 8 (24): e2102592. https://doi.org/10.1002/advs.202102592.

6 Lin, Z. et al. (2022). Language models of protein sequences at the scale of evolution enable accurate structure prediction. *bioRxiv* doi: 10.1101/2022.07.20.500902.

7 Chowdhury, R. et al. (2021). Single-sequence protein structure prediction using language models from deep learning. *bioRxiv* doi: 10.1101/2021.08.02.454840.

8 Kryshtafovych, A. et al. (2021). Critical assessment of methods of protein structure prediction (CASP)—round XIV. *Proteins: Structure, Function, and Bioinformatics* 89 (12): 1607–1617. https://doi.org/10.1002/prot.26237.

9 Burley, S.K. et al. (2019). RCSB protein data Bank: biological macromolecular structures enabling research and education in fundamental biology, biomedicine, biotechnology and energy. *Nucleic Acids Res.* 47 (D1): D464–D474. https://doi.org/10.1093/nar/gky1004.

10 Goodsell, D.S. et al. (2020). RCSB protein data Bank: enabling biomedical research and drug discovery. *Protein Sci.* 29 (1): 52–65. https://doi.org/10.1002/pro.3730.

11 Available from: https://github.com/deepmind/alphafold.

12 Varadi, M. et al. (2022). AlphaFold protein structure database: massively expanding the structural coverage of protein-sequence space with high-accuracy models. *Nucleic Acids Res.* 50 (D1): D439–D444. https://doi.org/10.1093/nar/gkab1061.

13 Xu, T. et al. (2022). Induced-fit docking enables accurate free energy perturbation calculations in homology models. *J. Chem. Theory Comput.* 18 (9): 5710–5724. https://doi.org/10.1021/acs.jctc.2c00371.

14 Masrati, G. et al. (2021). Integrative structural biology in the era of accurate structure prediction. *J. Mol. Biol.* 433 (20): 167127. https://doi.org/10.1016/j.jmb.2021.167127.

15 Van Drie, J.H. and Tong, L. (2020). Cryo-EM as a powerful tool for drug discovery. *Bioorg. Med. Chem. Lett.* 30 (22): 127524. https://doi.org/10.1016/j.bmcl.2020.127524.

16 Scapin, G., Potter, C.S., and Carragher, B. (2018). Cryo-EM for small molecules discovery, design, understanding, and application. *Cell. Chem. Biol.* 25 (11): 1318–1325. https://doi.org/10.1016/j.chembiol.2018.07.006.

17 de Oliveira, T.M. et al. (2021). Cryo-EM: the resolution revolution and drug discovery. *SLAS Discov* 26 (1): 17–31. https://doi.org/10.1177/2472555220960401.

18 Renaud, J.P. et al. (2018). Cryo-EM in drug discovery: achievements, limitations and prospects. *Nat. Rev. Drug Discov.* 17 (7): 471–492. https://doi.org/10.1038/nrd.2018.77.

19 Topf, M. et al. (2008). Protein structure fitting and refinement guided by Cryo-EM density. *Structure* 16 (2): 295–307. https://doi.org/10.1016/j.str.2007.11.016.

20 Palmer, C.M. and Aylett, C.H.S. (2022). Real space in cryo-EM: the future is local. *Acta Crystallogr. D Struct. Biol.* 78 (Pt. 2): 136–143. https://doi.org/10.1107/S2059798321012286.

21 Emsley, P. et al. (2010). Features and development of Coot. *Acta Crystallogr. D Biol. Crystallogr.* 66 (Pt 4): 486–501. https://doi.org/10.1107/S0907444910007493.

22 Kang, Y. and Chen, L. (2022). Structure and mechanism of NALCN-FAM155A-UNC79-UNC80 channel complex. *Nat. Commun.* 13 (1): 2639. https://doi.org/10.1038/s41467-022-30403-7.

23 Liebschner, D. et al. (2019). Macromolecular structure determination using X-rays, neutrons and electrons: recent developments in phenix. *Acta Crystallogr. D Struct. Biol.* 75 (Pt 10): 861–877. https://doi.org/10.1107/S2059798319011471.

24 Qiao, Z. et al. (2022). Cryo-EM structure of the entire FtsH-HflKC AAA protease complex. *Cell Rep.* 39 (9): 110890. https://doi.org/10.1016/j.celrep.2022.110890.

25 Tai, L. et al. (2022). 8 a structure of the outer rings of the Xenopus laevis nuclear pore complex obtained by cryo-EM and AI. *Protein Cell* https://doi.org/10.1007/s13238-021-00895-y.

26 Evans, R. et al. (2022). Protein complex prediction with AlphaFold-Multimer. *bioRxiv* https://doi.org/10.1101/2021.10.04.463034.

27 Xu, Z. et al. (2022). Pentameric assembly of the Kv2.1 tetramerization domain. *Acta Crystallogr. D Struct. Biol.* 78 (Pt 6): 792–802. https://doi.org/10.1107/S205979832200568X.

28 Yuan, Y. et al. (2022). Cryo-EM structure of human glucose transporter GLUT4. *Nat. Commun.* 13 (1): 2671. https://doi.org/10.1038/s41467-022-30235-5.

29 Sanchez-Garcia, R. et al. (2021). DeepEMhancer: a deep learning solution for cryo-EM volume post-processing. *Commun. Biol.* 4 (1): 874. https://doi.org/10.1038/s42003-021-02399-1.

30 Kim, H. et al. (2022). Structural basis for assembly and disassembly of the IGF/IGFBP/ALS ternary complex. *Nat. I.D.A.A. Commun.* 13 (1): https://doi.org/10.1038/s41467-022-32214-2.

31 Foucher, A.-E. et al. (2022). Structural analysis of Red1 as a conserved scaffold of the RNA-targeting MTREC/PAXT complex. *Nat. I.D.A.A. Commun.* 13 (1): https://doi.org/10.1038/s41467-022-32542-3.

32 Oeffner, R.D. et al. (2022). Putting AlphaFold models to work with phenix.process_predicted_model and ISOLDE. *Acta Crystallographica Section D* 78 (11): 1303–1314. https://doi.org/10.1107/S2059798322010026.

33 Hryc, C.F. and Baker, M.L. (2022). AlphaFold2 and CryoEM: revisiting CryoEM modeling in near-atomic resolution density maps. *iScience* https://doi.org/10.1016/j.isci.2022.104496.

34 Kryshtafovych, A. et al. (2021). Computational models in the service of X-ray and cryo-electron microscopy structure determination. *Proteins* 89 (12): 1633–1646. https://doi.org/10.1002/prot.26223.

35 Jumper, J. et al. (2021). Applying and improving AlphaFold at CASP14. *Proteins: Structure, Function, and Bioinformatics* 89 (12): 1711–1721. https://doi.org/10.1002/prot.26257.

36 Dias, D.M. and Ciulli, A. (2014). NMR approaches in structure-based lead discovery: recent developments and new frontiers for targeting multi-protein complexes. *Prog. Biophys. Mol. Biol.* 116 (2–3): 101–112. https://doi.org/10.1016/j.pbiomolbio.2014.08.012.

37 Sugiki, T. et al. (2018). Current NMR techniques for structure-based drug discovery. *Molecules* 23 (1): https://doi.org/10.3390/molecules23010148.

38 Zweckstetter, M. (2021). NMR hawk-eyed view of AlphaFold2 structures. *Protein Sci.* 30 (11): 2333–2337. https://doi.org/10.1002/pro.4175.

39 Robertson, A.J. et al. (2021). Concordance of X-ray and AlphaFold2 models of SARS-CoV-2 main protease with residual dipolar couplings measured in solution. *J. Am. Chem. Soc.* 143 (46): 19306–19310. https://doi.org/10.1021/jacs.1c10588.

40 Fowler, N.J., Sljoka, A., and Williamson, M.P. (2020). A method for validating the accuracy of NMR protein structures. *Nat. Commun.* 11 (1): 6321. https://doi.org/10.1038/s41467-020-20177-1.

41 Fowler, N.J. and Williamson, M.P. (2022). The accuracy of protein structures in solution determined by AlphaFold and NMR. *Structure* https://doi.org/10.1016/j.str.2022.04.005.

42 Tejero, R. et al. (2022). AlphaFold models of small proteins rival the accuracy of solution NMR structures. *Front. Mol. Biosci.* 9: 877000. https://doi.org/10.3389/fmolb.2022.877000.

43 Leney, A.C. and Heck, A.J. (2017). Native mass spectrometry: what is in the name? *J. Am. Soc. Mass Spectrom.* 28 (1): 5–13. https://doi.org/10.1007/s13361-016-1545-3.

44 Heck, A.J. (2008). Native mass spectrometry: a bridge between interactomics and structural biology. *Nat. Methods* 5 (11): 927–933. https://doi.org/10.1038/nmeth.1265.

45 Allison, T.M. et al. (2022). Complementing machine learning-based structure predictions with native mass spectrometry. *Protein Sci.* 31 (6): e4333. https://doi.org/10.1002/pro.4333.

46 Rohl, C.A. et al. (2004). Protein structure prediction using Rosetta. In: *Methods in Enzymology*, 66–93. Academic Press.

47 Sala, D. et al. (2022). Modeling of protein conformational changes with Rosetta guided by limited experimental data. *Structure* https://doi.org/10.1016/j.str.2022.04.013.

48 del Alamo, D. et al. (2022). Integrated AlphaFold2 and DEER investigation of the conformational dynamics of a pH-dependent APC antiporter. *Proc. Natl. Acad. Sci. U S A* 119 (34): e2206129119. https://doi.org/10.1073/pnas.2206129119.

49 del Alamo, D. et al. (2022). Sampling alternative conformational states of transporters and receptors with AlphaFold2. *eLife* 11: e75751. https://doi.org/10.7554/eLife.75751.

50 Edich, M. et al. (2022). The impact of AlphaFold on experimental structure solution. *bioRxiv* doi: 10.1101/2022.04.07.487522.

51 Derewenda, Z.S. and Vekilov, P.G. (2006). Entropy and surface engineering in protein crystallization. *Acta Crystallogr. Sec. D* 62 (1): 116–124. https://doi.org/10.1107/S0907444905035237.

52 Manfredi, M. et al. (2021). DeepREx-WS: a web server for characterising protein–solvent interaction starting from sequence. *Comput. Struct. Biotechnol. J.* 19: 5791–5799. https://doi.org/10.1016/j.csbj.2021.10.016.

53 Terwilliger, T.C. et al. (2022). AlphaFold predictions: great hypotheses but no match for experiment. *bioRxiv* doi: 10.1101/2022.11.21.517405.

54 Mulligan, V.K. (2021). Current directions in combining simulation-based macromolecular modeling approaches with deep learning. *Expert Opin. Drug Discov.* 16 (9): 1025–1044. https://doi.org/10.1080/17460441.2021.1918097.

55 Heo, L., Janson, G., and Feig, M. (2021). Physics-based protein structure refinement in the era of artificial intelligence. *Proteins* 89 (12): 1870–1887. https://doi.org/10.1002/prot.26161.

56 Heo, L. and Feig, M. (2020). High-accuracy protein structures by combining machine-learning with physics-based refinement. *Proteins* 88 (5): 637–642. https://doi.org/10.1002/prot.25847.

57 Steinegger, M. et al. (2019). HH-suite3 for fast remote homology detection and deep protein annotation. *BMC Bioinform.* 20 (1): 473. https://doi.org/10.1186/s12859-019-3019-7.

58 Jorgensen, W.L. et al. (1983). Comparison of simple potential functions for simulating liquid water. *J. Chem. Phys.* 79 (2): 926–935. https://doi.org/10.1063/1.445869.

59 Kryshtafovych, A. et al. (2018). Evaluation of the template-based modeling in CASP12. *Proteins: Struct. Funct. Bioinform.* 86 (S1): 321–334. https://doi.org/10.1002/prot.25425.

60 Zhang, Y. et al. (2022). Benchmarking refined and unrefined AlphaFold2 structures for hit discovery. *ChemRxiv* https://doi.org/10.26434/chemrxiv-2022-kcn0d-v2.

61 Mysinger, M.M. et al. (2012). Directory of useful decoys, enhanced (DUD-E): better ligands and decoys for better benchmarking. *J. Med. Chem.* 55 (14): 6582–6594. https://doi.org/10.1021/jm300687e.

62 Scardino, V., Di Filippo, J.I., and Cavasotto, C.N. (2022). How good are AlphaFold models for docking-based virtual screening? *iScience* 105920. https://doi.org/10.1016/j.isci.2022.105920.

63 Miller, E.B. et al. (2021). Reliable and accurate solution to the induced fit docking problem for protein–ligand binding. *J. Chem. Theory Comput.* 17 (4): 2630–2639. https://doi.org/10.1021/acs.jctc.1c00136.

64 Wang, L. et al. (2015). Accurate and reliable prediction of relative ligand binding potency in prospective drug discovery by way of a modern free-energy calculation protocol and force field. *J. Am. Chem. Soc.* 137 (7): 2695–2703. https://doi.org/10.1021/ja512751q.

65 Beuming, T. et al. (2022). Are deep learning structural models sufficiently accurate for free-energy calculations? Application of FEP+ to AlphaFold2-predicted

structures. *J. Chem. Inf. Model.* 62 (18): 4351–4360. https://doi.org/10.1021/acs.jcim.2c00796.

66 Steinegger, M., Mirdita, M., and Soding, J. (2019). Protein-level assembly increases protein sequence recovery from metagenomic samples manyfold. *Nat. Methods* 16 (7): 603–606. https://doi.org/10.1038/s41592-019-0437-4.

67 Mitchell, A.L. et al. (2020). MGnify: the microbiome analysis resource in 2020. *Nucl. Acids Res.* 48 (D1): D570–D578. https://doi.org/10.1093/nar/gkz1035.

68 Suzek, B.E. et al. (2007). UniRef: comprehensive and non-redundant UniProt reference clusters. *Bioinformatics* 23 (10): 1282–1288. https://doi.org/10.1093/bioinformatics/btm098.

69 Audagnotto, M. et al. (2022). Machine learning/molecular dynamic protein structure prediction approach to investigate the protein conformational ensemble. *Sci. Rep.* 12 (1): https://doi.org/10.1038/s41598-022-13714-z.

70 Zhang, C. et al. (2020). DeepMSA: constructing deep multiple sequence alignment to improve contact prediction and fold-recognition for distant-homology proteins. *Bioinformatics* 36 (7): 2105–2112. https://doi.org/10.1093/bioinformatics/btz863.

71 Davtyan, A. et al. (2012). AWSEM-MD: protein structure prediction using coarse-grained physical potentials and bioinformatically based local structure biasing. *J. Phys. Chem. B.* 116 (29): 8494–8503. https://doi.org/10.1021/jp212541y.

72 Schlessinger, A. and Bonomi, M. (2022). Exploring the conformational diversity of proteins. *eLife* 11: e78549. https://doi.org/10.7554/eLife.78549.

73 Steinegger, M. and Söding, J. (2017). MMseqs2 enables sensitive protein sequence searching for the analysis of massive data sets. *Nat. Biotechnol.* 35 (11): 1026–1028. https://doi.org/10.1038/nbt.3988.

74 He, X.H. et al. (2022). AlphaFold2 versus experimental structures: evaluation on G protein-coupled receptors. *Acta Pharmacol. Sin.* https://doi.org/10.1038/s41401-022-00938-y.

75 Nicoli, A. et al. (2022). Modeling the orthosteric binding site of the G protein-coupled odorant receptor OR5K1. *bioRxiv* doi: 10.1101/2022.06.01.494157.

76 Heo, L. and Feig, M. (2022). Multi-state modeling of G-protein coupled receptors at experimental accuracy. *Proteins: Struct. Funct. Bioinform.* n/a(n/a) https://doi.org/10.1002/prot.26382.

77 Jakubec, D. et al. (2022). PrankWeb 3: accelerated ligand-binding site predictions for experimental and modelled protein structures. *Nucl. Acids Res.* https://doi.org/10.1093/nar/gkac389.

78 Wang, S. et al. (2022). CavitySpace: a database of potential ligand binding sites in the human proteome. *Biomolecules* 12 (7): https://doi.org/10.3390/biom12070967.

79 Yuan, Q. et al. (2022). AlphaFold2-aware protein–DNA binding site prediction using graph transformer. *Brief. Bioinform.* 23 (2): https://doi.org/10.1093/bib/bbab564.

80 Konc, J. and Janežič, D. (2022). ProBiS-fold approach for annotation of human structures from the AlphaFold database with no corresponding structure in the

PDB to discover new druggable binding sites. *J. Chem. Inf. Model.* https://doi.org/10.1021/acs.jcim.2c00947.
81 Trudeau, S.J. et al. (2022). PrePCI: A structure- and chemical similarity-informed database of predicted protein compound interactions. *bioRxiv* doi: 10.1101/2022.09.17.508184.
82 Li, F. et al. (2022). DrugMAP: molecular atlas and pharma-information of all drugs. *Nucl. Acids Res.* https://doi.org/10.1093/nar/gkac813.
83 White, M.E.H., Gil, J., and Tate, E.W. (2022). Proteome-wide structure-based accessibility analysis of ligandable and detectable cysteines in chemoproteomic datasets. *bioRxiv* doi: 10.1101/2022.12.12.518491.
84 Wicky, B.I.M. et al. (2022). Hallucinating symmetric protein assemblies. *Science* 378 (6615): 56–61. https://doi.org/10.1126/science.add1964.
85 Anishchenko, I. et al. (2021). De novo protein design by deep network hallucination. *Nature* 600 (7889): 547–552. https://doi.org/10.1038/s41586-021-04184-w.
86 Ding, W., Nakai, K., and Gong, H. (2022). Protein design via deep learning. *Brief. Bioinform.* 23 (3): https://doi.org/10.1093/bib/bbac102.
87 Pozzati, G. et al. (2021). Limits and potential of combined folding and docking. *Bioinformatics* https://doi.org/10.1093/bioinformatics/btab760.
88 Bryant, P., Pozzati, G., and Elofsson, A. (2022). Improved prediction of protein-protein interactions using AlphaFold2. *Nat. Commun.* 13 (1): 1265. https://doi.org/10.1038/s41467-022-28865-w.
89 Gao, M. et al. (2022). AF2Complex predicts direct physical interactions in multimeric proteins with deep learning. *Nat. Commun.* 13 (1): 1744. https://doi.org/10.1038/s41467-022-29394-2.
90 Gao, M., Nakajima An, D., and Skolnick, J. (2022). Deep learning-driven insights into super protein complexes for outer membrane protein biogenesis in bacteria. *Elife* 11. https://doi.org/10.7554/eLife.82885.
91 Schlick, T. and Portillo-Ledesma, S. (2021). Biomolecular modeling thrives in the age of technology. *Nat. Comput. Sci.* 1 (5): 321–331. https://doi.org/10.1038/s43588-021-00060-9.
92 Mullard, A. (2021). What does AlphaFold mean for drug discovery? *Nat. Rev. Drug Discov.* 20 (10): 725–727. https://doi.org/10.1038/d41573-021-00161-0.
93 Hekkelman, M.L. et al. (2021). AlphaFill: enriching the AlphaFold models with ligands and co-factors. *bioRxiv* doi: 10.1101/2021.11.26.470110.
94 Hekkelman, M.L. et al. (2022). AlphaFill: enriching AlphaFold models with ligands and cofactors. *Nat. Methods* https://doi.org/10.1038/s41592-022-01685-y.
95 Wilson, C.J., Choy, W.Y., and Karttunen, M. (2022). AlphaFold2: a role for disordered protein/region prediction? *Int. J. Mol. Sci.* 23 (9): https://doi.org/10.3390/ijms23094591.
96 Strodel, B. (2021). Energy landscapes of protein aggregation and conformation switching in intrinsically disordered proteins. *J. Mol. Biol.* 433 (20): 167182. https://doi.org/10.1016/j.jmb.2021.167182.
97 Lee, C., Su, B.H., and Tseng, Y.J. (2022). Comparative studies of AlphaFold, RoseTTAFold and Modeller: a case study involving the use of G-protein-coupled receptors. *Brief. Bioinform.* https://doi.org/10.1093/bib/bbac308.

98 Chakravarty, D. and Porter, L.L. (2022). AlphaFold2 fails to predict protein fold switching. *Protein Sci.* 31 (6): e4353. https://doi.org/10.1002/pro.4353.

99 Bagdonas, H. et al. (2021). The case for post-predictional modifications in the AlphaFold protein structure database. *Nat. Struct. Mol. Biol.* 28 (11): 869–870. https://doi.org/10.1038/s41594-021-00680-9.

100 Outeiral, C., Nissley, D.A., and Deane, C.M. (2022). Current structure predictors are not learning the physics of protein folding. *Bioinformatics* https://doi.org/10.1093/bioinformatics/btab881.

101 Chen, S.J. et al. (2023). Opinion: protein folds vs. protein folding: differing questions, different challenges. *Proc. Natl. Acad. Sci. U. S. A.* 120 (1): e2214423119. https://doi.org/10.1073/pnas.2214423119.

102 Sen, N. et al. (2022). Characterizing and explaining the impact of disease-associated mutations in proteins without known structures or structural homologs. *Brief. Bioinform.* https://doi.org/10.1093/bib/bbac187.

103 Buel, G.R. and Walters, K.J. (2022). Can AlphaFold2 predict the impact of missense mutations on structure? *Nat. Struct. Mol. Biol.* 29 (1): 1–2. https://doi.org/10.1038/s41594-021-00714-2.

104 David, A. et al. (2022). The AlphaFold database of protein structures: a Biologist's guide. *J. Mol. Biol.* 434 (2): 167336. https://doi.org/10.1016/j.jmb.2021.167336.

105 Mirdita, M. et al. (2022). ColabFold: making protein folding accessible to all. *Nat. Methods* https://doi.org/10.1038/s41592-022-01488-1.

106 Ros-Lucas, A. et al. (2022). The use of AlphaFold for in silico exploration of drug targets in the parasite Trypanosoma cruzi. Frontiers in cellular and infection. *Microbiology* 12. https://doi.org/10.3389/fcimb.2022.944748.

107 Rossi Sebastiano, M. et al. (2021). AI-based protein structure databases have the potential to accelerate rare diseases research: AlphaFoldDB and the case of IAHSP/Alsin. *Drug Discov. Today* https://doi.org/10.1016/j.drudis.2021.12.018.

108 Burley, S.K. et al. (2022). RCSB protein data Bank: tools for visualizing and understanding biological macromolecules in 3D. *Protein Sci.* e4482. https://doi.org/10.1002/pro.4482.

109 Edich, M. et al. (2022). The impact of AlphaFold on experimental structure solution. *Faraday Discuss.* https://doi.org/10.1039/d2fd00072e.

11

Deep Learning for the Structure-Based Binding Free Energy Prediction of Small Molecule Ligands

Venkatesh Mysore, Nilkanth Patel, and Adegoke Ojewole

NVIDIA Corporation, 2788 San Tomas Expy, Santa Clara, CA 95051, USA

11.1 Introduction

The prediction of the binding affinity between a ligand and a protein is one of the hardest and most important problems in computational drug discovery. These predictions are employed at all stages of the pipeline, from hit identification, to hit-to-lead development and lead optimization. Throughput and speed become relevant while screening catalogs of billions of compounds to find viable hits, when accuracy becomes paramount when prioritizing analogs of a hit to be synthesized. Ultimately, any virtual screening method is only as good as its scoring function, which serves as an estimate of the ligand-binding free energy. Understanding its assumptions and factoring in its limitations will enable the design of a robust workflow, informing key decisions such as the number of compounds to be screened and the number of hits to be followed up experimentally.

At a high level, the binding free energy prediction approaches can be classified as ligand-based and structure-based. In ligand-based approaches, the two- or three-dimensional constraints that a ligand must satisfy are used to mine a large database for compounds. These requirements may be derived from a known active compound, the characteristics of the binding pocket, or other relevant aspects known a priori. The structure-based approach, on the other hand, is characterized by reasoning about the docked three-dimensional protein–ligand complex. The ligand has multiple conformations accessible through its rotatable bonds, while the protein has multiple conformations that could expose multiple binding sites, each with varying degrees of flexibility afforded by the backbone and side-chain atoms of the amino-acid chain. For the structure-based approach to be viable, a reliable model of the protein, specifically the binding site, is required. Additionally, a reliable protocol for predicting the bound pose of the ligand is required, which automatically necessitates a reliable predictor of the protein–ligand binding free energy as the bound pose typically minimizes this quantity. It is important to explicitly state the assumptions of this approach. The protein's binding site is

Computational Drug Discovery: Methods and Applications, First Edition.
Edited by Vasanthanathan Poongavanam and Vijayan Ramaswamy.
© 2024 WILEY-VCH GmbH. Published 2024 by WILEY-VCH GmbH.

presumed to be rigid, with alternate conformations typically handled by multiple rounds of docking and scoring to a representative ensemble. While ligand binding is also a dynamic process with the ligand sampling multiple poses and conformations even in its bound state, a single bound pose is usually what gets scored by the binding free energy estimator. Scores of multiple poses, potentially with multiple binding site conformations, can be used to account for the true nature of the binding process.

In this chapter, we focus on the use of deep learning in the structure-based prediction of binding free energy. We do not cover the precursor steps of protein structure prediction, binding site prediction, or bound pose prediction, which have been covered in some recent reviews [1]. We restrict our attention to the binding affinity prediction of an assembled complex, building on recent reviews on the topic [2–5].

Traditional approaches for addressing this problem have involved fitting a scoring function using known protein--ligand complex structures. Physics-inspired functional forms with pre-factors capturing the statistical distribution of protein--ligand atoms in complexes have proven to be remarkably useful in high-throughput virtual screening applications. However, such methods suffer from some inherent shortcomings: the functional form of the interaction terms has to be fixed ahead of time, and only experimentally determined co-crystal structures can be utilized for fitting. Further, the process of fitting and evaluating the scoring function involves substantial human inspection and ad hoc decision-making in many cases.

Deep-learning-based approaches have attempted to address these problems in the following ways. (1) A sufficiently deep and wide neural network can approximate arbitrary mathematical functions, so the functional form need not be hard-coded ahead of time. The model architecture, be it a 3D-convolutional network or graph convolutional one, can be conferred with as many parameters as necessary by increasing the number of features, nodes, or layers, enabling the complexity needed to capture the end-point. (2) The training procedure is less sensitive to low-quality training examples (with potentially incorrect poses) as long as there are a very large number of examples. Docked poses predicted by one or more methods can be used to expand the training set, increasing the number of training examples by several orders of magnitude. (3) Well-established procedures exist for the automated training and evaluation of models. Carefully splitting the available labeled data into training, validation, and test (or hold-out) can allow reliable and reproducible model development and unbiased evaluation. These three advantages have fueled the widespread development and application of deep learning for protein–ligand binding affinity prediction. The development of sophisticated software packages and libraries that encapsulate breakthrough technologies in deep learning has made it incredibly easy for researchers working on binding affinity to benefit from advances in image or object recognition, natural language processing, or network analysis. This has enabled rapid iterative experimentation, resulting in numerous successful applications of deep learning for structure-based binding affinity prediction. See Figure 11.1 for the approaches used to predict binding affinity.

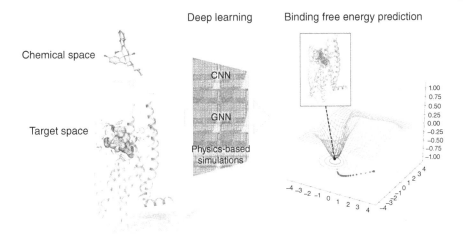

Figure 11.1 Schematic overview of the different models used for binding affinity prediction. Physics-based methods operate directly on ligand and target structures to predict binding affinity. Deep learning methods based on convolutional neural networks (CNNs) or graph neural networks (GNNs), for example, operate on compressed and learned representations of protein–ligand complexes to predict binding affinity.

In the following sections, we outline the current best practices in the application of these methods, providing a brief commentary on their evolution, limitations, and future.

11.2 Deep Learning Models for Reasoning About Protein–Ligand Complexes

The central challenge in adapting deep learning algorithms to the structure-based drug discovery domain is the representation of the protein–ligand complex in a format that can be digested as input by the deep learning framework. The protein–ligand interaction site features have to capture the 3-D structural and chemical aspects of the protein and ligand atoms involved. Specifically, it becomes crucial to capture the three-dimensional chemical neighborhood of a protein or ligand atom to facilitate the learning of a function that estimates binding free energy given the atoms of the protein–ligand complex. The decisions to be made center around whether explicit Cartesian coordinates, interatomic distances, or a coarser representation using grid points would capture this spatial context best. This representation of the input often dictates the model architecture capable of harnessing the locality information. An additional engineering challenge is incorporating information about the atoms, bonds, distances, and angles into the featurization of the atoms. In other words, all the information necessary to be able to make a reasonable prediction must be captured in the input representation and featurization. In this section, we go over the two broad classes of models – convolutional neural networks (CNNs) and graph neural networks (GNNs) – that have been successfully employed to predict protein–ligand binding affinity. We quickly introduce the datasets before describing the models.

11.2.1 Datasets

Predicting protein–ligand binding affinities requires a training set comprising the ligand binding affinity values, such as Ki, Kd, and IC50, against the target protein or a set of proteins. Several publicly available resources contain such information. For example, the Protein Data Bank and its derived datasets, PDBbind, BindingMOAD, BindingDB, and BindingDB subsets, are frequently used for CNN model development. Scientific community-driven standardized benchmarking initiatives, such as SAMPL, CASF, and CACHE, maintain datasets for model performance evaluations [6, 7]. While these datasets capture experimentally validated protein–ligand complexes, several data sources include experimental values of protein–ligand binding affinities without the 3-D details of the protein–ligand complexes (ChEMBL, PubChem, DUD-E, DOCKSTRING, ExCAPE, MoleculeNet, etc.). In the absence of experimental 3-D complex information, there are several approaches developed for utilizing such datasets for training a model. For example, it often includes employing computationally generated 3-D models of target proteins (using AlphaFold2 and RosettaFold) using protein sequence as an input in FASTA format, modeling the protein–ligand complex using docking (AutoDOCK), and more recently, using deep learning approaches for predicting the ligand binding sites in the proteins and predicting protein–ligand complexes [8–10].

11.2.2 Convolutional Neural Networks

11.2.2.1 Background

A CNN is a type of artificial neural network, mainly developed for computer vision tasks where the input features have locality information in their sequence. The reasoning to be performed should exploit the spatial proximity inherent in the input feature ordering. Identifying objects in an image would thus entail employing the same neural "filters" to smaller parts of the input. This is more parameter efficient than having a network where all input features had to interact with all others to enable the detection of some input object. As the depth of CNN layers increases, the model will be able to recognize increasingly complex objects.

11.2.2.2 Voxelized Grid Representation

Learnings from impactful CNN applications in computer vision were translated to tasks related to structure-based drug discovery by observing that the structure of the binding site and the compound could be treated as a three-dimensional grid of points where some are occupied by atoms [11]. The process of generating such features involved defining the ligand binding site as a 3-D box of a specific size and creating a grid within that box. An atom could be assigned to the nearest grid point or be spread out over a portion of the grid.

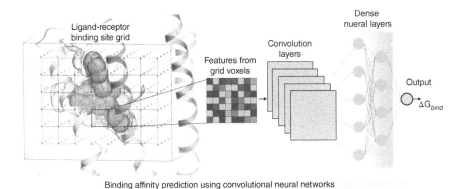

Figure 11.2 Featurization of a protein–ligand complex structure for binding affinity prediction using a convolutional neural network. CNN models for binding affinity can operate on 3-D tensors to capture the structural and physicochemical properties of a protein–ligand complex. The accuracy of 3-D CNNs is further enhanced by introducing random translations and rotations before applying convolutional filters.

While the initial CNN implementations for image recognition tasks used 2-D array representations of image segments as feature sets for input, advanced CNN models developed for drug discovery involve using 3-D tensors obtained from protein–ligand complexes (Figure 11.2). This representation also lends itself to data augmentation by random rotation and translation.

11.2.2.3 Descriptors

3D CNN models use various input features to represent the protein–ligand binding sites in a vectorized format similar to input tensors from images. These vectorized 3D grids of the binding site capture structural and/or physicochemical features. Such features include the presence or absence of atoms, atom types, and the electronic environment, including aromaticity, partial charge, protonation states, and atom interactions.

There are also approaches where the feature sets are generated from protein and ligand separately [12, 13]. Such feature sets include properties related to interaction surfaces – for example, solvent-accessible surface area, surface charge distribution, etc. Other surface representation approaches include utilizing and generating advanced geometric representations [14] and feeding them into a CNN model. Some recent models also have additional sources for feature calculations, such as the ions and solvents present or modeled at the ligand binding site and varying grid resolutions [15, 16].

11.2.2.4 Applications

Initial CNN-based models, such as AtomNet, GNINA, OnionNet, Kdeep, and MedusaNet, required an input protein–ligand binding pose to estimate the complex's binding affinity. While these models show improved performance over the traditional molecular docking scoring programs and functions in the benchmark datasets, their ability is often limited by the docking step required to estimate the protein–ligand binding pose. This step also impacts the throughput of the binding affinity prediction as the binding pose(s) generation step becomes the bottleneck compared to the interaction energy scoring step. A few models bypass the molecular docking step using amino acid sequences for proteins and SMILES format for ligands, followed by feature generation and binding affinity predictions. While bypassing the docking step may expedite the overall estimation process, it loses the input 3-D context of the interaction [6, 13, 15, 17–24].

11.2.3 Graph Neural Networks

11.2.3.1 Background

Graphs are represented by nodes (N) and edges (E) and a feature vector per node (f). Since the advent of social media and extensive data, graphs have become an important medium to capture and process such information efficiently. In the past decade, Neural Network architectures that operate on graph data structures have become helpful in drug discovery and molecular simulations.

11.2.3.2 Graph Representation

Protein–ligand interactions can be captured as a graph using various architectures. For example, a simple representation of a protein–ligand complex may have atoms from the binding pocket as nodes in the graph. In contrast, the edges represent the connectivity or vicinity of that atom. Graph representation captures the molecular features, along with its rotation and translation invariance features making it faster and more robust, while consuming much less memory than CNN.

11.2.3.3 Descriptors

While the choices for the atomic descriptors are the same as those for the CNNs, GNNs provide an additional avenue for encoding information about connectivity and neighborhood through the use of edge features. An edge between two atoms can carry information about the nature of the interaction (bonded or nonbonded), the distance, and other attributes. The graph convolutional operator aggregates information from all atoms connected to an atom; hence, the presence or absence of an edge can have a significant impact on the final prediction. The cut-off distance for deeming two atoms as interacting and the functional form of the distance-dependent term are some of the key design choices in a GNN. Unlike a CNN, whose input vector will be affected by rotation and translation of the complex, a GNN is invariant to these transformations as only internal distances are effectively used. Since there is no implicit order in which the edges of a node are to be traversed, chirality information is not automatically captured.

11.2.3.4 Applications

GNNs have been extensively applied to represent small molecules, for use in predicting compound properties and interactions. The first successful application of GNNs for protein–ligand binding affinity prediction was ?. Some of the initial GNN architectures for small molecules, such as PotentialNet, included features from the PLI site with atoms and bonds from the protein–ligand complex. While other works, such as D-MPNN and graphDelta, involved features calculated by combining pre-trained fingerprints from QM datasets and fitting them into the extracted features from the PLI sites [25–27]. Cao and Shen [28] reported energy-based graph convolutional network for predicting intra- and inter-molecular interactions and related energies. Alternatively, Lim et al. [29] applied a mixed approach by using GAN for pose prediction and CNN for interaction-based scoring. Fusion models of 3D-CNN and GNN have shown better performance [30].

11.2.3.5 Extension to Attention Based Models

In the past few years, the field of NLP has seen a major shift with the advent of Transformer architecture, enabling the model to pay attention during self-supervised learning in order to capture long-range context. Following up on these advances, Liu et al. [14] employed the transformer architecture to develop intramolecular graph transformer (IGT) for binding pose and activity prediction. Similarly, Yuan et al. [31] developed a graph attention network with added custom mechanisms designed to capture bond features and node-level features. With the field of physics-informed ML gaining traction in recent years, Moon et al. [32] developed a model to incorporate physics-informed equations in addition to neural networks for predicting the protein–ligand binding affinity.

11.2.3.6 Geometric Deep Learning and Other Approaches

GNNs have started replacing 3D-CNNs in applications where rotational and translational invariance hold. Going beyond, a series of developments such as the SE(3)-transformer have led to more mathematically sophisticated models for protein structure prediction, binding pose prediction, and ultimately binding free energy. Of note is TANKBind from [33], which predicts both the bound pose and the binding affinity given a protein and a compound separately. Attempts have also been made to incorporate other mathematical concepts from geometry and topology to characterize the connectivity of atoms beyond graphs [34–36]; their wider adoption will aid their evaluation on real-world problems.

11.3 Deep Learning Approaches Around Molecular Dynamics Simulations

Structural biologists have long understood that the rigid body "lock-and-key" hypothesis to binding is an oversimplification and that modeling the flexibility of both the receptor and ligand is necessary to accurately predict binding affinity. From this understanding, molecular dynamics (MD) simulations have emerged as a powerful tool to investigate the dynamical properties of protein–ligand complexes. Despite the explosive growth in computational power in the last 30 years [37], long-time scale trajectories are required to probe the conformational changes that are relevant to binding under different conditions. Such MD simulations are computationally expensive and only apply to a small fraction of ligand–receptor pairs of interest. In addition, classical molecular mechanics-based forcefields undersample biologically relevant high-energy transitions and binding states. Those states that are deeply sampled are explored in discrete steps. The theoretically correct approaches to computationally estimating ligand binding free energy typically employ explicit solvent MD simulations in some form [see other chapters on this topic].

Deep learning is beginning to permeate this sphere in multiple ways. Accurate force fields are essential for the MD simulations to be correct and for the quantities estimated to be realistic. This requires the computation of quantum mechanical potential energies for various ligand and peptide conformations. These computationally intensive calculations can now be performed using deep learning models trained on vast datasets of exact QM calculations. Going a step further, the neural networks developed to predict these potential energies can in fact drive the entire MD simulations. Machine learnt force-fields and their use in simulation are covered in a separate chapter. We highlight other important domains where deep learning is aiding MD simulation-based inference.

11.3.1 Enhanced Sampling

Approaches to marry deep learning and enhanced sampling are showing greater promise. Rufa et al. [38] describe a hybrid approach to perform alchemical free energy calculations correctly on a system where the ligand alone is simulated using neural potentials. Bertazzo et al. [39] describe a metadynamics-based approach for estimating binding free energy where machine learning is used to extract a free energy path from the initial steered-MD simulation.

11.3.2 Physics-inspired Neural Networks

Another family of methods incorporates force-field terms into their free energy prediction model. Hassan-Harrirou et al. [17] use the decomposition of MM/GBSA and other terms to featurize a voxelized 3D grid representing the ligand atoms. Guedes et al. [40] featurized a protein–ligand complex with force-field terms augmented with solvation, lipophilic, and ligand torsional terms, and then employed linear regression, random forests, and SVMs to predict the ligand binding free energy. Dong et al. [41] extract terms from MM / GBSA calculations, augmenting them with numerous descriptors before employing everything from linear regression to neural networks to predict free energy. Ji et al. [42] also employ MM /GBSA calculations and use the protein–ligand interaction energies as the featurization of the complex. Cain et al. [43] invoke the Generalized Born and Poisson–Boltzmann calculations for implicit solvent MD simulation, and propose a way of integrating those terms with a GNN representing the protein–ligand complex.

While machine-learnt force-fields concentrate on being able to replicate the potential energies employed in Hamiltonians of standard MD simulation force-fields, efforts to approximate the binding free energy directly using force-field-inspired neural networks have been underway in parallel. Zhu et al. [44] attempted to develop a single neural network model that could predict the contribution to the binding free energy of a protein–ligand atom pair, given the force-field's atom parameterization consisting of its partial charge and Lennard-Jones parameters. Cao and Shen [28] use a graph convolutional network modified to use operations inspired by the functional form of energy potentials. More recently, Moon et al. [32] used gated GNNs to learn atomic representations of a protein–ligand complex, subsequently feeding them pairwise into separate physics-inspired equations for different force-field terms.

11.3.3 Modeling Dynamics

Machine learning and deep learning are beginning to be evaluated for their ability to predict binding using conformational ensembles of protein–ligand complexes. These are based on approaches for predicting the course of an MD simulation, trying to reason over self-supervised representations of MD simulation frames. These exciting research areas are in the early stages of being adopted for industry-grade protein–ligand binding free energy prediction.

11.3.3.1 Applications

A variety of methods have been proposed to leverage protein–ligand conformational dynamics for binding affinity prediction. [45] began with 8888 initial candidate compounds and used a workflow consisting of physics-based flexible docking (Autodock Vina), followed by inference using the DeepBindBC ResNet binary classifier, and then a 100ns MD simulation step to predict 69 final candidate molecules. Of these, four were experimentally tested and shown to competitively inhibit TIPE2 (tumor necrosis factor-alpha-induced protein 8-like 2 protein) with µM affinity for target PIP2. While the docking and MD steps explicitly computed dynamics, the deep learning step did not.

In contrast, Yasuda et al. [46] took a different approach predicated on the observation that binding affinity is associated with energy differences between the unbound and ligand-bound conformational ensembles. Their method begins with MD simulations of different ligands with a range of binding affinities for a specific binding pocket. These simulations, called the local dynamics ensemble (LDE), are defined as an ensemble of short-term trajectories of atoms of interest in the binding site. The authors used a multi-layer perceptron to compute the difference of LDE distributions between a ligand-bound and ligand-free system based on the Wasserstein Distance for all N pairs of bound and unbound systems. The resulting $N \times N$ matrix was embedded into points in a lower-dimensional space, and principal component analysis was performed on the embedded points. The first principal component was used as a proxy for ligand-binding energies.

Wang et al. [15] also begin by featurizing short-trajectory MD simulations. However, the authors trained random forest and LSTM-based models to predict the impact of active site point mutations on binding affinity. Structures of protein–ligand complexes were obtained from experimentally determined Platinum database and X-ray crystal structures (3Å or finer) from the PDB. Frames of nanosecond-scale

MD trajectories of wild-type and receptor mutants in complex with the ligands were featurized by the following descriptors: shape and topology, differences in estimated free energy, and local geometry (closeness, local surface area, orientation, contacts, and interfacial hydrogen bonds). The authors found that LSTM models trained on MD trajectories were more accurate than predictors based on energy estimation or descriptors alone.

Another approach is to predict binding affinity from an ensemble of protein–ligand structures not computed from an MD simulation. Intuitively, cross-docking a ligand to an ensemble of receptor conformations should provide a more comprehensive set of binding poses for structure-based virtual screening than single-pose docking does. However, two problems remain: 1) there is still no agreement on how to aggregate individual docking results into an ensemble docking score rank a ligand, and 2) ranking from traditional ensemble docking typically yields only modest improvement over docking to a rigid receptor. Ricci-Lopez et al. [47] propose to use ML to determine ensemble docking scores for four proteins: CDK2, FXa, EGFR, and HSP90. Receptor ensembles were prepared by docking ligands from standard libraries to crystal structures in the PDB. The authors used these ensembles to train binary classifiers using logistic regression and gradient-boosted decision trees and showed that these models significantly outperformed standard consensus docking at predicting binders and non-binders. Following a similar idea, Stafford et al. [48] propose AtomNet PoseRanker (ANPR), a graph convolutional network that predicts binding affinity from a collection of ligand poses. The input of ANPR is an ensemble of protein–ligand complexes computed from RosettaCM, which was used to sample low-energy structures in the vicinity of an input structure from PDBbind v2019. These structures are augmented with structures computed from docking the ligand to alternative structures of the same target protein. From this cross-docked data set, ANPR learns to recognize distinct ligand poses as valid in different receptor conformations. Ultimately, learning from ligand and receptor conformational diversity helps ANPR recognize a multitude of valid binding modes, improving ANPR's binding predictions vs. Smina.

11.4 Modifying AlphaFold2 for Binding Affinity Prediction

AlphaFold2 [9] provided a step function improvement in protein structure prediction, generating unbound "apo" protein structures with near-experimental accuracy. This new capability created great interest among structural biologists. However, it is undetermined whether AlphaFold2 or similar models such as RosettaFold [8] can predict an arbitrary ligand binding site with accuracy sufficient to accommodate a variety of known binding partners. High accuracy in this task is necessary for computational drug discovery programs. The drug discovery community has, therefore, dedicated its efforts to determine how to modify AlphaFold2 to predict ligand binding.

11.4.1 Modifying AlphaFold2 Input Protein Database for Accurate Free Energy Predictions

Advances in classical force fields and sampling algorithms have made it possible to apply free energy perturbation (FEP) simulations to predict protein–ligand binding affinities with accuracies approaching those of biochemical and biophysical assays [49]. The improved accuracy of FEP methods shifts the burden of computational structure-based design to the availability of high-resolution protein–ligand complexes. It remains unanswered to what extent protein folds predicted by deep learning models such as AlphaFold2 can assist in free energy calculations. To partially address this knowledge gap, Beuming and colleagues [50] retrospectively substituted AlphaFold2 protein targets into Schrodinger FEP+ simulations [49] to reproduce previously validated studies. In this study, the AlphaFold2 database was modified to exclude template structures and sequences with more than 30% sequence identity to the query sequence at inference time. The modified model, called $AF2_{30}$, not only predicts protein active sites but also has substantial structural differences from the standard AlphaFold2 model. In a study limited to 14 protein targets, $AF2_{30}$ predicted relative changes in ligand affinities with an accuracy of 1.04 kcal/mol, which is statistically comparable to FEP+ calculations obtained using crystal structure complexes.

11.4.2 Modifying Multiple Sequence Alignment for AlphaFold2-Based Docking

An alternative is to modify AlphaFold2 to predict homo- and hetero-dimeric protein–protein interactions. The first approach to this end is AlphaFold-Multimer [51], an AlphaFold2 variant expressly trained on multimeric inputs. AlphaFold-Multimer incorporates additional changes to AlphaFold2, including modified losses to avoid superposing structures of different chains, to reduce steric clashes, and to account for permutation symmetry among homomeric chains. Other changes include paired multiple sequence alignments (MSAs) to capture interchain coevolutionary information and minor adjustments to the AlphaFold2 model architecture. On both small and large datasets of protein complexes, AlphaFold-Multimer exhibited superior accuracy to AlphaFold2 (input adapted with a flexible linker) per DockQ scores [52].

Bryant et al. [53] developed FoldDock, which uses a "fold and dock" approach with AlphaFold2 to predict binding between two different proteins each of length greater than 50 amino acids. A MSA is computed for the sequence of each binding partner, where each MSA is computed from a distinct database. A third MSA is computed by concatenating the first two MSAs. And a fourth "paired" MSA is constructed from the first two MSAs to determine inter-sequence coevolutionary information. In the FoldDock study, AlphaFold2 consumed these MSAs and predicted complexes with acceptable quality (DockQ $\geq$ 0.23) for 63% of 216 heterodimers. The predicted DockQ scores, used as a proxy for binding strength, also predicted interacting protein pairs from noninteracting pairs with a 1% false positive rate. However, FoldDock has a greater empirical success rate with proteins from bacteria and archaea than those from eukaryotes and viruses.

An alternative approach by [54] is ColAttn. This method uses the MSA transformer to estimate a column attention score from the MSAs corresponding to the putative binding partners. AlphaFold-Multimer consumes this input to predict the complex structure. Overall, ColAttn may yield better complex structure prediction than AlphaFold-Multimer, particularly in eukaryotes.

11.5 Conclusion

11.5.1 New Models for Binding Affinity Prediction

Language-based models have recently been proposed to predict binding affinity. For example, Vielhaben et al. [55] use USMPep, an RNN architecture, to predict neopeptide binding affinity for Class 1 and Class 2 major histocompatibility complex (MHC) binding pockets. Vielhaben et al. [56] propose applying USMPep to predict viral peptides binding affinity to MHC. Cheng et al. [57] developed BERTMHC, a transformer-based multi-instance learning model to predict peptide-MHC Class 2 binding. Recently, Bachas et al. [58] proposed a BERT-style language model on antibody sequence data and binding affinity labels to quantitatively predict binding of unseen antibody sequence variants. Language-based approaches to binding affinity prediction promise to improve the productivity of early stages of drug discovery. Excitingly, transformer language models appear to improve in predictive performance on primary and downstream tasks as the architectures are scaled in parameter size from tens of millions to hundreds of billions [59]. This observation has not been tested for language models trained for binding affinity prediction. Doing so will require pretraining and fine-tuning large domain-specific models for DNA, RNA, proteins, and small molecules. Fortunately, recent advances in computing hardware, training frameworks, and inference frameworks have made the challenge tractable.

11.5.2 Retrospective from the Compute Industry

The foundational methods that have enabled the advancements in DL for binding affinity are GPU-accelerated algorithms from computational chemistry and popular machine learning architectures. The broad availability of application programming interfaces (APIs) such as CUDA [60] and OpenCL [61] has enabled these algorithms and architectures to leverage advances in GPU computing hardware. Over the last ten years, GPU acceleration has increased the size and complexity of MD-simulated systems by 100-fold [37]. The increased computational power afforded by GPUs has enabled more accurate ensemble computation in protein–ligand systems using density functional theory (DFT) [62] and hybrid quantum mechanics-molecular mechanics [63]. GPU-accelerated free-energy perturbation simulations [49] are routinely used to compute accurate relative binding affinities for biomedically and industrially relevant protein–ligand systems. Machine-learned force-fields such as ANI [64] and AIMNet [65] have the potential to further accelerate these free energy simulations.

Graphics processing unit (GPU)-accelerated docking algorithms can also rapidly generate structural ensembles used to train DL models to predict binding affinity. Synthesizable molecules now number in the tens of billions, vastly expanding the scope of virtual screening and increasing the utility of massively parallel docking calculations [62]. For example, AutodockGPU was used to screen a billion molecules against a SARS-CoV-2 spike protein using the Summit supercomputer by parallelizing pose searching and scoring [66].

11.5.2.1 Future DL-Based Binding Affinity Computation will Require Massive Scalability

Despite impressive innovations in GPU architectures [handwiki.com], CPU codes that are amenable to GPU acceleration are precisely those that can be decomposed into a large number of thread blocks that are executed concurrently on GPU hardware. When optimization for a single GPU is insufficient to meet compute requirements, computation must either be scaled to multiple GPUs in the same compute node or distributed among multiple compute nodes with low-latency interconnects. Powerful yet simplified distributed training frameworks are necessary to make distributed supercomputing architectures routinely accessible for DL. The following discussion describes the emerging computational challenge and discusses the software and hardware frameworks that have been proposed to address the problem.

11.5.2.2 Single GPU Optimizations for DL

To take advantage of exponentially increasing datasets, the DL models for binding described in this chapter will need to radically expand in size. As a result, the GPU platforms and APIs that run these larger models must also increase in speed, complexity, and scalability. Owing to Moore's Law, individual GPUs will continue to scale in transistor count [handwiki.com], allowing for larger computational graphs to be mapped onto individual devices. GPUs will also be optimized at the hardware level to accelerate computational bottlenecks within DL architectures. For example, the recently announced Hopper GPU [67] will substantially accelerate transformer models by automatically altering tensor precision to maintain model accuracy while reaping the hardware performance benefits of representing data in smaller, faster numerical formats such as FP8.

11.5.2.3 Distributed DL Training and Inference

Data and Model Parallelism As data and model sizes have increased dramatically, it is common that DL models must be trained using multiple GPUs, which necessitates data or model parallelism. In the former case, different data batches, along with an identical set of model weights, are distributed to each GPU. Gradients are computed concurrently on each GPU and are returned to the main compute node, where they are aggregated and redistributed, along with new data batches, for the next iteration. Model parallelism is necessary when the model itself is so large that it cannot be stored in the memory of a single GPU. In this case, the high memory requirement is managed by storing different layers of the model on different GPUs and transmitting the results of forward and backward propagation to the appropriate GPU(s). Efficient data and model parallelism require fast intra-node GPU interconnection technologies and fast internode networking, along with scalable, feature-rich, and user-friendly APIs such as Pytorch Lightning [68] and NeMo [69]. Inference frameworks such as Triton [70, 71] and associated SDKs such as TensorRT [72] are increasingly becoming necessary to deploy these large DL models with the requisite scalability and response time of a research and development environment.

Federated Learning A logical extension to data and model parallelism is Federated Learning (FL). FL frameworks, such as NVIDIA FLARE [73], apply the concepts of distributed learning to geographically dispersed sites. By design, FL has the potential to obviate the barriers to sharing proprietary data, leading to more accurate and robust models. In FL, participating sites perform localized training using their respective unshared data. Gradients and losses from these local models are subsequently aggregated using the previously described principles of data parallelism. And since no data leave their respective silos, the resulting global model can be shared beyond the individual sites.

References

1 Dhakal, A., McKay, C., Tanner, J.J., and Cheng, J. (2022). Artificial intelligence in the prediction of protein–ligand interactions: recent advances and future directions. *Brief. Bioinform.* 23 (1): bbab476. https://doi.org/10.1093/bib/bbab476.
2 Qin, T., Zhu, Z., Wang, X.S. et al. (2021). Computational representations of protein–ligand interfaces for structure-based virtual screening. *Expert Opin. Drug Discovery* 16 (10): 1175–1192. https://doi.org/10.1080/17460441.2021.1929921.
3 Anighoro, A. (2022). Deep learning in structure-based drug design. *Methods Mol. Biol.* 2390: 261–271. https://doi.org/10.1007/978-1-0716-1787-8_11. PMID: 34731473.
4 Kimber, T.B., Chen, Y., and Volkamer, A. (2021). Deep learning in virtual screening: recent applications and developments. *Int. J. Mol. Sci.* 22: 4435. https://doi.org/10.3390/ijms22094435.
5 Li, H., Sze, K.-H., Lu, G., and Ballester, P. (2020). Machine-learning scoring functions for structure-based drug lead optimization. *Wiley Interdiscip. Rev.: Comput. Mol. Sci.* 10: e1465. https://doi.org/10.1002/wcms.1465.
6 Zhang, H., Liao, L., Saravanan, K.M. et al. (2019). DeepBindRG: A deep learning based method for estimating effective protein-ligand affinity. *Peer J* 7: 2019. https://doi.org/10.7717/peerj.7362.
7 Ackloo, S., Al-Awar, R., Amaro, R.E. et al. (2022). CACHE (Critical Assessment of Computational Hit-finding Experiments): a public–private partnership benchmarking initiative to enable the development of computational methods for

hit-finding. *Nature Rev. Chem.* 6 (4) Nature Research: 287–295. https://doi.org/10.1038/s41570-022-00363-z.

8 Baek, M., DiMaio, F., Anishchenko, I. et al. (2021). Accurate prediction of protein structures and interactions using a three-track neural network. *Science (1979)* 373 (6557): 871–876. https://doi.org/10.1126/science.abj8754.

9 Jumper, J., Evans, R., Pritzel, A. et al. (2021). Highly accurate protein structure prediction with AlphaFold. *Nature* 596: 583. https://doi.org/10.1038/s41586-021-03819-2.

10 Stärk, H., Ganea, O., Pattanaik, L. et al. (2022). Geometric deep learning for drug binding structure prediction. In *Proceedings of the 39th International Conference on Machine Learning*, pp. 20503–20521.

11 Wallach, I., Dzamba, M., and Heifets, A. (2015). AtomNet: a deep convolutional neural network for bioactivity prediction in structure-based drug discovery. http://arxiv.org/abs/1510.02855.

12 Ahmed, A., Mam, B., and Sowdhamini, R. (2021). DEELIG: a deep learning approach to predict protein-ligand binding affinity. *Bioinform. Biol. Insights* 15. https://doi.org/10.1177/11779322211030364.

13 Jiménez, J., Škalič, M., Martínez-Rosell, G., and de Fabritiis, G. (2018). KDEEP: protein–ligand absolute binding affinity prediction via 3D-convolutional neural networks. *J. Chem. Inf. Model.* 58 (2): 287–296. https://doi.org/10.1021/acs.jcim.7b00650.

14 Liu, Q., Wang, P.-S., Zhu, C. et al. (2021). OctSurf: Efficient hierarchical voxel-based molecular surface representation for protein-ligand affinity prediction. *J. Mol. Graph. Model.* 105: 107865. https://doi.org/10.1016/j.jmgm.2021.107865.

15 Wang, Z., Zheng, L., Liu, Y. et al. (2021). OnionNet-2: a convolutional neural network model for predicting protein-ligand binding affinity based on residue-atom contacting shells. *Front. Chem.* 9. https://doi.org/10.3389/fchem.2021.753002.

16 Gao, A. and Remsing, R.C. (2022). Self-consistent determination of long-range electrostatics in neural network potentials. *Nat. Commun.* 13 (1): 1–11. https://doi.org/10.1038/s41467-022-29243-2.

17 Hassan-Harrirou, H., Zhang, C., and Lemmin, T. (2020). RosENet: improving binding affinity prediction by leveraging molecular mechanics energies with an ensemble of 3D convolutional neural networks. *J. Chem. Inf. Model.* 60 (6): 2791–2802. https://doi.org/10.1021/acs.jcim.0c00075.

18 Sunseri, J., King, J.E., Francoeur, P.G., and Koes, D.R. (2019). Convolutional neural network scoring and minimization in the D3R 2017 community challenge. *J. Comput. Aided Mol. Des.* 33 (1): 19–34. https://doi.org/10.1007/s10822-018-0133-y.

19 Wu, Z., Ramsundar, B., Feinberg, E.N. et al. (2017). MoleculeNet: a benchmark for molecular machine learning. http://arxiv.org/abs/1703.00564.

20 Cang, Z. and Wei, G. (2017). TopologyNet: Topology based deep convolutional and multi-task neural networks for biomolecular property predictions. *PLoS Comput. Biol.* 13 (7). https://doi.org/10.1371/journal.pcbi.1005690.

21 Stepniewska-Dziubinska, M.M., Zielenkiewicz, P., and Siedlecki, P. (2018). Development and evaluation of a deep learning model for protein–ligand binding affinity prediction. *Bioinformatics* 34 (21): 3666–3674. https://doi.org/10.1093/bioinformatics/bty374.

22 McNutt, A.T., Francoeur, P., Aggarwal, R. et al. (2021). GNINA 1.0: molecular docking with deep learning. *J. Cheminform* 13: 43. https://doi.org/10.1186/s13321-021-00522-2.

23 Öztürk, H., Özgür, A., and Ozkirimli, E. (2018). DeepDTA: deep drug–target binding affinity prediction. *Bioinformatics* 34 (17): i821–i829. https://doi.org/10.1093/bioinformatics/bty593.

24 Li, Y., Rezaei, M.A., Li, C., and Li, X. (2019). DeepAtom: a framework for protein-ligand binding affinity prediction. In: *Proceedings – 2019 IEEE International Conference on Bioinformatics and Biomedicine, BIBM 2019*, 303–310. https://doi.org/10.1109/BIBM47256.2019.8982964.

25 Yang, K., Swanson, K., Jin, W. et al. (2019). Analyzing learned molecular representations for property prediction. *J. Chem. Inf. Model.* 59: 3370–3388. https://doi.org/10.1021/acs.jcim.9b00237.

26 Feinberg, E.N., Sur, D., Wu, Z. et al. (2018). PotentialNet for molecular property prediction. *ACS Cent. Sci.* 4 (11): 1520–1530. https://doi.org/10.1021/acscentsci.8b00507.

27 Karlov, D.S., Sosnin, S., Fedorov, M.V., and Popov, P. (Mar. 2020). GraphDelta: MPNN scoring function for the affinity prediction of protein-ligand complexes. *ACS Omega* 5 (10): 5150–5159. https://doi.org/10.1021/acsomega.9b04162.

28 Cao, Y. and Shen, Y. (2020). Energy-based graph convolutional networks for scoring protein docking models. *Proteins: Struct. Function Bioinform.* 88. https://doi.org/10.1002/prot.25888.

29 Lim, J., Ryu, S., Park, K. et al. (2019). Predicting drug-target interaction using 3D structure-embedded graph representations from graph neural networks. http://arxiv.org/abs/1904.08144.

30 Knutson, C., Bontha, M., Bilbrey, J.A., and Kumar, N. (2022). Decoding the protein–ligand interactions using parallel graph neural networks. *Sci. Rep.* 12 (1). https://doi.org/10.1038/s41598-022-10418-2.

31 Yuan, H., Huang, J., and Li, J. (2021). Protein-ligand binding affinity prediction model based on graph attention network. *Math. Biosci. Eng.* 18 (6): 9148–9162. https://doi.org/10.3934/mbe.2021451.

32 Moon, S., Zhung, W., Yang, S. et al. (2022). PIGNet: a physics-informed deep learning model toward generalized drug-target interaction predictions. *Chem. Sci.* 13 (13): 3661–3673. https://doi.org/10.1039/d1sc06946b.

33 Lu, W., Wu, Q., Zhang, J. et al. (2022). TANKBind: trigonometry-aware neural networks for drug-protein binding structure prediction. *bioRxiv*. https://doi.org/10.1101/2022.06.06.495043.

34 Nguyen, D.Q., Nguyen, T.D., and Phung, D. (2019). Universal graph transformer self-attention networks. https://arxiv.org/abs/1909.11855v1.

35 Meng, Z. and Xia, K. (2021). Persistent spectral–based machine learning (PerSpect ML) for protein-ligand binding affinity prediction. *Sci. Adv.* 7 (19). https://doi.org/10.1126/sciadv.abc5329.

36 Wee, J. and Xia, K. (2021). Ollivier persistent Ricci curvature-based machine learning for the protein–ligand binding affinity prediction. *J. Chem. Inf. and Model.* 61 (4): 1617–1626.

37 Pandey, M., Fernandez, M., Gentile, F. et al. (2022). The transformational role of GPU computing and deep learning in drug discovery. *Nature Mach. Intell.* 4 (3): 211–221.

38 Rufa, D.A., Bruce Macdonald, H.E., Fass, J. et al. (2020). Towards chemical accuracy for alchemical free energy calculations with hybrid physics-based machine learning/molecular mechanics potentials. *bioRxiv*. https://doi.org/10.1101/2020.07.29.227959.

39 Bertazzo, M., Gobbo, D., Decherchi, S., and Cavalli, A. (2021). Machine learning and enhanced sampling simulations for computing the potential of mean force and standard binding free energy. *J. Chem. Theory Comput.* 17 (8): 5287–5300. https://doi.org/10.1021/acs.jctc.1c00177.

40 Guedes, I.A., Barreto, A.M.S., Marinho, D. et al. (2021). New machine learning and physics-based scoring functions for drug discovery. *Sci. Rep.* 11: 3198. https://doi.org/10.1038/s41598-021-82410-1.

41 Dong, L., Qu, X., Zhao, Y., and Wang, B. (2021). Prediction of binding free energy of protein–ligand complexes with a hybrid molecular mechanics/generalized born surface area and machine learning method. *ACS Omega* 6 (48): 32938–32947. https://doi.org/10.1021/acsomega.1c04996.

42 Ji, B., He, X., Zhai, J. et al. (2021). Machine learning on ligand-residue interaction profiles to significantly improve binding affinity prediction. *Brief. Bioinf.* 22 (5). https://doi.org/10.1093/bib/bbab054.

43 Cain, S., Risheh, A., and Forouzesh, N. (2022). A physics-guided neural network for predicting protein–ligand binding free energy: from host–guest systems to the PDBbind database. *Biomolecules* 12: 919. https://doi.org/10.3390/biom12070919.

44 Zhu, F., Zhang, X., Allen, J.E. et al. (2020). Binding affinity prediction by pairwise function based on neural network. *J. Chem. Inf. Model.* 60 (6): 2766–2772. https://doi.org/10.1021/acs.jcim.0c00026.

45 Zhang, H., Li, J., Saravanan, K.M. et al. (2021). An integrated deep learning and molecular dynamics simulation-based screening pipeline identifies inhibitors of a new cancer drug target TIPE2. *Front. Pharmacol.* 12.

46 Yasuda, I., Endo, K., Yamamoto, E. et al. (2022). Differences in ligand-induced protein dynamics extracted from an unsupervised deep learning approach correlate with protein–ligand binding affinities. *Commun. Biol.* 5 (1): 1–9.

47 Ricci-Lopez, J., Aguila, S.A., Gilson, M.K., and Brizuela, C.A. (2021). Improving structure-based virtual screening with ensemble docking and machine learning. *J. Chem. Inf. Model.* 61 (11): 5362–5376.

48 Stafford, K., Anderson, B.M., Sorenson, J., and van den Bedem, H. (2022). AtomNet PoseRanker: enriching ligand pose quality for dynamic proteins in virtual high-throughput screens. *J. Chem. Inf. Model.* 62 (5): 1178–1189.

49 Abel, R., Wang, L., Harder, E.D. et al. (2017). Advancing drug discovery through enhanced free energy calculations. *Acc. Chem. Res.* 50: 1625–1632.

50 Beuming, T., Martín, H., Díaz-Rovira, A.M. et al. (2022). Are deep learning structural models sufficiently accurate for free-energy calculations? Application

of FEP+ to AlphaFold2-predicted structures. *J. Chem. Inf. Model.* 62 (18): 4351–4360.
51 Evans, R., O'Neill, M., Pritzel, A. et al. (2021). Protein complex prediction with AlphaFold-Multimer. *BioRxiv*.
52 Basu, S. and Wallner, B. (2016). DockQ: a quality measure for protein-protein docking models. *PloS One* 11 (8): 1–9.
53 Bryant, P., Pozzati, G., and Elofsson, A. (2022). Improved prediction of protein-protein interactions using AlphaFold2. *Nat. Commun.* 13: 1265.
54 Chen, B., Xie, Z., Xu, J. et al. (2022). Improve the protein complex prediction with protein language models. *BioRxiv*.
55 Vielhaben, J., Wenzel, M., Samek, W., and Strodthoff, N. (2020). USMPep: universal sequence models for major histocompatibility complex binding affinity prediction. *BMC Bioinform.* 21 (1): 1–16.
56 Vielhaben, J., Wenzel, M., Weicken, E., and Strodthoff, N. (2021). Predicting the binding of SARS-CoV-2 peptides to the major histocompatibility complex with recurrent neural networks. arXiv preprint arXiv:2104.08237.
57 Cheng, J., Bendjama, K., Rittner, K., and Malone, B. (2021). BERTMHC: improved MHC–peptide class II interaction prediction with transformer and multiple instance learning. *Bioinformatics* 37 (22): 4172–4179.
58 Bachas, S., Rakocevic, G., Spencer, D. et al. (2022). Antibody optimization enabled by artificial intelligence predictions of binding affinity and naturalness. *bioRxiv*.
59 Rae, J.W., Borgeaud, S., Cai, T. et al. (2021). Scaling language models: methods, analysis & insights from training gopher. arXiv preprint arXiv:2112.11446.
60 Vingelmann, P. and Fitzek, F.H.P. (2020). CUDA release 10.2.89, NVIDIA.
61 Stone, J.E., Gohara, D., and Shi, G. (2010). OpenCL: a parallel programming standard for heterogeneous computing systems. *Comput. Sci. Eng.* 12: 66–72.
62 Grygorenko, O.O., Radchenko, D.S., Dziuba, I. et al. (2020). Generating multibillion chemical space of readily accessible screening compounds. *iScience* 23: 101681.
63 Yu, J.K., Liang, R., Liu, F., and Martínez, T.J. (2019). First-principles characterization of the elusive I fluorescent state and the structural evolution of retinal protonated Schif base in bacteriorhodopsin. *J. Am. Chem. Soc.* 141: 18193–18203.
64 Yoo, P., Sakano, M., Desai, S. et al. (2021). Neural network reactive force field for C, H, N, and O systems. *NPJ Comput. Mater.* 7: 9.
65 Zubatyuk, R., Smith, J.S., Leszczynski, J., and Isayev, O. (2021). Accurate and transferable multitask prediction of chemical properties with an atoms-in-molecules neural network. *Sci. Adv.* 5, eaav6490.
66 LeGrand, S., Scheinberg, A., Tillack, A.F. et al. (2020). GPU-accelerated drug discovery with docking on the summit supercomputer: porting, optimization, and application to COVID-19 research. *Proceedings of 11th ACM International Conference on Bioinformatics, Computational Biology and Health Informatics*. https://doi.org/10.1145/3388440.3412472.
67 Salvator, D. (2022). H100 Transformer Engine Supercharges AI Training, Delivering Up to 6x Higher Performance Without Losing Accuracy, NVIDIA.

68 Falcon, W. (2019). Pytorch lightning. GitHub. Note: https://github.com/PyTorchLightning/pytorch-lightning 3.6.
69 Kuchaiev, O., Li, J., Nguyen, H. et al. (2019). Nemo: a toolkit for building ai applications using neural modules. arXiv preprint arXiv:1909.09577.
70 Jahanshahi, A., Sabzi, H.Z., Lau, C., and Wong, D. (2020). Gpu-nest: Characterizing energy efficiency of multi-gpu inference servers. *IEEE Comput. Architect. Lett.* 19 (2): 139–142.
71 de Souza Pereira Moreira, G., Rabhi, S., Ak, R., and Schifferer, B. (2021). End-to-end session-based recommendation on GPU. In: *Fifteenth ACM Conference on Recommender Systems*, 831–833.
72 Jeong, E., Kim, J., Tan, S. et al. (2022). Deep learning inference parallelization on heterogeneous processors with TensorRT. *IEEE Embedd. Syst. Lett.* 14 (1): 15–18.
73 Han, W., Mawhirter, D., Wu, B. et al. (2019). FLARE: flexibly sharing commodity GPUs to enforce QoS and improve utilization. In: *International Workshop on Languages and Compilers for Parallel Computing*. Cham: Springer.

12

Using Artificial Intelligence for *de novo* Drug Design and Retrosynthesis

Rohit Arora[1], Nicolas Brosse[2], Clarisse Descamps[2], Nicolas Devaux[2], Nicolas Do Huu[2], Philippe Gendreau[2], Yann Gaston-Mathé[2], Maud Parrot[2], Quentin Perron[2], and Hamza Tajmouati[2]

[1] *Iktos Inc, 50 Milk St, Boston, MA 02110, USA*
[2] *Iktos SAS, 65 Rue de Prony, 75017, Paris, France*

12.1 Introduction

12.1.1 Traditional Drug Design and Discovery Process Is Slow and Expensive

The drug discovery and development process is notably complex and iterative and is therefore time-consuming, arduous, and expensive. Typically, the process is initiated with the identification and validation of a potential therapeutic target (e.g. a protein) that is functionally implicated in a disease or multiple diseases [1]. Once the target has been identified and validated, the small molecules that can potentially interact with this target protein to either inhibit or activate its function (directly or otherwise) that has an impact on the disease state, must be identified. This launches the early stages of the drug design and discovery process (see Figure 12.1).

Initial hits against the biological target may be identified using a number of established methods, focusing primarily on compound activity against the target. These include (but are not limited to) high-throughput screening of chemical libraries, focused screening approaches, fragment-based drug design, virtual screening, and knowledge-based (both computer-based and medicinal chemistry-based from known chemical matter) drug design. Once hit molecules are identified, the process of identifying and prioritizing promising chemical series begins. Analogues of the original hit compounds may be synthesized and tested to investigate structure–activity relationships (SAR; physicochemical and absorption, distribution, metabolism, and excretion [ADME] properties in addition to activity) and identify lead compound series. Finally, lead optimization entails maintaining favorable activity, absorption, distribution, metabolism, excretion, and toxicity (ADMET) and physicochemical properties while tweaking the lead structure to ensure that the compound is successful in the downstream pre-clinical and clinical phases [2, 3].

Computational Drug Discovery: Methods and Applications, First Edition.
Edited by Vasanthanathan Poongavanam and Vijayan Ramaswamy.
© 2024 WILEY-VCH GmbH. Published 2024 by WILEY-VCH GmbH.

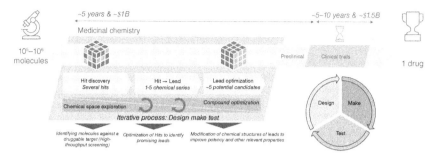

Figure 12.1 Early stages of drug design and discovery process.

The entire process follows the DMTA (Design-Make-Test-Analyze) iterative cycle at every phase to optimize the pharmacological and drug-like properties of potential drug candidates (Figure 12.1). Traditionally, in drug discovery campaigns, compound design and synthesis decisions are made with input from medicinal and synthetic chemistry teams and require multiple iterations. Owing to the iterative nature and challenges of each component of this cycle, bringing a new drug from the discovery stage to the market can take up to a decade and can cost well over a billion US dollars. This process can be accelerated by addressing the underlying challenges and streamlining the individual components of the DMTA cycle. This is a crucial issue that must be tackled at the discovery and optimization stages, before significant time and resources have been invested. Recently, some of the industry focus for *de novo* drug design [4] has shifted from lower-hanging fruits for chemical matter and toward using new modalities aimed toward drugging novel, first-in-class, hard-to-drug, or "undruggable" targets (e.g. PROTACs, molecular glues, RNA therapeutics, small-molecule modulators of protein–protein interactions, etc.) [5], that may go beyond Lipinski's Rule of Five [6–8]. These new modalities introduce additional variables and therefore pose additional optimization challenges for each component of the DMTA cycle that are harder to tackle by medicinal chemists. It is essential to address them to accelerate the entire discovery process.

12.1.2 Success and Limitations of Standard Computational Methods

Over the last three decades or so, computational methods have been a mainstay of the drug discovery process and have helped in the development of a number of therapeutically relevant drug-like molecules. This is true for both small-molecule drugs and (more recently) peptide therapeutics [9, 10]. These methods include structure-based approaches like protein–ligand or protein–protein docking, virtual screening, molecular dynamics simulations, etc., and ligand-based approaches like quantitative structure-activity relationship (QSAR) and pharmacophore modeling.

From the perspective of *de novo* drug design, virtual screening of large libraries of molecules has been successfully deployed in a number of drug discovery campaigns [11]. Virtual screening involves either docking a large library of compounds to a known target (structure-based) or evaluating the similarity between compounds in the large library and known actives against a target, especially in the absence of target structure information (ligand-based) [12]. The scale of these screening campaigns in terms of the compound library size has grown by orders of magnitude over the last few years – from $10^5 - 10^6$ to $10^8 - 10^9$ – in order to sample a larger chunk of the chemical space. This trend has been aided and accelerated by the advent of ultra-large chemical libraries of virtual compounds – e.g. Enamine REAL Space, Merck MASSIV library, GSKChemspace, and WuXi Apptec's GalaXi [13]. This has also encouraged development of a number of data analysis and predictive machine learning methods to manage these data, and to complete screening campaigns in a reasonable amount of time [14].

The impact of virtually screening ultra-large compound libraries has been significant – it has been shown to lead to a marked increase in the enrichment factor [15, 16]. This, however, does not address the inherent limitations of such methods. For *de novo* drug design, screening campaigns are limited to what is available in these compound libraries. Even the largest libraries ($\approx 10^9$ or larger) represent a fraction of the total size of the drug-like chemical space, which is estimated to be on the order of 10^{60} [17], and may not represent the overall diversity of this space. This drastically reduces the probability of finding the optimal compound for the target in question. Importantly, this approach does not address the issue of multi-parametric optimization (MPO) where the optimal drug candidate must be optimized across multiple objectives (solubility, potency, permeability, etc.). Prior knowledge of known actives can certainly improve this likelihood, but even with predictive methods, this problem can be fairly intractable. Screening methods – both ligand-based and structure-based – can evaluate existing ideas but cannot propose novel ideas. Put differently, these approaches can tell you what not to do, but they cannot tell you what to do. This is where artificial intelligence (AI)-based generative approaches can excel.

12.1.3 AI-Based Methods can Accelerate Medicinal Chemistry

AI-based workflows and pipelines are being adopted and deployed across a number of industries, including medical, pharmaceuticals, and life sciences, and are poised to have a significant impact on productivity, output, and the economy [18–20]. Over the last few years, the drug design and discovery sector, led by research and development by pharmaceutical industry and academic institutions, has also employed AI-based strategies to optimize a number of components in the pipeline (Figure 12.1). Specifically, this has impacted drug design, repurposing, screening, poly-pharmacology, and chemical synthesis [21].

Here we highlight the efforts that address the issues impacting the efficiency of the aforementioned DMTA cycle – drug design and chemical synthesis. Generative AI-based strategies are able to fulfill the promise of truly *de novo* drug design by being able to explore and sample the chemical space far more efficiently and comprehensively, subject to any medicinal chemistry constraints (target product profile, substructure or descriptor constraints, 3D scores, etc.), compared to traditional methods. Such generative approaches have found successful applications in tasks such as image synthesis, language translation, text generation, and music generation [22–24]. Now these approaches have also found their way in chemistry. The methods based on these approaches generally do not rely solely on structural similarity, instead they learn the property similarity in the latent space, and are therefore able to design diverse set of ideas for the molecules that may be close in the physicochemical property space [25]. This approach has been extensively validated [26–28] and has yielded positive results in real-life case studies [29]. A key component of the generative AI in *de novo* drug design is synthetic accessibility, which has come in to a sharp focus more recently [30, 31]. Compounds obtained from generative chemistry can satisfy multiple objectives *in silico*, but at the same time can be difficult to synthesize [32]. Therefore, it is essential to ensure that these AI systems can optimize synthetic accessibility of the generated compounds and allow chemists to more efficiently choose the compounds to be synthesized and tested.

In this chapter, various components of the generative AI and synthetic accessibility that are implemented in *de novo* drug design and have been impacting drug discovery projects are discussed. Specifically we explore the state-of-the-art methods and algorithms that power these AI systems and platforms that have been commonly deployed across a variety of drug discovery and design programs. The aim of this chapter is to introduce the reader to the technical and functional aspects of these systems, and pave the way to further diversify the ecosystem that has been developing at the intersection of AI/ML, other computational techniques, chemistry, and biology, for the benefit of drug discovery.

12.2 Quantitative Structure-Activity Relationship Models

12.2.1 Introduction to QSAR Models

QSAR models are developed using one or more statistical model building tools, which may be broadly categorized into regression- and classification-based approaches. Like other regression models, QSAR regression models relate a set of "predictor" variables to the potency of the response variable, while classification QSAR models relate the predictor variables to a categorical value of the response variable. The mathematical form of a QSAR model is:

$$\text{Activity} = f(\text{physiochemical properties and/or structural properties}) + \text{error}$$

In the following Sections (12.2.2 and 12.2.3), we distinguish between the traditional machine learning-based techniques for QSAR models and the newer techniques based on deep neural networks (DNN).

12.2.2 QSAR Machine Learning Methods

Building statistical models to predict chemical properties is of paramount importance as it would allow to discard compounds without having to synthesize and test them [33]. However, applying statistical modeling techniques to molecules face many hurdles. First, the molecules must be represented as fingerprints (vectors) to be processed by a statistical model [34]. Then, since the chemical space is very sparse, considerations about the applicability domain of the models are crucial. Finally, these models will make mistakes – no statistical model is perfect – and furthermore, the expert using them (e.g. medicinal or computational chemist) needs to be convinced of their usefulness. Therefore, explaining the results obtained through interpretability and not having black-box QSAR model is essential. Various fingerprints have been developed [35–39] not only for QSAR models but also for other operations on molecules, like similarity computation [40, 41] or clustering. The first class of fingerprints consists of listing molecular descriptors and building a vector from them, which in some cases shows good performance. These fingerprints relying on descriptors do not encode local features of the structure, but instead provide global information of the molecule. On the other hand, extended-connectivity fingerprints [42] enumerate the molecular features and directly encode the structure of the molecule. These fingerprints are fast-to-compute and generally make a good baseline for building QSAR models [43]. Finally, with the development of deep-learning methods a new category of fingerprints called Learned Representation has emerged [44]. It allows to encode molecular graphs or SMILES strings, and shows competitive results in QSAR model benchmarks [45]. These three kinds of molecular representations can be computed with the molecule alone (whether it be a small molecule or a peptide). It is worth mentioning that some fingerprints include the interaction between the molecule and a protein [46], which allows to build 3D QSAR models, especially useful in scaffold-hopping tasks [47].

Once molecules are converted to fingerprints, a statistical model is used to predict the property of interest. Depending on the model, varying performances can be obtained on the same task, but it is generally hard to know in advance which one to use. Use of multiple solutions like linear models [48, 49], random forests [50–52], support vector machines (SVM) [53, 54], neural networks, etc. [55] have been studied. Benchmarks exist to compare these models [49], [56–59]. However, these comparisons are limited to specific use cases and test sets, and generally do not allow to intrinsically rank the statistical models for QSAR. A summary diagram can be found in [37, Figure 1]. Aside from the model-measured performances, its applicability domain is critical to estimate. Though it is hard and is subject to many biases, efforts have been developed to measure the applicability domain and quantify the model's errors [60]. A simple yet efficient method to estimate the applicability domain is to evaluate the similarity of the estimated compound with the training dataset [61]. More sophisticated techniques take into account the defaults of the models aimed at improving these applicability domain metrics [62, 63]. The restrained applicability domain of a QSAR is the reason why it is generally very risky to use these models for scaffold hopping in very diverse chemical spaces. Efforts have been made to extend the applicability domain, thanks to federated learning [64], which consists of training a model on multiple private datasets while maintaining the privacy of the data involved [65]. This kind of approach is very promising, especially in the pharmaceutical industry, where data privacy is critical. However, there are, to-date, few demonstrations [66] that the performance of the QSAR model is significantly improved.

Finally, the interpretability of the QSAR model is key for proper usage and understanding [67]. Most models are indeed black boxes, either because of the model (neural network) or the features (learned representations). Adding interpretability (for instance, using SHAP values [68–70]) allows an expert user to spot the biases in the model and results that stem from spurious correlations rather than causality.

12.2.3 QSAR Deep Neural Networks Methods

Numerous research studies have demonstrated that graph neural networks (GNN) may be a more effective modeling technique for predicting chemical properties than descriptor-based techniques [71–78]. Due to their remarkable ability to learn complex and often non-linear relationships between structures and properties, deep learning (DL) algorithms have more recently revolutionized the classic cheminformatics activity presented in Section 12.2.2.

The DL-based models largely fall into two main categories – descriptor-based models and graph-based models. In descriptor-based DL models, molecular descriptors and/or fingerprints commonly used in traditional QSAR models are used as the input, and then a specific DL architecture is deployed to train a model [79]. While in the graph-based DL models, the basic chemical information encoded by molecular graphs is used as the input, and then a graph-based DL algorithm, such as GNN, is used to train a model.

The GNN generalizes the convolution technique to the irregular molecular graph, which is a natural representation of chemical structures, in a manner similar to how convolutions are applied to regular data such as pictures and texts. A graph $G = (V, E)$ can be defined as the connectivity relationships between a collection of nodes (V) and a collection of edges (E). Naturally, one may also think of a molecule as a graph made up of a number of atoms (nodes) and a number of bonds (edges). In essence, GNN aims to learn the representations of each atom by aggregating the information from its nearby atoms encoded by the atom feature vector and the information of the connected bonds encoded by the bond feature vector through messages passing across the molecular graph repeatedly. This is done after the state updating of the central atoms and read-out operation. Through the read-out phase, the learned atom representations can then be applied to the prediction of molecular attributes. A key feature of GNN is its ability to automatically learn task-specific representations using graph convolutions while not requiring conventional manually created descriptors and/or fingerprints. An explanatory diagram is presented in [[80], Figure 1].

As of 2022, the most common Python libraries to manipulate GNNs are PyTorch Geometric, Deep Graph library, DIG, Spektral, and TensorFlow GNNS.

12.3 Modes of Generative AI in Chemistry

12.3.1 General Introduction

Generative modeling is a challenging problem to solve. It is fundamentally harder than discriminative modeling (e.g. statistical classification) because it models objects from arbitrary distributions of arbitrary nature and complexity (images, texts, graphs, audios, etc.). Being unsupervised, generative models benefit from large unlabeled datasets especially when designed as deep neural networks.

Generative models are practical not only for sampling novel data points from a given distribution but also for characterizing the likelihood of the sampled data points. When coupled with an optimization method for a given fitness score, they model the distribution of good-scored data points. This is the desired behavior of generative models applied to drug discovery – generate novel molecules optimizing a defined fitness score.

Generative models for molecular design can be characterized by three main features:

1. **Which molecular representation they use**: It can be either text (SMILES [81], SELFIES [82]), a graph [83], a set of fragments [84], or a synthesis tree [85].
2. **How they generate molecules**: The generation strategy can use a simple policy [86], add or remove atoms or bonds, for example. The most successful generative models in drug design in the literature rely on DNN from two main families of models.
   - **Deep autoregressive networks [87]**: Trained to retrieve a missing part of a piece of data knowing the rest of this piece of data and are compatible with reinforcement learning optimization.
   - **Autoencoder models [88]**: Trained to reconstruct the desired piece of data starting from an initial input (same piece of data or another type of piece of data). It is compatible with optimization in an Euclidean latent space and evolutionary algorithms.
3. **How they perform property optimization**: The property optimization strategy can be based on reinforcement learning [89], Bayesian optimization [90], or evolutionary algorithms [91].

12.3.2 Generative AI in Lead Optimization

MPO is required in a drug discovery project especially in the lead optimization stage in order to identify the rare compounds satisfying all the objectives of the project, such as activity across multiple biological assays, selectivity, (lack of) toxicity, pharmacokinetics, synthetic accessibility, and novelty, to name a few [92]. In the last five years, the development of AI approaches to drug discovery, and more specifically *de novo* drug design through the use of deep generative models, has triggered a lot of interest in the computer-aided drug design (CADD) community. In this chapter, we briefly discuss the successful implementation of this approach.

The easiest way to generate molecules is to use a deep recurrent neural network (RNN), and more precisely, a deep long short-term memory (LSTM) [93], to generate molecules represented as SMILES [94]. The LSTM should first be trained on a big generic database such as ChEMBL or ZINC databases, using teacher forcing [95], to build a character-based language model for generating SMILES strings. Recall that the role of a language model p is to model the next character probability distribution given the sequence of previous characters:

$$p(x_{t+1}|x_1 x_2 x_t) = LSTM(x_{t+1}|x_1 x_2 x_t) \tag{12.1}$$

SMILES are generated by iteratively sampling the next character from its inferred past conditioned distribution $p(x_{t+1}|x_1 x_2 x_t)$ generating a SMILES starts and ends, respectively, with the special tokens of the vocabulary "START" and "END." Molecules in ChEMBL database are transformed into their canonical achiral RDKIT version. No data augmentation is performed either by enumerating the different ways of writing a SMILES, or by enumerating the tautomeric forms of the same compound. LSTM language model trained with this approach generates achiral SMILES. Identical compounds can be generated with different writings of their SMILES. Tautomers of the same compound are generated as distinct molecules. After being trained on ChEMBL, the LSTM language model had a 94% SMILES chemical validity rate, which implies that the LSTM trained on ChEMBL database has learnt to generate molecules belonging to ChEMBL chemical space.

Crucially, generated molecules should stay near the chemical space of the lead optimization series. Thus, the previous LSTM model is retrained by teacher forcing on the lead optimization dataset. This second training allows to zoom in on the chemical space studied to generate molecules similar to the lead optimization chemical series. The simplest molecule optimization strategy that can be used along with a SMILES LSTM generator is the hill climbing strategy [94]. It is an iterative process where the LSTM generative model is fine-tuned in teacher forcing on an optimal set of SMILES that evolves over time as follow: step after step, this set of SMILES is updated by retaining only the top-scored generated compounds (10% for example) since the first step.

An MPO lead optimization dataset is a list of molecules with experimental bioactivity measurements on multiple biological assays. In order to score novel generated molecules, QSAR models are trained on each assay measurements. We recommend using binary classification models after binarizing the data using the desired thresholds of the lead optimization project. Indeed, binary classification models better handle unbalanced datasets and can better predict the minority class. The reward (fitness) score used for ranking molecules in hill climbing combines the predicted probabilities from QSAR models (p_i), a measure of similarity to the initial dataset, and any other physical or chemical properties of the project (Molecular Weight for example). An aggregation function that works pretty well is the geometric mean of

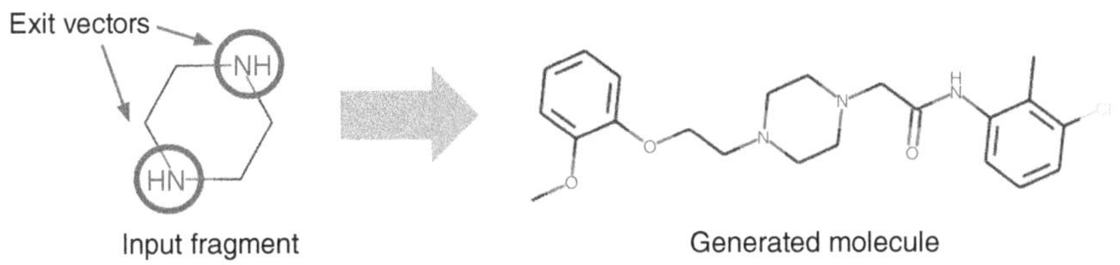

Figure 12.2 Growing from an initial fragment with defined exit vectors.

scaled scores (between 0 and 1), which allows us to transform our problem from multi-objective optimization to mono-objective optimization.

12.3.3 Fragment Growing

Generative models can be used to address different challenges in drug design, and one of them is fragment growing. The goal of fragment growing is to start from an initial molecular structure with a pre-defined exit vector, Figure 12.2, then choose new fragments to be plugged into each exit vector to obtain a new molecule. The fragments are selected from a dataset of commercial compounds and are added to the molecule through chemical reactions. One fragment and the associated reaction are predicted for each exit vector of the initial molecular structure.

This process can be learned by a deep learning model. The model relies on three architectures: a feed-forward neural network to predict the reactions, another feed-forward to choose the building blocks, and a recurrent neural network to capture the sequential aspect of the fragment growing process, Figure 12.3. With this architecture, it is possible to model the likelihood of the sequences with the model's parameters. Modeling the likelihood enables distribution learning and the use of reinforcement learning. With reinforcement learning, the selection of the building blocks plugged into the initial molecular structure is optimized for a given score such as docking score, quantitative estimate of druglikeness (QED), or similarity. The application of fragment growing is not limited to 2D approaches. It has been successfully implemented in structure-based drug design approaches, especially like fragment-based drug design wherein a hit fragment in the binding pocket of the target protein must be grown into a lead-like and (potentially) potent compound with good binding affinity in the binding pocket [96–99].

12.3.4 Novelty Generation

12.3.4.1 The Model

The novelty generator is a deep learning algorithm based on a transformer architecture [100], which was trained to jump from patent to patent creating, by design, highly novel molecules (Figure 12.4). The transformer model is used in a sequence-to-sequence way, very similar to machine translation tasks. The supervision dataset is composed of pairs of molecules (Molecule$_A$, Molecule$_B$) where:

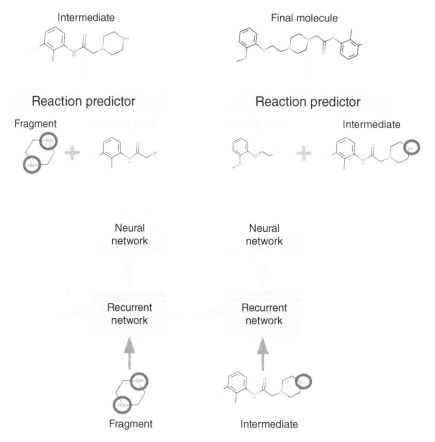

Figure 12.3 Model architecture for fragment growing. The model takes as input the fragment and its exit vectors. The model selects which building block among a dataset should react with the fragment provided.

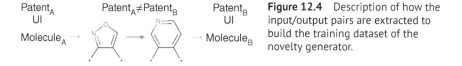

Figure 12.4 Description of how the input/output pairs are extracted to build the training dataset of the novelty generator.

- The molecules come from two different patents
- Molecule$_B$ can be retrieved using an MMP (Matched Molecular Pair) rule from Molecule$_A$

The two challenges are first building this dataset, and then training the model to perform the desired transformation. To build the supervised dataset, the input dataset is a cleaned sample of SureChEMBL [101], this leads to $N=18$ million pairs. Extracting all the MMP rules on this dataset is fairly untractable – for k elements, there are k^2 pairs, hence $>10^{12}$ rules to compute. To reduce the complexity to $\frac{N^2}{k}$, the SureChEMBL dataset was clustered into $k=1000$ clusters. The entire pipeline is summed up in Figure 12.5.

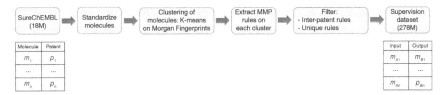

Figure 12.5 Pipeline to build the training dataset of the novelty generator.

As mentioned above, the model used for this task is a Transformer model, which is an encoder–decoder model with several attention mechanisms. The model was trained with the teacher forcing method [102], meaning that at each iteration the model samples a character, which is only used to contribute to the value of the loss function, but the truth character is added to the sequence. The model can be long and computationally expensive to train. As an estimate, training this model at Iktos took around 1 month on an 11 GB GPU (NVIDiA 2080 ti). The inference method is different from the training method. In inference, the model doesn't have access to the ground truth, so it samples character by character following the decoder probability distribution until an "end of sentence" character is sampled. If sampling occurs several times from the same input SMILES, the stochasticity of the decoding outputs different results with various probabilities. To further augment the solutions given by the model, we observed that enumerating the input SMILES and applying the inference to those SMILES increased the diversity of the generated molecules.

12.3.4.2 Optimization of the Novelty Generator

A reward function-based optimization method can be deployed for novelty generators. Given a defined reward function, a loop on the novelty generator can be created so that at every iteration the generator gets back the best-scored output molecules to be used in the next iteration, and therefore climb the reward function. In this optimization method, the model is considered as a black box, and its weights are not updated. The reward is optimized by selecting which molecules will be the input molecules of the next iteration. This optimization method is part of hill climbing optimization methods. An optimization method for the continuous latent space of the model have been tried [103]. A major limitation here is that the dimension of the encoded SMILES depends on the length of the SMILES, so the optimization is limited at each time step to the latent space corresponding to one SMILES length. The experiment to invalidate the [103] method was to compare the score for the same number of SMILES decoded with this method vs. decoding with random noise.

12.4 Importance of Synthetic Accessibility

12.4.1 Overview

In small molecule drug discovery projects, generative models can be used to design massive libraries of molecules with specific properties [29, 94]. The optimization of an AI-molecular generator to explore a given chemical space and propose new well-scored molecules in an MPO project is mostly based on molecular properties and fingerprints [75, 94, 103–106]. However, one of the major challenges in any CADD project is that the molecules need to be synthesized. The question of whether or not the generated molecules can be synthesized is not systematically taken into account during generation, even though the synthesizability of the generated molecules is a fundamental requirement for such methods to be useful in practice. Generative models are known to sample lots of non accessible molecules [107, 108], and few synthesizability scores are known in the literature to be used in the pipeline of molecular generation [32, 85, 109, 110].

12.4.2 Synthetic Scores

No chemical rule is able to completely answer the question of whether a molecule with a valid SMILES can be synthesized or not. Moreover, the evaluation of such scores is challenging, particularly due to the difficulty in interpreting the values. A simple way to define synthesizability is with a binary score denoting "synthesizable" or "not synthesizable." While a binary score is useful, it has limits, as it does not allow the prioritization of molecules of the same score. Also, a continuous score gives more signal when used as a reward for a *de novo* drug design algorithm. With the recent efforts of the community, some continuous scores were recently developed to describe synthetic accessibility [111–114]. These can be based on chemical substructures, domain expertise, or output of models fitting expert scores. However, as two very similar molecules may have different synthetic routes due to a single functional group, it may be difficult to find a proxy for a true retrosynthetic analysis. The RA score, for retrosynthetic accessibility score [113], is a predictor of the binary score given by the AiZynthFinder retrosynthesis tool [114]. Its value range is between 0 and 1, and according to the score, the higher the value the more optimistic the algorithm is about the synthesis of the molecule. The SC score, for synthetic complexity score [111], ranks the molecules and scores them from 1 to 5. Based on the criteria that products are more complex than their reactants, a neural network trained on a corpus of reactions was used to build the score. Molecules with lower values have a more optimistic synthesizability profile. Finally, the SA score for synthetic accessibility score [112], is based on a heuristics where molecular complexity and fragment contributions are used to evaluate synthetic tractability. Low scores indicate less complex molecules and therefore more feasible compounds. We believe that the features taken into account to compute these scores are not sufficient to encapsulate all of the information about synthesizability.

More recently, AI has been introduced to address some of these challenges and to aid synthetic, medicinal, and computational chemists. Iktos has developed Spaya [115], a template-based retrosynthesis AI software that computes synthetic routes and ranks them based on a synthesizability score. The Retro-Score (RScore) is a synthetic feasibility score derived from the output of a full Spaya retrosynthetic analysis for a given molecule. As highlighted below, conducting a full retrosynthetic analysis to determine synthesizability is essential. The RScore can be used:

(1) to evaluate the synthesizability of molecules given by generative models,
(2) to guide the molecular generator to an area of the chemical space where molecules are synthesizable.

Due to the high computational costs associated with the computation of a full retrosynthetic analysis needed to obtain the RScore, an easier way to compute score is RSPred, obtained by training a neural network on the output of the Spaya RScore and performs comparably to the RScore in a variety of tasks, but can be computed orders of magnitude faster. Further, in order to simplify the use of the algorithm on large batches of molecules (libraries), Spaya-API has been developed [115]. It is an API running on Spaya's algorithmic engine for library scoring purposes, which has been used herein to evaluate the synthetic accessibility of newly generated molecules. For a given molecule (m), the RScore is derived from routes proposed by Spaya, but handled in a high-throughput manner by Spaya-API. The worst RScore value is 0 (when no route is found in a given time interval), and the best score is 1 (when the route is a one-step retrosynthesis matching exactly a reaction in the literature). To score a molecule and obtain its RScore value, Spaya-API performs a retrosynthetic analysis with an early stopping process. The early stopping mode stops the Spaya run when a route with a score above the predefined threshold (set to 0.6 by default) is found, or after the defined timeout (set to one minute by default) has elapsed. The RScore of a molecule is defined as:

$$RScore(m) = \max_{\substack{\text{routes given by Spaya} \\ \text{with early stopping}}} (score(route(m))) \tag{12.2}$$

The score is rounded to 1 decimal, and hence can take 11 different values (from 0.0 to 1.0). Spaya-API also returns the number of steps for the best synthetic route found for each input molecule. The list of commercial compounds used for the retrosynthesis is a catalog of 60M commercially available starting materials provided by Mcule [116], Chemspace [117], eMolecules [118], and Key Organics [119].

Compute time is an essential attribute of a score as it may limit its usage on large-scale data sets. In Table 12.1 compute time estimates of the different synthetic scores are shown. The RScore, obtained through a full retrosynthesis (with a one minute timeout), is by far the most time-consuming score. Due to its scalability,

Table 12.1 Compute time per molecule for the different synthetic scores.

Synthetic score	Time per molecule (ms)
RA score	28
SC score	241
SA score	2
RScore	40 000
RSPred	1

Spaya-API accelerates RScore computation on large batch of molecules. The prediction of the latter, RSPred, is the fastest score to compute, only 1 ms per molecule, 40 000 times faster than the RScore. The SA score closely follows with 2 ms per molecule, the RA score is one order of magnitude slower while the SC score is two orders of magnitude slower.

12.4.3 Integration of Synthetic Scores in Generative AI

In the context of generative AI, the generator's weights are optimized in order to maximize a reward function. The various synthetic scores described above can be integrated in the reward function, so that the generated molecules get easier to synthesize. Figure 12.6 illustrates the pipeline on the use case presented below.

12.4.3.1 An Example of a Lead Optimization Use Case

This task is a generation around a library of 463 structurally homogeneous PI3K and mTOR inhibitors. The details of this experiment can be found in reference [120]. The dataset used here serves as a simplified proxy for a real-life MPO in a lead optimization project with four objectives to be optimized : QED, Pi3K, mTOR, and similarity to initial dataset. Six generations were run based on this dataset — one without any synthetic score constraint, and five with synthetic score constraints (RA, SC, SA, RScore, and RSPred). RScore1min and RScore3min correspond to RScore calculated at one and three minute timeouts, respectively.

The main metric to evaluate the quality of a generation method is the number of generated molecules validating all the constraints, which also have a good RScore. The right graph in Figure 12.6 shows for each of the five generations how many molecules validate the thresholds, and their RScore range. To summarize the obtained results in the PI3K/mTOR experiment, RScore appeared to be the best synthetic feasibility score to use as a synthetizability constraint integrated in an MPO generation, since it outperformed other methods in generating a high number of compounds in the defined blueprint with a good synthetic feasibility score. It is worth noting that RSPred is a good proxy of the RScore metric with a much lower computational cost. The SA score has some correlation to the RScore, but the generation under SA constraint outputs less than half as many interesting molecules as the generation under RScore1min or RSPred constraint. The other synthetic constraints were not very useful in this experiment — the RA score has poor precision, meaning that among the molecules well scored by RA score, very few actually have a good RScore3min, and when included in the reward of a generation, almost all molecules get a high reward and the generator can't be optimized toward easier to make molecules; the SC score has no correlation to the RScore3min, so it comes as no surprise that the generation under SC score constraint fails to optimize the RScore3min during the generation, and gives poor results.

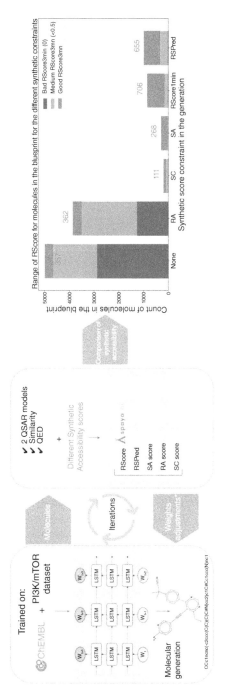

Figure 12.6 Pipeline of AI generative models with a synthetic accessibility constraint.

12.5 The Road Ahead

AI and machine learning-based tools are poised to become *must-haves* for every pharmaceutical company and drug discovery laboratory over the next decade. The methodologies and processes are maturing, and outstanding issues are being resolved. The innovation in both algorithmic and computational capacity is rapidly accelerating because of the desire to overcome longstanding issues in this field of study, and it is evident by the rapid increase in the scientific meetings and publications in this field of study. Caution is however warranted, and it is worth re-emphasizing that AI is not magic [121, 122]. While AI and machine learning algorithms are valuable in their ability to learn, it is the data generated by chemists in a myriad of experiments involving drug-like molecules that forms the backbone of this approach. Thus, both the quality of the AI framework (e.g. neural net architecture, deep learning algorithms) and the training data are of utmost importance [123]. It is also imperative to remember that the best AI system in drug discovery is the one that works alongside chemist's human intelligence — a chemist's domain and project knowledge cannot be replicated by AI.

Depending on the complexity of the problem, significant amounts of training data may be required, and it is unreasonable to expect one source to generate all or even most of it. Individual data sources (e.g. pharmaceutical companies) fiercely guard and seldom publicly disclose the data they generate in drug discovery pipelines for business confidentiality reasons. Data sharing remains a critical issue in the community, but crucial attempts are being made to use data from multiple sources to train model without actually making the data public (Federated Learning) [124, 125].

As the understanding of the benefits of using AI across the value chain of the drug discovery process continues to improve, some of the skepticism held by major stakeholders has been alleviated [126–128]. This understanding has been further augmented by examples of AI adding value to challenging projects [29]. Furthermore, there has been growing interest in developing tools for the explainability and interpretability of the models underneath the AI and machine learning systems used in the drug design and discovery space [129–131]. These methods can reveal the rationale behind the predictions made by models, offering insight in to the *black box*, thereby helping chemists devise useful strategies. Taken together, these methods can go a long way in further establishing the credentials of AI-based methods as a useful component in the drug discovery value chain.

References

1 Ha, J., Park, H., Park, J., and Park, S.B. (2021). Recent advances in identifying protein targets in drug discovery. *Cell Chemical Biology* 28 (3): 394–423.
2 Hughes, J.P., Rees, S., Kalindjian, S.B., and Philpott, K.L. (2011). Principles of early drug discovery. *British Journal of Pharmacology* 162 (6): 1239–1249.
3 Keserű, G.M. and Makara, G.M. (2006). Hit discovery and hit-to-lead approaches. *Drug Discovery Today* 11 (15–16): 741–748.

4 Mouchlis, V.D., Afantitis, A., Serra, A. et al. (2021). Advances in de novo drug design: from conventional to machine learning methods. *International Journal of Molecular Sciences* 22 (4): 1676.

5 Dang, C.V., Reddy, E.P., Shokat, K.M., and Soucek, L. (2017). Drugging the'undruggable'cancer targets. *Nature Reviews Cancer* 17 (8): 502–508.

6 An, S. and Fu, L. (2018). Small-molecule PROTACs: an emerging and promising approach for the development of targeted therapy drugs. *eBioMedicine* 36: 553–562.

7 Müller, C.E., Hansen, F.K., Gütschow, M. et al. (2021). New drug modalities in medicinal chemistry, pharmacology, and translational science: joint virtual special issue by *Journal of Medicinal Chemistry*, *ACS Medicinal Chemistry Letters*, and *ACS Pharmacology & Translational Science*. *Journal of Medicinal Chemistry* 64 (19): 13935–13936.

8 Yang, W., Gadgil, P., Krishnamurthy, V.R. et al. (2020). The evolving druggability and developability space: chemically modified new modalities and emerging small molecules. *The AAPS Journal* 22 (2): 1–14.

9 Maurya, N.S., Kushwaha, S., and Mani, A. (2019). Recent advances and computational approaches in peptide drug discovery. *Current Pharmaceutical Design* 25 (31): 3358-3366.

10 Sliwoski, G., Kothiwale, S., Meiler, J., and Lowe, E.W. (2014). Computational methods in drug discovery. *Pharmacological Reviews* 66 (1): 334–395.

11 Lionta, E., Spyrou, G., Vassilatis, D.K., and Cournia, Z. (2014). Structure-based virtual screening for drug discovery: principles, applications and recent advances. *Current Topics in Medicinal Chemistry* 14 (16): 1923–1938.

12 Hamza, A., Wei, N.-N., and Zhan, C.-G. (2012). Ligand-based virtual screening approach using a new scoring function. *Journal of Chemical Information and Modeling* 52 (4): 963–974.

13 Hoffmann, T. and Gastreich, M. (2019). The next level in chemical space navigation: going far beyond enumerable compound libraries. *Drug Discovery Today* 24 (5): 1148–1156.

14 Walters, W.P. and Wang, R. (2020). New trends in virtual screening. *Journal of Chemical Information and Modeling* 60 (9): 4109–4111.

15 Fresnais, L. and Ballester, P.J. (2021). The impact of compound library size on the performance of scoring functions for structure-based virtual screening. *Briefings in Bioinformatics* 22 (3): bbaa095.

16 Gentile, F., Yaacoub, J.C., Gleave, J. et al. (2022). Artificial intelligence–enabled virtual screening of ultra-large chemical libraries with deep docking. *Nature Protocols* 17 (3): 672–697.

17 Reymond, J.-L. (2015). The chemical space project. *Accounts of Chemical Research* 48 (3): 722–730.

18 Furman, J. and Seamans, R. (2019). Ai and the economy. *Innovation Policy and the Economy* 19 (1): 161–191.

19 Woo, M. (2019). An ai boost for clinical trials. *Nature* 573 (7775): S100–S100.

20 Muehlematter, U.J., Daniore, P., and Vokinger, K.N. (2021). Approval of artificial intelligence and machine learning-based medical devices in the USA and

EUROPE (2015–20): a comparative analysis. *The Lancet Digital Health* 3 (3): e195–e203.
21 Paul, D., Sanap, G., Shenoy, S. et al. (2021). Artificial intelligence in drug discovery and development. *Drug Discovery Today* 26 (1): 80.
22 Park, S.-W., Ko, J.-S., Huh, J.-H., and Kim, J.-C. (2021). Review on generative adversarial networks: focusing on computer vision and its applications. *Electronics* 10 (10): 1216.
23 Reed, S., Akata, Z., Yan, X. et al. (2016). Generative adversarial text to image synthesis. *International Conference on Machine Learning*, 1060–1069. PMLR.
24 Wang, L., Chen, W., Yang, W. et al. (2020). A state-of-the-art review on image synthesis with generative adversarial networks. *IEEE Access* 8: 63514–63537.
25 Vogt, M. (2022). Using deep neural networks to explore chemical space. *Expert Opinion on Drug Discovery* 17 (3): 297–304.
26 Wang, M., Wang, Z., Sun, H. et al. (2022). Deep learning approaches for de novo drug design: an overview. *Current Opinion in Structural Biology* 72: 135–144.
27 Schneider, G. and Clark, D.E. (2019). Automated de novo drug design: are we nearly there yet? *Angewandte Chemie International Edition* 58 (32): 10792–10803.
28 Blaschke, T., Arús-Pous, J., Chen, H. et al. (2020). REINVENT 2.0: an AI tool for de novo drug design. *Journal of Chemical Information and Modeling* 60 (12): 5918–5922.
29 Perron, Q., Mirguet, O., Tajmouati, H. et al. (2022). Deep generative models for ligand-based de novo design applied to multi-parametric optimization. *Journal of Computational Chemistry* 43 (10): 692–703.
30 Makara, G.M., Kovács, L., Szabó, I., and Põcze, G. (2021). Derivatization design of synthetically accessible space for optimization: *in silico* synthesis vs deep generative design. *ACS Medicinal Chemistry Letters* 12 (2): 185–194.
31 Miljković, F., Rodríguez-Pérez, R., and Bajorath, J. (2021). Impact of artificial intelligence on compound discovery, design, and synthesis. *ACS Omega* 6 (49): 33293–33299.
32 Gao, W. and Coley, C.W. (2020). The synthesizability of molecules proposed by generative models. *Journal of Chemical Information and Modeling* 60 (12): 5714–5723.
33 Kar, S., Roy, K., and Leszczynski, J. (2018). Impact of pharmaceuticals on the environment: risk assessment using QSAR modeling approach. In: *Computational Toxicology. Methods in Molecular Biology*, vol. 1800 (ed. O. Nicolotti), 395–443. New York: Springer.
34 Zagidullin, B., Wang, Z., Guan, Y. et al. (2021). Comparative analysis of molecular fingerprints in prediction of drug combination effects. *Briefings in Bioinformatics* 22 (6): bbab291.
35 Wigh, D.S., Goodman, J.M., and Lapkin, A.A. (2022). A review of molecular representation in the age of machine learning. *WIREs Computational Molecular Science* 12 (5): e1603.

36 Capecchi, A., Probst, D., and Reymond, J.-L. (2020). One molecular fingerprint to rule them all: drugs, biomolecules, and the metabolome. *Journal of Cheminformatics* 12: 43.

37 Pattanaik, L. and Coley, C.W. (2020). Molecular representation: going long on fingerprints. *Chem* 6 (6): 1204–1207.

38 Orosz, Á., Héberger, K., and Rácz, A. (2022). Comparison of descriptor- and fingerprint sets in machine learning models for ADME-Tox targets. *Frontiers in Chemistry* 10: 852893.

39 Sandfort, F., Strieth-Kalthoff, F., Kühnemund, M. et al. (2019). A structure-based platform for predicting chemical reactivity. *ChemRxiv*.

40 Venkatraman, V., Gaiser, J., Roy, A., and Wheeler, T.J. (2022). Molecular fingerprints are not useful in large-scale search for similarly active compounds†. *bioRxiv*.

41 O'Boyle, N.M. and Sayle, R.A. (2016). Comparing structural fingerprints using a literature-based similarity benchmark. *Journal of Cheminformatics* 8: 36.

42 Rogers, D. and Hahn, M. (2010). Extended-connectivity fingerprints. *Journal of Chemical Information and Modeling* 50 (5): 742–754.

43 Mittal, R.R., McKinnon, R.A., and Sorich, M.J. (2009). Comparison data sets for benchmarking QSAR methodologies in lead optimization. *Journal of Chemical Information and Modeling* 49 (7): 1810–1820.

44 Preuer, K., Renz, P., Unterthiner, T. et al. (2018). Fréchet ChemNet distance: a metric for generative models for molecules in drug discovery. *Journal of Chemical Information and Modeling* 58 (9): 1736–1741.

45 Yang, K., Swanson, K., Jin, W. et al. (2019). Are learned molecular representations ready for prime time? *ChemRxiv*.

46 Salentin, S., Schreiber, S., Haupt, V.J. et al. (2015). PLIP: fully automated protein-ligand interaction profiler. *Nucleic Acids Research* 43 (W1): W443–W447.

47 Laufkötter, O., Sturm, N., Bajorath, J. et al. (2019). Combining structural and bioactivity-based fingerprints improves prediction performance and scaffold hopping capability. *Journal of Cheminformatics* 11 (1): 54.

48 Duchowicz, P.R. (2018). Linear regression QSAR models for polo-like kinase-1 inhibitors. *Cells* 7 (2): 13.

49 Konovalov, D.A., Llewellyn, L.E., Heyden, Y.V., and Coomans, D. (2008). Robust cross-validation of linear regression QSAR models. *Journal of Chemical Information and Modeling* 48 (10): 2081–2094.

50 Svetnik, V., Liaw, A., Tong, C. et al. (2003). Random forest: a classification and regression tool for compound classification and QSAR modeling. *Journal of Chemical Information and Computer Sciences* 43 (6): 1947–1958.

51 Lee, K., Lee, M., and Kim, D. (2017). Utilizing random forest QSAR models with optimized parameters for target identification and its application to target-fishing server. *BMC Bioinformatics* 18 (16): 567.

52 Trinh, T.X., Seo, M., Yoon, T.H., and Kim, J. (2022). Developing random forest based QSAR models for predicting the mixture toxicity of TiO_2 based nano-mixtures to *Daphnia magna*. *NanoImpact* 25: 100383.

53 Shi, Y. (2021). Support vector regression-based QSAR models for prediction of antioxidant activity of phenolic compounds. *Scientific Reports* 11: 8806.

54 Mei, H., Zhou, Y., Liang, G., and Li, Z. (2005). Support vector machine applied in QSAR modelling. *Chinese Science Bulletin* 50: 2291–2296.

55 Darnag, R., Minaoui, B., and Fakir, M. (2017). QSAR models for prediction study of HIV protease inhibitors using support vector machines, neural networks and multiple linear regression. *Arabian Journal of Chemistry* 10: S600–S608.

56 Wu, Z., Ramsundar, B., Feinberg, E.N. et al. (2018). MoleculeNet: a benchmark for molecular machine learning. *Chemical Science* 9: 513–530.

57 Kokabi, M., Donnelly, M., and Xu, G. (2020). Benchmarking small-dataset structure-activity-relationship models for prediction of wnt signaling inhibition. *IEEE Access* 8: 228831–228840.

58 Arshadi, A.K., Salem, M., Firouzbakht, A., and Yuan, J.S. (2022). MolData, a molecular benchmark for disease and target based machine learning. *Journal of Cheminformatics* 14 (1): 10.

59 Czub, N., Pacławski, A., Szlek, J., and Mendyk, A. (2021). Curated database and preliminary AutoML QSAR model for 5-HT1A receptor. *Pharmaceutics* 13 (10): 1711.

60 Norinder, U., Carlsson, L., Boyer, S., and Eklund, M. (2014). Introducing conformal prediction in predictive modeling. A transparent and flexible alternative to applicability domain determination. *Journal of Chemical Information and Modeling* 54 (6): 1596–1603.

61 Liu, R. and Wallqvist, A. (2019). Molecular similarity-based domain applicability metric efficiently identifies out-of-domain compounds. *Journal of Chemical Information and Modeling* 59 (1): 181–189.

62 Sahigara, F., Ballabio, D., Todeschini, R., and Consonni, V. Defining a novel k-nearest neighbours approach to assess the applicability domain of a QSAR model for reliable predictions. *Journal of Cheminformatics* 5 (1): 27.

63 Aniceto, N., Freitas, A.A., Bender, A., and Ghafourian, T. (2016). A novel applicability domain technique for mapping predictive reliability across the chemical space of a QSAR: reliability-density neighbourhood. *Journal of Cheminformatics* 8: 69.

64 McMahan, H.B., Moore, E., Ramage, D., and y Arcas, B.A. (2016). Federated learning of deep networks using model averaging. *arXiv, 2, 2016*.

65 Pejó, B. (2020). The good, the bad, and the ugly: quality inference in federated learning. *arXiv*, abs/2007.06236.

66 Davies, R., Fowkes, A., Williams, R., and Johnston, L. (2020). Consortium-led federated QSAR models for secondary pharmacology - preparing the data. Granary Wharf House, 2 Canal Wharf, Leeds, LS11 5PS.

67 Matveieva, M. and Polishchuk, P. (2021). Benchmarks for interpretation of QSAR models. *Journal of Cheminformatics* 13: 41.

68 Lundberg, S.M. and Lee, S.-I. (2017). A unified approach to interpreting model predictions. *Advances in Neural Information Processing Systems* 30 (NIPS 2017).

69 Rodríguez-Pérez, R. and Bajorath, J. (2019). Interpretation of compound activity predictions from complex machine learning models using local approximations and shapley values. *Journal of Medicinal Chemistry* 63 (16): 8761–8777.

70 Wojtuch, A., Jankowski, R., and Podlewska, S. (2021). How can SHAP values help to shape metabolic stability of chemical compounds? *Journal of Cheminformatics* 13: 74.

71 Dahl, G.E., Jaitly, N., and Salakhutdinov, R. (2014). Multi-task neural networks for QSAR predictions. *arXiv*.

72 Xu, Y., Dai, Z., Chen, F. et al. Deep learning for drug-induced liver injury. *Journal of Chemical Information and Modeling* 55: 2085–2093.

73 Gawehn, E., Hiss, J.A., and Schneider, G. (2016). Deep learning in drug discovery. *Molecular Informatics* 35: 3–14.

74 Zhang, L., Tan, J., Han, D., and Zhu, H. (2017). From machine learning to deep learning: progress in machine intelligence for rational drug discovery. *Drug Discovery Today* 22: 1680–1685.

75 Chen, H., Engkvist, O., Wang, Y. et al. (2018). The rise of deep learning in drug discovery. 23 (6): 1241–1250.

76 Li, X., Xu, Y., Lai, L., and Pei, J. Prediction of human cytochrome P450 inhibition using a multitask deep autoencoder neural network. *Molecular Pharmaceutics* 15: 4336–4345.

77 Bhhatarai, B., Walters, W.P., Hop, C.E.C.A. et al. (2019). Opportunities and challenges using artificial intelligence in ADME/Tox. *Nature Materials* 18: 418–422.

78 Sun, M., Zhao, S., Gilvary, C. et al. Graph convolutional networks for computational drug development and discovery. *Briefings in Bioinformatics* 21 (3): 919–935.

79 Ma, J., Sheridan, R.P., Liaw, A. et al. (2015). Deep neural nets as a method for quantitative structure-activity relationships. *Journal of Chemical Information and Modeling* 55: 263–274.

80 Jiang, D., Wu, Z., Hsieh, C.Y. et al. (2021). Could graph neural networks learn better molecular representation for drug discovery? A comparison study of descriptor-based and graph-based models. *Journal of Cheminformatics* 13: 1–23.

81 Weininger, D. (1988). SMILES, a chemical language and information system. 1. Introduction to methodology and encoding rules. *Journal of Chemical Information and Computer Sciences* 28 (1): 31–36.

82 Krenn, M., Häse, F., Nigam, A.K. et al. (2020). Self-referencing embedded strings (SELFIES): a 100 robust molecular string representation. *Machine Learning: Science and Technology* 1 (4): 045024.

83 Mercado, R., Rastemo, T., Lindelöf, E. et al. (2020). Practical notes on building molecular graph generative models. *Applied AI Letters* 1 (2): https://doi.org/10.1002/ail2.18.

84 Chen, B., Fu, X., Barzilay, R., and Jaakkola, T. (2021). Fragment-based sequential translation for molecular optimization.

85 Bradshaw, J., Paige, B., Kusner, M.J. et al. (2020). Barking up the right tree: an approach to search over molecule synthesis dags. *CoRR*, abs/2012.11522.

86 Zhou, Z., Kearns, S., Li, L. et al. (2018). Optimization of molecules via deep reinforcement learning. *CoRR*, abs/1810.08678.

87 Gregor, K., Danihelka, I., Mnih, A. et al. (2014). Deep autoregressive networks. *Proceedings of Machine Learning Research* 32 (2): 1242–1250.

88 Bank, D., Koenigstein, N., and Giryes, R. (2020). Autoencoders. *CoRR*, abs/2003.05991

89 Kaelbling, L.P., Littman, M.L., and Moore, A.W. (1996). Reinforcement learning: a survey. *CoRR*, cs.AI/9605103.

90 Frazier, P.I. (2018). A tutorial on Bayesian optimization.

91 Bartz-Beielstein, T., Branke, J., Mehnen, J., and Mersmann, O. (2014). Evolutionary algorithms. *WIREs Data Mining and Knowledge Discovery* 4 (3): 178–195.

92 Nicolaou, C.A. and Brown, N. (2013). Multi-objective optimization methods in drug design. *Drug Discovery Today: Technologies* 10 (3): e427–e435.

93 Greff, K., Srivastava, R.K., Koutník, J. et al. (2016). LSTM: a search space odyssey. *IEEE Transactions on Neural Networks and Learning Systems* 28 (10): 2222–2232.

94 Segler, M.H.S., Kogej, T., Tyrchan, C., and Waller, M.P. (2018). Generating focused molecule libraries for drug discovery with recurrent neural networks. *ACS Central Science* 4 (1): 120–131.

95 Williams, R.J. and Zipser, D. (1989). A learning algorithm for continually running fully recurrent neural networks. *Neural Computation* 1 (2): 270–280.

96 de Souza Neto, L.R., Moreira-Filho, J.T., Neves, B.J. et al. (2020). In silico strategies to support fragment-to-lead optimization in drug discovery. *Frontiers in Chemistry* 8: 93.

97 Li, Q. (2020). Application of fragment-based drug discovery to versatile targets. *Frontiers in Molecular Biosciences* 7: 180.

98 Zhang, G., Zhang, J., Gao, Y. et al. (2022). Strategies for targeting undruggable targets. *Expert Opinion on Drug Discovery* 17 (1): 55–69.

99 Penner, P., Martiny, V., Gohier, A. et al. (2020). Shape-based descriptors for efficient structure-based fragment growing. *Journal of Chemical Information and Modeling* 60 (12): 6269–6281.

100 Vaswani, A., Shazeer, N., Parmar, N. et al. (2017). Attention is all you need. *Advances in Neural Information Processing Systems 30 (NIPS 2017)*.

101 Papadatos, G., Davies, M., Dedman, N. et al. (2015). SureChEMBL: a large-scale, chemically annotated patent document database. *Nucleic Acids Research* 44 (D1): D1220–D1228.

102 Lamb, A.M., ALIAS PARTH GOYAL, A.G., Zhang, Y. et al. (2016). Professor forcing: a new algorithm for training recurrent networks. *Advances in Neural Information Processing Systems 29 (NIPS 2016)*.

103 Winter, R., Montanari, F., Steffen, A. et al. (2019). Efficient multi-objective molecular optimization in a continuous latent space. *Chemical Science* 10: 8016–8024.

104 Gómez-Bombarelli, R., Wei, J.N., Duvenaud, D. et al. (2018). Automatic chemical design using a data-driven continuous representation of molecules. *ACS Central Science* 4 (2): 268–276.

105 Sattarov, B., Baskin, I.I., Horvath, D. et al. (2019). De novo molecular design by combining deep autoencoder recurrent neural networks with generative topographic mapping. *Journal of Chemical Information and Modeling* 59 (3): 1182–1196.

106 Gao, K., Nguyen, D.D., Tu, M., and Wei, G.-W. (2020). Generative network complex for the automated generation of drug-like molecules. *Journal of Chemical Information and Modeling* 60 (12): 5682–5698.

107 Renz, P., Van Rompaey, D., Wegner, J.K. et al. (2019). On failure modes in molecule generation and optimization. *Drug Discovery Today: Technologies* 32–33: 55–63.

108 Brown, N., Fiscato, M., Segler, M.H.S., and Vaucher, A.C. (2019). GuacaMol: benchmarking models for de novo molecular design. *Journal of Chemical Information and Modeling* 59 (3): 1096–1108.

109 Bradshaw, J., Paige, B., Kusner, M.J. et al. (2019). A model to search for synthesizable molecules. *CoRR*, abs/1906.05221.

110 Liu, C.-H., Korablyov, M., Jastrzebski, S. et al. (2020). RetroGNN: approximating retrosynthesis by graph neural networks for de novo drug design. *CoRR*, abs/2011.13042.

111 Coley, C.W., Rogers, L., Green, W.H., and Jensen, K.F. (2018). SCScore: synthetic complexity learned from a reaction corpus. *Journal of Chemical Information and Modeling* 58 (2): 252–261.

112 Ertl, P. and Schuffenhauer, A. (2009). Estimation of synthetic accessibility score of drug-like molecules based on molecular complexity and fragment contributions. *Journal of Cheminformatics* 1 (1): 1–11.

113 Thakkar, A., Chadimová, V., Bjerrum, E.J. et al. (2021). Retrosynthetic accessibility score (RAscore)–rapid machine learned synthesizability classification from AI driven retrosynthetic planning. *Chemical Science* 12: 3339–3349.

114 Genheden, S., Thakkar, A., Chadimová, V. et al. (2020). AiZynthFinder: a fast, robust and flexible open-source software for retrosynthetic planning. *Journal of Cheminformatics* 12 (1): 1–9.

115 Spaya. https://spaya.ai/ (accessed 26 August 2023).

116 Mcule database. https://mcule.com/database/ (accessed 26 August 2023).

117 Chem-space. https://chem-space.com/ (accessed 26 August 2023).

118 eMolecules. https://www.emolecules.com/ (accessed 26 August 2023).

119 Key Organics. https://www.keyorganics.net/ (accessed 26 August 2023).

120 Parrot, M., Tajmouati, H., da Silva, V.B.R. et al. (2021). Integrating synthetic accessibility with AI-based generative drug design. *ChemRxiv*.

121 Marcus, G. and Davis, E. (2019). *Rebooting AI: Building Artificial Intelligence We Can Trust*. Vintage.

122 Collins, H. (2021). The science of artificial intelligence and its critics. *Interdisciplinary Science Reviews* 46 (1–2): 53–70.

123 Turk, J.-A., Gendreau, P., Drizard, N., and Gaston-Mathé, Y. (2022). A molecular assays simulator to unravel predictors hacking in goal-directed molecular generations. *ChemRxiv*.

124 Wise, J., de Barron, A.G., Splendiani, A. et al. (2019). Implementation and relevance of fair data principles in biopharmaceutical r&d. *Drug Discovery Today* 24 (4): 933–938.

125 Lhuillier-Akakpo, M., Hoffmann, B., Huu, N.D. et al. (2021). Preparing a public dataset for drug discovery. https://www.melloddy.eu/blog/preparing-public-dataset/ (accessed 26 August 2023).

126 Smalley, E. (2017). Ai-powered drug discovery captures pharma interest. *Nature Biotechnology* 35 (7): 604–606.

127 Jiménez-Luna, J., Grisoni, F., Weskamp, N., and Schneider, G. (2021). Artificial intelligence in drug discovery: recent advances and future perspectives. *Expert Opinion on Drug Discovery* 16 (9): 949–959.

128 Vijayan, R.S.K., Kihlberg, J., Cross, J.B., and Poongavanam, V. (2021). Enhancing preclinical drug discovery with artificial intelligence. *Drug Discovery Today* 27 (4): 967–984.

129 Jiménez-Luna, J., Grisoni, F., and Schneider, G. (2020). Drug discovery with explainable artificial intelligence. *Nature Machine Intelligence* 2 (10): 573–584.

130 Preuer, K., Klambauer, G., Rippmann, F. et al. (2019). Interpretable deep learning in drug discovery. In: *Explainable AI: Interpreting, Explaining and Visualizing Deep Learning*, Lecture Notes in Computer Science, vol. 11700 (ed. W. Samek, G. Montavon, A. Vedaldi, et al.), 331–345. Cham: Springer.

131 Luo, Y., Peng, J., and Ma, J. (2022). Next Decade's AI-based drug development features tight integration of data and computation. *Health Data Science* 2022: 9816939.

13

Reliability and Applicability Assessment for Machine Learning Models

Fabio Urbina and Sean Ekins

Collaborations Pharmaceuticals, Inc., 840 Main Campus Drive, Raleigh, NC 27606, USA

13.1 Introduction

Techniques for using small molecule structures and related physicochemical property or bioactivity data to generate computational models of different types have existed for decades (and are outlined elsewhere in this book). Over the last 10 years, we have observed an increased use of machine learning and quantitative structure-activity relationship (QSAR) across the pharmaceutical industry for a range of property predictions and virtual screening for drug discovery, lead optimization, and toxicity prediction [1, 2], which can in turn accelerate the production of new hits and drug lead candidates [3]. At the same time, there is now a wide array of accessible databases containing thousands of structure-activity datasets available for physicochemical properties, molecules screened against drug targets, or phenotypic screens in public resources like ChEMBL [4], PubChem [5, 6], or others [7]. These provide the starting points for demonstrating the application of a diverse number of machine learning methods with many classic algorithms such as k-Nearest Neighbors (kNN) [8], naïve Bayesian [9–13], decision trees [14], support vector machines (SVMs) [15–21], and others [22, 23], as well as newer algorithms such as deep neural networks (DNNs) [24–32], long short term memory (LSTM) [33], and transformers [34]. These efforts have enabled several large-scale analyses of datasets with different machine learning methods and molecular descriptors [35–42]. Some of the largest comparisons of machine learning models have used over 1000 models [43–47]. Most recently, we have described extracting over 5000 datasets from CHEMBL (endpoints such as IC_{50}, K_i, and MIC) for use with the ECFP6 fingerprint descriptor and comparing random forest, k-Nearest Neighbors, support vector classification, naïve Bayesian, AdaBoosted decision trees, and deep neural networks. The model performance was assessed using fivefold cross-validation metrics that were generated, including area-under-the-curve, F1 score, Cohen's kappa, and Matthews correlation coefficient. We demonstrated using ranked normalized scores for the metrics for all methods that they appeared comparable, while the distance from the top metric suggested our implementation of the

Computational Drug Discovery: Methods and Applications, First Edition.
Edited by Vasanthanathan Poongavanam and Vijayan Ramaswamy.
© 2024 WILEY-VCH GmbH. Published 2024 by WILEY-VCH GmbH.

Bayesian algorithm and support vector classification were essentially comparable. This work represents one of the largest-scale comparisons of machine learning algorithms [48]. ECFP6 represents just one fingerprint of many that could be potentially accessed in this manner [49]. Another large-scale evaluation performed by the Novartis Institute for Biomedical Research used 8558 proprietary Novartis assays to generate Random Forest Regressor models [50].

13.2 Challenges for Modeling

Such machine learning models are not perfect, they are "an approximation of reality." The models may have errors inherent in the data, as biological data that are generated *in vitro* will have experimental errors resulting from dispensing, analytical or other errors [51]. Other issues may include small dataset size, lack of diversity, poor activity distribution of the data, or other biases [7]. These experimental errors will impact the dataset that is being modeled and could affect the resulting predictions. Also, the training set for the model may be of a limited size, which will impact the diversity of the chemical property space covered, the data distribution, and the utility of the model to make predictions outside of the training set space. An early concept was that more diverse models could be considered as "global," whereas those with a very limited focus on a narrow SAR were likely to describe "local" properties [2]. Hence, a machine learning model's prediction confidence will be impacted by its domain of applicability (also called domain extrapolation) outside of this training domain [52]. An early demonstration of the calculation of prediction confidence and domain extrapolation used the estrogen receptor binding activity datasets and decision forest models. In this case, the prediction accuracy was related to the ratio of correct predictions to the total number of molecules in the domain and showed that accuracy declined sharply with domain extrapolation [52] (Table 13.1). Other groups have suggested how the initial poor performance of ADME-Tox models could have been related to their applicability domain. Several methods were described early on that related to the calculation of the applicability domain using either the descriptor space (missing fragments approach) or methods based on the similarity of molecules in descriptor space. The latter uses different methods such as Euclidean, city block, Tanimoto, Mahalanobis, hotelling T2, and leverage to measure the distance from a training set to a test set [56, 65, 66] to compare the chemical space of the datasets [2]. The applicability domain was also proposed early on as an important problem for QSAR studies, and a review described several limitations of the methods employed [67]. Pharmaceutical companies have applied applicability domains early on. For example, Bayer and Schering used several approaches, including Bayesian Gaussian process models, distance-based methods, and ensembles for regression models of lipophilicity [57] and solubility [68] that showed that the mean absolute error decreased as compounds were binned with higher confidence. For a visual appreciation of the applicability domain concept, the reader is also referred to these published articles and others, such as Aniceto et al. [53].

Table 13.1 Selected methods for applicability domain, error, and confidence predictions. See also additional articles in the text Adapted from Aniceto et al., 2016; Rakhimbekova et al., 2020; Sushko 2011.

Method	Reference
Domain extrapolation	[52]
Prediction confidence	[52]
Euclidean distance	[56]
City block	[56]
Mahalanobis	[56]
Hotelling T^2	[56]
Leverage	[56]
Bayesian Gaussian process models, distance-based methods, and ensembles	[57]
Optimal assignment kernel, flexible optional assignment kernel, marginalized graph kernel	[58]
Number of fingerprint features	[59]
A reliability-density neighborhood	[53]
Sum of distance-weighted contributions	[60]
Conformal prediction	[61]
Test time dropout	[62]
Rivality index	[63]
Entropy, Monte Carlo dropout, Multi-Initial, FPsDist, and LatentDist	[64]

As SVMs have been widely used for QSAR, one study has proposed three applicability domain approaches for kernel-based QSAR relying on similarity: the optimal assignment kernel, the flexible optional assignment kernel, and the marginalized graph kernel. Using three different virtual screening examples, these showed the models performed better inside the domains [58]. As molecular fingerprints such as ECFP are widely used in machine learning, they have also been used to define applicability domains. One study used the nearest neighbor or random forests in combination to provide an applicability domain for an Ames mutagenicity model. The number of ECFP_4 or ECFP_2 features for a test compound that is not present in a training set was used as an indicator of the applicability domain [59]. Several different methods have also been used to assess machine learning model reliability with 20 regression model datasets. These reliability estimates included Mahalanobis distance to nearest neighbors, Mahalanobis distance to the data set center, sensitivity analysis scores, bootstrap variance, local cross-validation error, local prediction error modeling, and combination of bootstrap variance and local prediction error score. Error-based estimation methods outperformed or were on a par with similarity-based methods, while performance did not depend on global or local model or descriptor type [69]. A reliability-density neighborhood approach was used as an applicability domain for P-gp, Ames, and CYP450 models and was proposed to take into account sparse regions by mapping data density and local precision and bias [53]. A new applicability domain metric that considered the contribution of every training sample weighted by its distance to the molecule being predicted (called sum of distance-weighted contributions) was demonstrated with several toxicities and physicochemical property datasets and correlated more strongly with prediction error than other methods like distance to model or ensemble variance measures [60]. This approach has also been used with a melting point dataset and showed that it outperformed the other methods utilized [70]. Several approaches have been proposed to compute the uncertainty of predictions [71], and one is conformal prediction (also see later section), which can provide confidence regions and was used with deep neural networks and benchmarked on 24 regression datasets from ChEMBL as well as against random forest-based conformal predictions. The confidence intervals for the deep confidence approach had a smaller spread than for the random forest approach [61]. The test time dropout and conformal prediction approaches have been used to reliably compute errors for neural network models created for the same 24 bioactivity datasets [62]. Conformal prediction has also been the subject of a minireview, which also applied the approach with three transporter models [72].

The rivality index has been proposed as another method for assessing the reliability of predictions or applicability domains by generating a normalized distance measurement between each molecule and its nearest neighbor belonging to the same class and the nearest neighbor belonging to a different class. This approach was tested with four classification datasets across 12 algorithms [63]. Uncertainty estimation using five methods, including Entropy, Monte Carlo dropout, Multi-Initial, FPsDist, and LatentDist, was used with a BBB dataset and several different machine learning approaches. The combination of Entropy and Monte Carlo dropout to predict uncertainty was used for the GROVER BBB model [64].

While these represent just a snapshot of some of the many efforts to address the applicability domain or confidence in prediction, these areas are not always covered in exhaustive reviews describing machine learning or QSAR methods [73]. It is for this reason that they should perhaps be given more exposure. We now provide several examples from our own work to explore this further.

13.3 Example 1: BBB Applicability Domain Comparison

The blood brain barrier (BBB) represents a significant challenge for drug delivery to the central nervous system (CNS). Molecules (such as antidepressants and antipsychotics, for example) must cross the BBB to act within the CNS or the brain. We recently published a BBB model based on a binary dataset of 2358 published compounds that either crossed the brain barrier or did not [74, 75].To illustrate the use of applicability domains, we have investigated a modified reliability-density neighborhood approach. We compare this approach to simple training set distance metrics as baselines.

Machine learning: Our software, Assay Central, uses multiple algorithms integrated into web-based software to build models, as described previously in detail [76]. The machine learning model validation was performed using a nested fivefold cross-validation with an external test set. This optimized model is then used to predict the initial 20% hold-out set. The final nested fivefold cross-validation scores are an average of each of the holdout set metrics. We chose a random forest model to investigate as it was among the top-performing models, and a standard deviation can be extracted by aggregating the predictions from the individual trees.

Modified reliability-density (RDE) neighborhood estimation: We adopted the approach from Aniceto et al. [53]. As our model uses ECFP6 descriptors, we simplified the AD score to the following: After the random forest model is built on the training data, it is then used to predict on the same training set to extract bias = $abs(\hat{Y}i - Yi)$, as well as the standard deviation of the individual tree predictions. Upon inference with a new molecule, the applicability domain is applied using the following equation:

$$AD\ score = w_T * w_{AD}$$

where w_T is the average Tanimoto similarity between the top-*n* most similar molecule(s) in the training dataset and the molecule for inference and $w_{AD} = \left(\frac{\sum(1-STD_i)}{n}\right) * \left(\frac{\sum 1 - abs(\hat{y}_i - y_i)}{n}\right)$ for the top-*n* most similar molecules in the model. As the Tanimoto distance is performed on the same ECFP6 1024-bit vector that informs the model, the maximum Tanimoto similarities highlight how informed the model is of the input features while w_{ADi} penalizing the similarity score based on the bias and decision tree agreement of the model on the most similar molecules. We performed two iterations, using *n* of 1 and *n* of 5 for the modified RDE calculation. We compare this method by using the average Tanimoto similarity to the training data, the maximum Tanimoto similarity to the training data, and the top-5 average Tanimoto similarities to the dataset as AD score baselines. We perform this comparison using stratified fivefold cross-validation on the BBB dataset and fit a locally estimated scatterplot smoothing (LOESS) regression model.

Results: While there is no correlation between the average Tanimoto similarity to the dataset and the absolute error, both the maximum Tanimoto and the top-5 Tanimoto distance to the training set show a small but non-robust correlation (Figure 13.1). The modified RDE method of weighting the maximum or top-5 maximum Tanimoto similarity shows a stronger correlation to the absolute error that is more consistent between folds and closer to a strictly-decreasing function. Evaluating the top *X*% of AD scores shows that all methods enrich correct predictions while the average Tanimoto lags significantly (Figure 13.2). The RDE top-5 method shows the most consistent enrichment, suggesting the corrective weighting factor helps reduce the influence of model bias.

13.4 Example 2: Models for Uncertainty Estimation for Multitask Toxicity Predictions

Many applicability domain scores have restrictions, such as being only applicable to classification tasks or specific feature inputs. While applicability domains are usually

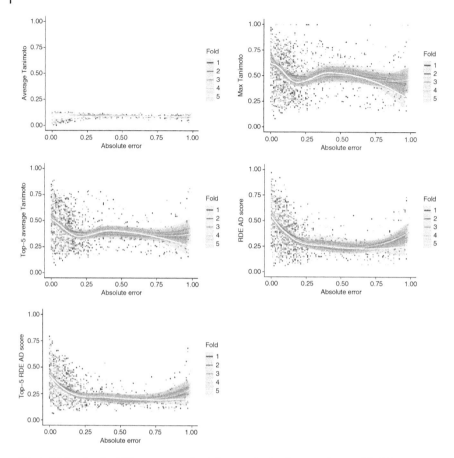

Figure 13.1 Applicability score vs. absolute error test set scores for stratified 5-cross-validation of a binary classification blood brain barrier random forest model (8850 actives/ 2580 inactives). Each fold's test set had 2286 molecules. Average Tanimoto: average Tanimoto against the training set, Max Tanimoto: Maximum Tanimoto against the training set. Top-5 Average: Top 5 Tanimoto distances against the training set averaged. RDE AD: Top-1 Reliability-density neighborhood estimation. Top-5 RDE: Top-5 Reliability-density neighborhood estimation.

investigated for single-endpoint models, the advent of larger and more complex deep learning models may require a new set of applicability domain algorithms to better represent their predictive performance. Here, we investigate the use of Monte Carlo (MC) dropout for uncertainty estimation in the predictions of a deep learning end-to-end multitask regression model.

Overview: The requirements for a candidate molecule to become a drug at a simplistic level include efficacy and specificity against a target of interest, as off-target interactions can lead to undesirable side effects and pose a potential safety hazard. Commercial *in vitro* safety profiling screens are often used to search for critical off-targets, which could lead to adverse drug reactions [77]. In silico approaches offer an appealing substitute, as they are comparably inexpensive and inference is

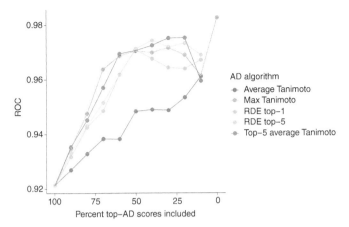

Figure 13.2 Comparison of ROC of applicability domain (AD) score inclusion. The top X% of molecules have AD scores and are evaluated at each interval.

significantly faster than *in vitro* assays. Thus, much research has been performed over several decades into predicting toxicity.

Especially crucial to toxicity models is how trustworthy predictions are, as false negatives are not tolerated in safety profiling. We recently built and described a multi-task neural network model to predict IC_{50}'s of 42 of the 44 endpoints used in the SafetyScreen44™, a commercial *in vitro* safety profiling assay for which we could find sufficient publicly available datasets [78]. The purpose of this machine learning model was to increase inference speed by utilizing SMILES as the only input, removing the temporal bottleneck of feature creation (e.g. generating Morgan fingerprints). As a further useful case study for applicability domains, we now revisit this model and investigate using MC dropout to approximate uncertainty estimation for the model.

Datasets: The data was curated as described in Ref. [78]. Briefly, IC_{50}-only target-activity data for 42 toxicity targets were downloaded from ChEMBL 30 and standardized (salts removed, charges neutralized, and canonical SMILES generated). The datasets were split randomly at 70%/15%/15% and stratified for each target. Seventy percent of the data was used for training, 15% for validation, and 15% for test results.

Machine Learning: We used a convolutional long-short term memory (ConvLSTM)-based model with an embedding layer (size 50), 1-D convolutional layer (size 256), a batch-norm layer, a bidirectional LSTM layer (size 1024), and three fully-connected layers with dropout (25%) followed by rectified linear unit (ReLU) layer of size 2048, 1024, and a final 42 for the output layer.

Results: During model training, dropout layers are used for model regularization [79]. These layers are generally turned off during inference so that the full model can be utilized in a deterministic manner. Dropout layers can be utilized during inference, however, to approximate Bayesian model uncertainty without alterations to the final model by using dropout layers during inference [80]. Running multiple predictions with different neurons due to dropout is equivalent to an ensemble of models performing inference. The predictions are then averaged for

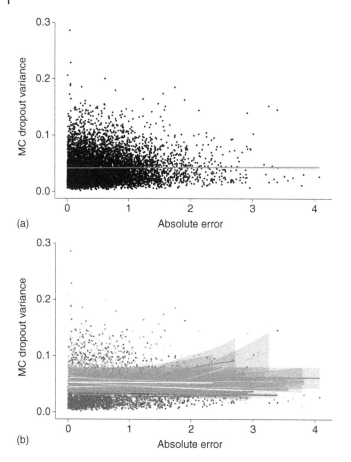

Figure 13.3 MC dropout variance vs. Absolute error for prediction on a test set using a multitask regression model (LSTM model using 42 toxicity models). Data fit using Generative additive models (GAM). (a) MC Dropout with a GAM fit on the entire test set. (b) MC Dropout with a GAM fit on each of the 42 individual target endpoints.

a final prediction. The variation of predictions gives an estimation of uncertainty: Predictions with high variation can be inferred as high uncertainty. Applying this technique, we performed MC dropout inference (25 predictions per test set datapoint) and calculated the standard deviation of each set of predictions (Figure 13.3). Surprisingly, while the variance correlated with several independent target predictions, the variance of predictions did not correlate with a significant number of endpoints or the predictive error of the entire test set (Figure 13.3). Altering the dropout rate (20–50%), number of neural network layers with dropout (1–3), or number of predictions (25–100) did not change the relationship between variance and predictive performance. This suggests that MC dropout may not be generally applicable to all scenarios, and care must be taken when selecting an applicability domain for more complex models.

13.5 Example 3: Class-Conditional Conformal Predictors

Conformal predictors are a framework that can be applied to any model that outputs prediction scores and have been used for various drug discovery and environmental QSAR applications [72, 81, 82]. A recent extensive review [83] describes how, in simple terms, the framework uses a calibration dataset to determine the optimal prediction score threshold for each class (1, 0) such that prediction scores that exceed the class-specific threshold have the following validity: $1 - \alpha \leq P\left(Y_{test} \in C(X_{test}) \leq 1 - \alpha + \frac{1}{n-1}\right)$ for a user-chosen value of $\alpha \in [0, 1]$.

More formally, let k be the number of classes predicted by a model $\hat{f}(x) \in [0,1]^k$. Let $(x_i, y_i)_i^n$, be an independent calibration set, assumed I.I.D., and from the same distribution as the training data. Define a conformal score to be $S_i = 1 - \hat{f}(x_i)_{y_i}$ of the calibration set and $\hat{q}_k = \frac{\lceil (n+1)(1-\alpha) \rceil}{n}$ as the empirical quantile of the conformal scores $S_{k_1}, \ldots, S_{k_i}$ for each k. For any new predictions, return the set of possible classes $C(X_{test}) = \{y : \hat{f}(X_{test})_k \geq 1 - \hat{q}_k\}$. In the binary case, return the prediction score for class 0 if $1 - \hat{f}(X_{test}) \geq 1 - \hat{q}_0$ and the prediction score for class 1 if $\hat{f}(X_{test}) \geq 1 - \hat{q}_1$. As a heuristic, we may treat the return of no prediction scores or prediction scores for both classes to be inconclusive and "trust" only single-class predictions that are returned.

Evaluation: Using a two-class biodegradation dataset as an example. The two classes are "not-readily biodegradable" and "readily biodegradable." Following the concatenation, these datasets were cleaned (subjected to charge neutralization, salt removal, and standardization via custom software using open-source RDkit functions. Following these, the duplicates were then removed after a check for unambiguous classification) prior to model building. The final dataset contained 3428 unique compounds (inorganic compounds removed), with 962 classified as "readily biodegradable." Using a class-stratified random 70/20/10% split for the training, test, and calibration sets, respectively, we then built a Random Forest model with 1000 trees. The initial model had a recall of 0.72 and a ROC of 0.86; however, it had a lower precision of 0.63 (Figure 13.4a). We next calculated the thresholds at different levels of α using the calibration dataset. Predictions that did not meet the required threshold of either class (no predictions returned) or were predicted to belong to both classes (both prediction scores returned) were labeled "inconclusive." Only conclusive predictions were used for metric calculations. Stricter levels of alpha classified more of the test set as "inconclusive" (Figure 13.4b), but enriched correct predictions, up to an $\alpha = 0.2$, where the ROC rose to 0.9, Recall to 0.89, and Precision to 0.86, coinciding with the validity guarantee of $1 - \alpha$ accuracy for each class. As outlined in this example and as has been shown earlier by others, this methodology therefore allows for stricter confidence in returned predictions and rejection of prediction scores that do not meet the designated threshold. This therefore has utility as a method for applicability domain prediction.

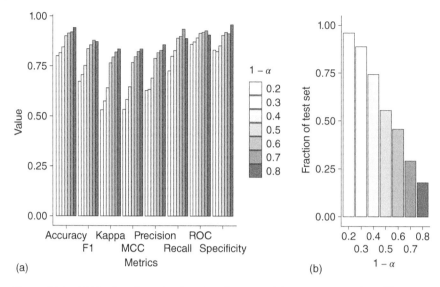

Figure 13.4 (a) Metrics of a trained random forest model on a test set using different α thresholds to classify test molecules as either not-readily biodegradable, readily biodegradable, or inconclusive. Metrics were calculated on non-inconclusive data points. (b) The fraction of the test set that was considered not inconclusive.

13.6 Conclusions

Applicability domains are becoming a de facto requirement before taking action on "black box" model predictions. Future challenges include that generative models that include property predictions to derive molecules with ideal properties may be limited by the applicability domain or model confidence in predictions [7, 84]. This is not an area that has been broadly addressed to date. Many applicability domains rely on a similarity measure to the training data, while generative models are often attempting to discover novel molecules, putting the use of such applicability domains at odds with molecule generation [84]. Another challenge is that while many applicability domains have been introduced and used effectively, their performance is not always significantly better than using the model's prediction scores alone [85]. When coupled with the sometimes-strict requirements and complexity to calculate AD scores, a need for more effective AD algorithms with less strict requirements and more efficacy, is therefore highly desirable. Further, as we show, many AD algorithms are not readily applicable to more complex models such as multitask models or models with distinct inputs. There is clearly still considerable scope to both evaluate existing AD methods and develop new ones for the reliability and applicability assessment of machine learning models. Beyond applicability, the next important area will be in the interpreting or explainability of the model predictions, and there have already been recent discussions on this topic [86–88]. Perhaps in 20 years, model interpretability/explainability will be at a similar level of acceptance as a model applicability domain is today. Ultimately, the key will be the transition of these concepts into commercial or widely used software products, by which point they will be regarded as standard.

Funding

We kindly acknowledge NIH funding from R44GM122196-02A1 from NIGMS and 2R44ES031038-02A1 from NIEHS. Research reported in this publication was supported by the National Institute of Environmental Health Sciences of the National Institutes of Health under Award Number 2R44ES031038-02A1. The content is solely the responsibility of the authors and does not necessarily represent the official views of the National Institutes of Health."

Competing Interests

S.E. is owner, and F.U. is an employee of Collaborations Pharmaceuticals, Inc.

References

1 Ekins, S., Puhl, A.C., Zorn, K.M. et al. (2019). Exploiting machine learning for end-to-end drug discovery and development. *Nat. Mater.* 18 (5): 435–441.
2 Cheng, F., Li, W., Liu, G., and Tang, Y. (2013). In silico ADMET prediction: recent advances, current challenges and future trends. *Curr. Top. Med. Chem.* 13 (11): 1273–1289.
3 Zhavoronkov, A., Ivanenkov, Y.A., Aliper, A. et al. (2019). Deep learning enables rapid identification of potent DDR1 kinase inhibitors. *Nat. Biotechnol.* 37 (9): 1038–1040.
4 Gaulton, A., Bellis, L.J., Bento, A.P. et al. (2012). ChEMBL: a large-scale bioactivity database for drug discovery. *Nucleic Acids Res.* 40 (Database issue): D1100–D1107.
5 Kim, S., Thiessen, P.A., Bolton, E.E. et al. (2016). PubChem substance and compound databases. *Nucleic Acids Res.* 44 (D1): D1202–D1213.
6 Anon *The PubChem Database.* http://pubchem.ncbi.nlm.nih.gov.
7 Nigam, A., Pollice, R., Hurley, M.F.D. et al. (2021). Assigning confidence to molecular property prediction. *Expert Opin. Drug Discovery* 16 (9): 1009–1023.
8 Shen, M., Xiao, Y., Golbraikh, A. et al. (2003). Development and validation of k-nearest neighbour QSPR models of metabolic stability of drug candidates. *J. Med. Chem.* 46: 3013–3020.
9 Wang, S., Sun, H., Liu, H. et al. (2016). ADMET evaluation in drug discovery. 16. Predicting hERG blockers by combining multiple pharmacophores and machine learning approaches. *Mol. Pharmaceutics* 13 (8): 2855–2866.
10 Li, D., Chen, L., Li, Y. et al. (2014). ADMET evaluation in drug discovery. 13. Development of in silico prediction models for P-glycoprotein substrates. *Mol. Pharmaceutics* 11 (3): 716–726.

11 Nidhi, Glick, M., Davies, J.W., and Jenkins, J.L. (2006). Prediction of biological targets for compounds using multiple-category Bayesian models trained on chemogenomics databases. *J. Chem. Inf. Model.* 46 (3): 1124–1133.

12 Azzaoui, K., Hamon, J., Faller, B. et al. (2007). Modeling promiscuity based on in vitro safety pharmacology profiling data. *ChemMedChem* 2 (6): 874–880.

13 Bender, A., Scheiber, J., Glick, M. et al. (2007). Analysis of pharmacology data and the prediction of adverse drug reactions and off-target effects from chemical structure. *ChemMedChem* 2 (6): 861–873.

14 Susnow, R.G. and Dixon, S.L. (2003). Use of robust classification techniques for the prediction of human cytochrome P450 2D6 inhibition. *J. Chem. Inf. Comput. Sci.* 43 (4): 1308–1315.

15 Bennet, K.P. and Campbell, C. (2000). Support vector machines: hype or hallelujah? *SIGKDD Explor.* 2: 1–13.

16 Christianini, N. and Shawe-Taylor, J. (2000). *Support Vector Machines and Other Kernel-Based Learning Methods*. Cambridge, MA: Cambridge University Press.

17 Chang, C.C. and Lin, C.J. (2011). LIBSVM: a library for support vector machines. *ACM Trans. Intell. Syst. Technol.* 2 (3): 1–27.

18 Lei, T., Chen, F., Liu, H. et al. (2017). ADMET evaluation in drug discovery. Part 17: development of quantitative and qualitative prediction models for chemical-induced respiratory toxicity. *Mol. Pharmaceutics* 14 (7): 2407–2421.

19 Kriegl, J.M., Arnhold, T., Beck, B., and Fox, T. (2005). A support vector machine approach to classify human cytochrome P450 3A4 inhibitors. *J. Comput.-Aided Mol. Des.* 19 (3): 189–201.

20 Guangli, M. and Yiyu, C. (2006). Predicting Caco-2 permeability using support vector machine and chemistry development kit. *J. Pharm. Pharm. Sci.* 9 (2): 210–221.

21 Kortagere, S., Chekmarev, D., Welsh, W.J., and Ekins, S. (2009). Hybrid scoring and classification approaches to predict human pregnane X receptor activators. *Pharm. Res.* 26 (4): 1001–1011.

22 Mitchell, J.B. (2014). Machine learning methods in chemoinformatics. *Wiley Interdiscip. Rev. Comput. Mol. Sci.* 4 (5): 468–481.

23 Wacker, S. and Noskov, S.Y. (2018). Performance of machine learning algorithms for qualitative and quantitative prediction drug blockade of hERG1 channel. *Comput. Toxicol.* 6: 55–63.

24 Schmidhuber, J. (2015). Deep learning in neural networks: an overview. *Neural Netw.* 61: 85–117.

25 Capuzzi, S.J., Politi, R., Isayev, O. et al. (2016). QSAR modeling of Tox21 challenge stress response and nuclear receptor signaling toxicity assays. *Front. Environ. Sci.* 4 (3).

26 Russakovsky, O., Deng, J., Su, H., et al. (2015) ImageNet Large Scale Visual Recognition Challenge. https://arxiv.org/pdf/1409.0575.pdf.

27 Zhu, H., Zhang, J., Kim, M.T. et al. (2014). Big data in chemical toxicity research: the use of high-throughput screening assays to identify potential toxicants. *Chem. Res. Toxicol.* 27 (10): 1643–1651.

28 Clark, A.M. and Ekins, S. (2015). Open source Bayesian models: 2. Mining a "big dataset" to create and validate models with ChEMBL. *J. Chem. Inf. Model.* 55: 1246–1260.
29 Ekins, S., Clark, A.M., Swamidass, S.J. et al. (2014). Bigger data, collaborative tools and the future of predictive drug discovery. *J. Comput.-Aided Mol. Des.* 28 (10): 997–1008.
30 Ekins, S., Freundlich, J.S., and Reynolds, R.C. (2014). Are bigger data sets better for machine learning? Fusing single-point and dual-event dose response data for *Mycobacterium tuberculosis. J. Chem. Inf. Model.* 54: 2157–2165.
31 Ekins, S. (2016). The next era: deep learning in pharmaceutical research. *Pharm. Res.* 33 (11): 2594–2603.
32 Baskin, I.I., Winkler, D., and Tetko, I.V. (2016). A renaissance of neural networks in drug discovery. *Expert Opin. Drug Discovery* 11: 785–795.
33 Greff, K., Srivastava, R.K., Koutník, J. et al. (2017). LSTM: a search space odyssey. *IEEE Trans. Neural Netw. Learn. Syst.* 28 (10): 2222–2232.
34 Devlin, J., Chang, M.-W., Lee, K., and Toutanova, K. (2018). BERT: Pre-training of Deep Bidirectional Transformers for Language Understanding. arXiv, 1810.04805.
35 Wang, L., Ma, C., Wipf, P. et al. (2013). TargetHunter: an in silico target identification tool for predicting therapeutic potential of small organic molecules based on chemogenomic database. *AAPS J.* 15 (2): 395–406.
36 Koutsoukas, A., Lowe, R., Kalantarmotamedi, Y. et al. (2013). In silico target predictions: defining a benchmarking data set and comparison of performance of the multiclass Naive Bayes and Parzen-Rosenblatt window. *J. Chem. Inf. Model.* 53 (8): 1957–1966.
37 Cortes-Ciriano, I. (2016). Benchmarking the predictive power of ligand efficiency indices in QSAR. *J. Chem. Inf. Model.* 56 (8): 1576–1587.
38 Qureshi, A., Kaur, G., and Kumar, M. (2017). AVCpred: an integrated web server for prediction and design of antiviral compounds. *Chem. Biol. Drug Des.* 89 (1): 74–83.
39 Bieler, M., Reutlinger, M., Rodrigues, T. et al. (2016). Designing multi-target compound libraries with Gaussian process models. *Mol. Inf.* 35 (5): 192–198.
40 Huang, T., Mi, H., Lin, C.Y. et al., andfor MZRW Group(2017). MOST: most-similar ligand based approach to target prediction. *BMC Bioinf.* 18 (1): 165.
41 Cortes-Ciriano, I., Firth, N.C., Bender, A., and Watson, O. (2018). Discovering highly potent molecules from an initial set of inactives using iterative screening. *J. Chem. Inf. Model.* 58 (9): 2000–2014.
42 Bosc, N., Atkinson, F., Felix, E. et al. (2019). Large scale comparison of QSAR and conformal prediction methods and their applications in drug discovery. *J. Cheminf.* 11 (1): 4.
43 Lenselink, E.B., Ten Dijke, N., Bongers, B. et al. (2017). Beyond the hype: deep neural networks outperform established methods using a ChEMBL bioactivity benchmark set. *J. Cheminf.* 9 (1): 45.

44 Mayr, A., Klambauer, G., Unterthiner, T. et al. (2018). Large-scale comparison of machine learning methods for drug target prediction on ChEMBL. *Chem. Sci.* 9 (24): 5441–5451.

45 Lee, K. and Kim, D. (2019). In-silico molecular binding prediction for human drug targets using deep neural multi-task learning. *Genes (Basel)* 10 (11): 906.

46 Awale, M. and Reymond, J.L. (2019). Polypharmacology browser PPB2: target prediction combining nearest neighbors with machine learning. *J. Chem. Inf. Model.* 59 (1): 10–17.

47 Škuta, C., Cortés-Ciriano, I., Dehaen, W. et al. (2020). QSAR-derived affinity fingerprints (part 1): fingerprint construction and modeling performance for similarity searching, bioactivity classification and scaffold hopping. *J. Cheminf.* 12 (1): 39.

48 Lane, T.R., Foil, D.H., Minerali, E. et al. (2021). Bioactivity comparison across multiple machine learning algorithms using over 5000 datasets for drug discovery. *Mol. Pharmaceutics* 18 (1): 403–415.

49 Clark, A.M., Dole, K., Coulon-Spektor, A. et al. (2015). Open source Bayesian models. 1. Application to ADME/Tox and drug discovery datasets. *J. Chem. Inf. Model.* 55 (6): 1231–1245.

50 Martin, E.J., Polyakov, V.R., Zhu, X.W. et al. (2019). All-assay-Max2 pQSAR: activity predictions as accurate as four-concentration IC50s for 8558 Novartis assays. *J. Chem. Inf. Model.* 59 (10): 4450–4459.

51 Ekins, S., Olechno, J., and Williams, A.J. (2013). Dispensing processes impact apparent biological activity as determined by computational and statistical analyses. *PLoS One* 8 (5): e62325.

52 Tong, W., Xie, Q., Hong, H. et al. (2004). Assessment of prediction confidence and domain extrapolation of two structure-activity relationship models for predicting estrogen receptor binding activity. *Environ. Health Perspect.* 112 (12): 1249–1254.

53 Aniceto, N., Freitas, A.A., Bender, A., and Ghafourian, T. (2016). A novel applicability domain technique for mapping predictive reliability across the chemical space of a QSAR: reliability-density neighbourhood. *J. Cheminf.* 8 (1): 69.

54 Rakhimbekova, A., Madzhidov, T.I., Nugmanov, R.I. et al. (2020). Comprehensive analysis of applicability domains of QSPR models for chemical reactions. *Int. J. Mol. Sci.* 21 (15): 5542.

55 Sushko, I. (2011). *Applicability Domain of QSAR Models*. Technische Universität München.

56 Tetko, I.V., Bruneau, P., Mewes, H.W. et al. (2006). Can we estimate the accuracy of ADME-Tox predictions? *Drug Discovery Today* 11 (15–16): 700–707.

57 Schroeter, T., Schwaighofer, A., Mika, S. et al. (2007). Machine learning models for lipophilicity and their domain of applicability. *Mol. Pharmaceutics* 4 (4): 524–538.

58 Fechner, N., Jahn, A., Hinselmann, G., and Zell, A. (2010). Estimation of the applicability domain of kernel-based machine learning models for virtual screening. *J. Cheminf.* 2 (1): 2.

59 Liu, R. and Wallqvist, A. (2014). Merging applicability domains for in silico assessment of chemical mutagenicity. *J. Chem. Inf. Model.* 54 (3): 793–800.

60 Liu, R., Glover, K.P., Feasel, M.G., and Wallqvist, A. (2018). General approach to estimate error bars for quantitative structure-activity relationship predictions of molecular activity. *J. Chem. Inf. Model.* 58 (8): 1561–1575.

61 Cortes-Ciriano, I. and Bender, A. (2019). Deep confidence: a computationally efficient framework for calculating reliable prediction errors for deep neural networks. *J. Chem. Inf. Model.* 59 (3): 1269–1281.

62 Cortes-Ciriano, I. and Bender, A. (2019). Reliable prediction errors for deep neural networks using test-time dropout. *J. Chem. Inf. Model.* 59 (7): 3330–3339.

63 Luque Ruiz, I. and Gomez-Nieto, M.A. (2019). Building of robust and interpretable QSAR classification models by means of the rivality index. *J. Chem. Inf. Model.* 59 (6): 2785–2804.

64 Tong, X., Wang, D., Ding, X. et al. (2022). Blood-brain barrier penetration prediction enhanced by uncertainty estimation. *J. Cheminf.* 14 (1): 44.

65 Nikolova-Jeliazkova, N. and Jaworska, J. (2005). An approach to determining applicability domains for QSAR group contribution models: an analysis of SRC KOWWIN. *Altern. Lab Anim.* 33 (5): 461–470.

66 Jaworska, J., Nikolova-Jeliazkova, N., and Aldenberg, T. (2005). QSAR applicability domain estimation by projection of the training set descriptor space: a review. *Altern. Lab Anim.* 33 (5): 445–459.

67 Tropsha, A. and Golbraikh, A. (2007). Predictive QSAR modeling workflow, model applicability domains, and virtual screening. *Curr. Pharm. Des.* 13 (34): 3494–3504.

68 Schroeter, T.S., Schwaighofer, A., Mika, S. et al. (2007). Estimating the domain of applicability for machine learning QSAR models: a study on aqueous solubility of drug discovery molecules. *J. Comput.-Aided Mol. Des.* 21 (12): 651–664.

69 Toplak, M., Mocnik, R., Polajnar, M. et al. (2014). Assessment of machine learning reliability methods for quantifying the applicability domain of QSAR regression models. *J. Chem. Inf. Model.* 54 (2): 431–441.

70 Liu, R. and Wallqvist, A. (2019). Molecular similarity-based domain applicability metric efficiently identifies out-of-domain compounds. *J. Chem. Inf. Model.* 59 (1): 181–189.

71 Mervin, L.H., Johansson, S., Semenova, E. et al. (2021). Uncertainty quantification in drug design. *Drug Discovery Today* 26 (2): 474–489.

72 Alvarsson, J., Arvidsson McShane, S., Norinder, U., and Spjuth, O. (2021). Predicting with confidence: using conformal prediction in drug discovery. *J. Pharm. Sci.* 110 (1): 42–49.

73 Mao, J., Akhtar, J., Zhang, X. et al. (2021). Comprehensive strategies of machine-learning-based quantitative structure-activity relationship models. *iScience* 24 (9): 103052.

74 Wang, Z., Yang, H., Wu, Z. et al. (2018). In silico prediction of blood-brain barrier permeability of compounds by machine learning and resampling methods. *ChemMedChem* 13 (20): 2189–2201.

75 Urbina, F., Zorn, K.M., Brunner, D., and Ekins, S. (2021). Comparing the Pfizer central nervous system multiparameter optimization calculator and a BBB machine learning model. *ACS Chem. Neurosci.* 12 (12): 2247–2253.

76 Lane, T., Russo, D.P., Zorn, K.M. et al. (2018). Comparing and validating machine learning models for *Mycobacterium tuberculosis* drug discovery. *Mol. Pharmaceutics* 15 (10): 4346–4360.

77 Bowes, J., Brown, A.J., Hamon, J. et al. (2012). Reducing safety-related drug attrition: the use of in vitro pharmacological profiling. *Nat. Rev. Drug Discovery* 11 (12): 909–922.

78 Blay, V., Li, X., Gerlach, J. et al. (2022). Combining DELs and machine learning for toxicology prediction. *Drug Discovery Today* 27 (11): 103351.

79 Srivastava, N., Hinton, G., Krizhevsky, A. et al. (2014). Dropout: a simple way to prevent neural networks from overfitting. *J. Mach. Learn. Res.* 15: 1929–1958.

80 Gal, Y. and Ghahramani, Z. (2015). Dropout as a Bayesian Approximation: Representing Model Uncertainty in Deep Learning.

81 Norinder, U. and Boyer, S. (2016). Conformal prediction classification of a large data set of environmental chemicals from ToxCast and Tox21 estrogen receptor assays. *Chem. Res. Toxicol.* 29 (6): 1003–1010.

82 Fagerholm, U., Hellberg, S., Alvarsson, J. et al. (2021). In silico prediction of volume of distribution of drugs in man using conformal prediction performs on par with animal data-based models. *Xenobiotica* 51 (12): 1366–1371.

83 Angelopoulou, A.N. and Bates, S. (2021). *A Gentle Introduction to Conformal Prediction and Distribution-Free Uncertainty Quantification*. arXiv:2107.07511.

84 Langevin, M., Grebner, C., Guessregen, S. et al. (2022). Impact of applicability domains to generative artificial intelligence. *ChemRxiv*.

85 Klingspohn, W., Mathea, M., Ter Laak, A. et al. (2017). Efficiency of different measures for defining the applicability domain of classification models. *J. Cheminf.* 9 (1): 44.

86 Lundberg, S.M. and Lee, S.-I. (2017). A unified approach to interpreting model predictions. In: *Advances in Neural Information Processing Systems*.

87 Murdoch, W.J., Singh, C., Kumbier, K. et al. (2019). Definitions, methods, and applications in interpretable machine learning. *Proc. Natl. Acad. Sci. U. S. A.* 116 (44): 22071–22080.

88 Jiménez-Luna, J., Grisoni, F., and Schneider, G. (2020). Drug discovery with explainable artificial intelligence. *Nat. Mach. Intell.* 2 (10): 573–584.

Printed in the USA/Agawam, MA
January 31, 2024